IET MANUFACTURING SERIES 23

Optimal Design Exploiting 3D Printing and Metamaterials

Optimal Design Exploiting 3D Printing and Metamaterials

Edited by
Paolo Di Barba and Sławomir Wiak

The Institution of Engineering and Technology

Published by The Institution of Engineering and Technology, London, United Kingdom

The Institution of Engineering and Technology is registered as a Charity in England & Wales (no. 211014) and Scotland (no. SC038698).

First published 2021

The Institution of Engineering and Technology
Michael Faraday House
Six Hills Way, Stevenage
Herts, SG1 2AY, United Kingdom

www.theiet.org

British Library Cataloguing in Publication Data
A catalogue record for this product is available from the British Library

ISBN 978-1-83953-351-8 (hardback)
ISBN 978-1-83953-352-5 (PDF)

Typeset in India by MPS Limited

Contents

About the editors

Paolo Di Barba, DSc PhD was appointed as a full professor of electrical engineering in tenure track position at the University of Pavia in the year 2002, where he currently teaches various courses at both BSc and MSc levels in the following areas:

- circuit theory;
- field theory;
- principles and applications of electrical engineering;
- numerical models for electricity and magnetism.

For the time being, he is the Head of the Doctoral School in Electronics, Computer Science and Electrical Engineering, University of Pavia.

Moreover, he is the dean of BSc degree and MSc degrees in Industrial Engineering.

He also leads of the Research Centre for the History of Electrical Technology, University of Pavia.

He is responsible of the Electromagnetic Device CAD Laboratory at the Department of Electrical, Computer and Biomedical Engineering, University of Pavia, where he currently leads a group of young scientists active in the area of computational electromagnetics.

At the Lodz University of Technology (Poland), after receiving the DSc degree, he has got a position of visiting professor at the International Faculty of Engineering since the year 2013.

His scientific interests are focused on the analysis and synthesis of electric and magnetic fields, with special emphasis on evolutionary algorithms for inverse problems and optimisation. Main applications include computer-aided design of electromagnetic and electromechanical devices. In this area, he has been one of the pioneering researchers who contributed to transfer the theory of Pareto optimality from classical microeconomics to computational electromagnetics.

The core activity is focused on advanced numerical methods for field analysis and field synthesis. The former includes finite-element method (FEM) for magnetic analysis and virtual-element method which extends the capabilities of FEM to polygonal and polyhedral elements. The latter include multi-objective evolutionary computing for solving non-linear inverse problems and optimization methods for higher dimensionality objective space. The study of coupled-field inverse problems, characterized by magnetic and thermal domains is another subject of research. Applications cover the automated optimal design of induction heating devices for electromagnetic processing of materials, permanent-magnet motors for electric vehicles, antennas for wireless body area networks.

For his scientific achievements, in 2021 the President of Poland conferred him the title of Professor, the highest academic award.

As far as publications are concerned, he is an author or co-author of more than 200 papers, presented to international conferences or published in international journals.

In particular, he is a co-author of the monograph entitled "MEMS Field Models and Optimal Design" (Springer, 2019), the author of the monograph entitled "Multiobjective Shape Design in Electricity and Magnetism" (Springer, 2010) and a co-author of the textbook entitled "Field Models in Electricity and Magnetism" (Springer, 2008).

He is an associate editor of the IEEE Transactions on Magnetics – Conferences.

He has been a guest editor of special issues of international journals and also an invited speaker of international conferences.

He is a member of the steering committee of international symposia in the area of computational electromagnetism, in particular,

- International Symposium on Electromagnetic Fields in Electrical Engineering (ISEF),
- Workshop on Optimization and Inverse Problems in Electromagnetism (OIPE).

He is a member of the editorial board of COMPUMAG (Conference on the Computation of Electromagnetic Fields), a major international forum of the electromagnetic community.

He contributes to research projects in cooperation with various international institutions, in particular,

- Institute of Mechatronics and Information Systems, Lodz University of Technology;
- Institute of Electronics, Lodz University of Technology;
- Dipartimento di Ingegneria Industriale, Università degli Studi di Padova.

He is a scientific consultant of former Infolytica Corporation (Montreal, Canada), now a part of Siemens, for the development of a code devoted to the automated optimal design of electromagnetic devices, based on the finite-element method for the field analysis. This was a major project of know-how transfer from university to industrial world.

He is a scientific consultant of the ABB Corporate Research Centre (Baden, Switzerland) in the area of thermal and dielectric design of power transformers.

Sławomir Wiak Professor Sławomir Wiak DSc, PhD, MEng, Doctor Honoris Causa University d'Atrois, Arras, France and Yaroslav-the-Wise Novgorod State University, Russia.

Rector of Lodz University of Technology (2016–2020), Vice Rector of Lodz University of Technology (2012–2016), Dean of the Faculty Electrical, Electronic, Computer and Control Engineering, Lodz University of Technology (2008–2012), Director of the Institute of Mechatronics and Information Systems Lodz University of Technology (since 2008).

Scientific specialization: Electrical Engineering and Computer Science (leading topics: numerical methods, expert systems, e-learning systems, MEMS, mechatronic systems, engineering software).

Visiting Professor to University of Pavia (4 years), Italy and Universite d'Artois, Arras, France (8 years).

Invited lectures to foreign Universities – over 30.

Referee Reports: Total number of review reports 620 *(papers in Journals, PhD, DSc, books, professor titles, papers of the Conferences, grants)*. Referee reports of DSc: 18, Referee reports of PhD degree: 24 (3 for University of Skopije, 1 for University of Vigo, 2 for INP-ENSEEIHT-LAPLACE Touluse), Referee report of Professor title: 14 (1 for University of Pavia, Italy, 1 for École Polytechnique Fédérale de Lausanne – EPFL, 1 for University of Ghent, Belgium).

Referee of 5 Doctor Honoris Causa degree for eminent professors, among them – Professor Kay Hameyer, University RWTH Aachen.

Promoter of DHC: Professor Sir Jim McDonald, Principal and Vice-Chancellor, University of Strathclyde, Glasgow, and Dr Börje Ekholm, President and CEO of Ericsson.

Chairman of the Session – 100 sessions at the Conferences (Conference examples: ICEM, ISEF, ACEMP, EHE, OIPE, ICAISC, etc.).

Member of International Steering Committees – 17, author and co-author of over 400 publications (19 monographs and text books), member and manager of 21 national and 24 international grants.

Recent monograph: Paolo Di Barba and Sławomir Wiak – MEMS: Field Models and Optimal Design, Springer, 2020.

Promotor of 12 PhD projects. General Chairman of SEF-International Conference on Electromagnetic Fields in Mechatronic, Electrical and Electronic Engineering.

Cooperation with the Universities: University of Southampton, University of West Bohemia – Pilzno, University of Prague, University of Arras, University of Maribor, University of Pavia, University of Vigo, Czech Academy of Sciences, Middle East Technical University in Ankara (Turkey), University of Valencia, Spain, University d'Atrois, Arras, France, Nowgorod State University, Russia, University of Skopije, University of Université de Lorraine, Nancy, France.

Member of the Advisory Boards (selected): National Board for FORESIGHT POLAND 2020 (2012–2016) – Nomination by Polish Ministry of Science and Higher Education, Member of Advisory Board of Ministry of Education and Science for Industry PhD projects – Nomination by Polish Ministry of Science and Higher Education, Advisory Committee of Lodz province governor, Advisory Committee of Marshal of Lodz voivodship, Chairman of the Scientific Committee of Lodz Center for Teachers Competences Development and Practical Education, Advisor for the strategy of Rector of Lodz University of Technology, Member of European University Association Expert Group for European Innovation Ecosystem, Advisor of Lukasiewicz Net of Polish Research Institutes, Managing Board Member – Bionanopark Ltd. Lodz (Science and Technology Park), Member of National Board for FORESIGHT POLAND 2020, Member of Advisory Committee of Lodz Province Governor, Vice Chairman of KRASP (Conference of Rectors of Polish Universities) Commission for IT Infrastructure of Polish Universities, Member of Advisory Board of Ministry of Education and Science for Industry PhD projects.

Honorary distinctions: Award of Scientific Secretary of Polish Academy of Sciences, Individual Award of Scientific Secretary of IV Branch of Polish Academy of Sciences, Silver Cross of Merit, Gold Cross of Merit, Silver Medal of Allesandro Volta of the University of Pavia, Italy, Medal Pro-Patria, Honorary Medal of Lodz City Council, Medal of 100-Anniversary of Polish freedom, Medal of the 100th anniversary of Poland regaining independence, Knight's Cross of the Rebirth of Poland (awarded by the President of the Republic of Poland in 2020), for outstanding merits in research, teaching and organizational work.

Introduction

This book addresses significant advances in the area of automated optimal design. Recent progress in this field has enabled the development of innovative electronic and mechatronic devices, but there is a risk that design uncertainties and manufacturing tolerances inherent in old technologies often limit the practical realisation of the novel designs. The solution might be found in new manufacturing opportunities offered by 3D printing technologies and techniques for the fabrication of conductive layers in low- and high-frequency applications.

The monograph approaches the topic from several perspectives, including the design of 3D fields, advances in shape synthesis, the role of additive manufacturing in synthesising metamaterials and manipulating ferromagnetic materials, as well as the steps from numerical models to printed mechatronic devices. The final chapter discusses design challenges and opportunities in the industrial environment. Benefiting from several contributions by authors with a range of backgrounds across academia and industrial research, the book provides key information for researchers, high courses students and industry professionals working in the field of advanced manufacturing, mechatronics, electrical and electronic engineering.

Throughout the book each theoretical concept is illustrated with the help of case studies, following a problem-solving approach and never forgetting that the engineering task is to formulate and solve problems using computational tools. Bearing this in mind, the authors have collected the experiences they have accumulated while teaching computer engineering electromagnetics and electromechanics at various levels and in different countries.

Specifically, Chapter 1 develops a historical preamble: it highlights the role of iron and machining in the first industrial revolution, steel and copper in the second revolution, finally the role of plastics and solid-state semiconductors in the third one. In all these cases, all the implications and disruptive effects of the new materials and the related technologies could hardly be envisaged when they emerged.

Chapter 2 reviews the state-of-the-art in the design optimisation of electromechanical and electromagnetic devices with particular emphasis on new trends and possible future directions. The underlying concept of 'hierarchical design' is explained with focus on different stages from preliminary to final and relevant associated modelling and simulation methodologies. The structural optimisation and its versions, size, shape and topology optimisation, is defined. The 'no-free-lunch' theorem is recalled. It is argued that the main challenge in electromagnetic design is a high computational cost of function evaluation. Finally, the concept of 'robust optimisation' is elaborated upon.

In Chapter 3, after reviewing the methods of optimal shape design, an algorithm of freeform optimisation with an application to dielectric design is presented; then, the results of a 3D printing experiment – which allows us to create a prototype of an electromagnetic lens based on the concept of dielectric meta-material – are presented. The underlying rationale is as follows: while classical (i.e. subtractive) production found its main constraint in the tolerances of the operating machines, the limits of additive manufacturing have yet to be fully explored and in turn may generate problems related to the final tolerances allowed by the printing machines. Nevertheless, the huge conceptual advantage, even more than the practical one, is to have a technology capable of creating surfaces of any shape: this unique feature has greatly enhanced the class of achievable prototypes, so allowing for the implementation of shapes and geometries identified through numerical methods of design optimisation. In particular, non-parametric – or freeform – optimisation appears to be an ever closer practical objective.

The core of Chapter 4 is the methodology proposed for the shape synthesis of electric motors. It is primarily based on the design of an existing motor, where it is possible to implement non-complex changes in certain parts of the design and/or materials, in order to achieve better performance characteristics. Generally, an optimal design of an electrical motor is a constrained optimisation problem, usually with a large number of design parameters, but also a variety of constraints to be fulfilled during the optimisation process. The priority is given to evolutionary algorithms and techniques, as well as to nature-mimicking algorithms.

The aim of Chapter 5 is to review the status and current developments of new ferromagnetic materials which form the vital parts of many devices, mainly electric motors applied in a wide range of products, from toys to rockets. The application of electric motors is continuously expanding thanks to ferromagnetic materials which are used in machines and devices as magnetic cores and permanent magnets. Their properties influence the performance of motors and devices with potential benefits from the environmental point of view in terms of a better efficiency.

Chapter 6 presents an overview of the fundamental metamaterial concepts and design as well as a review of selected emerging state-of-the-art metamaterial structures and fabrication methods. It will be of interest to scientists and engineers working on practical electromagnetic metamaterial technologies and applications.

Finally, the purpose of Chapter 7 is to provide an overview of the current and future industrial design combining metamaterials and 3D printing processes. Practical considerations concerning the adaptability of 3D printing for a variety of metamaterial applications are explored, as are prospective future developments in industry. In particular, a review of the state-of-the-art research progress in integrating additive manufacturing with metamaterials is provided, aiming at stimulating further research in the area of synthesis of metamaterials.

Because of the multidisciplinary nature of the covered topics, the skills and expertise of the contributing authors were exploited in preparing the manuscript, with the aim to attract an international audience. The book is particularly recommended and addressed to senior undergraduate and graduate students of departments of electrical and computer engineering, but it will also be helpful for students preparing their PhD dissertation, as well as for researchers and engineers working in the broad area of mechatronics.

Pavia, Łódź

Paolo Di Barba, Sławomir Wiak

Chapter 1

Innovative materials, computational methods and their disruptive effects

Massimo Guarnieri[1]

1.1 Materials that changed the world

The history of man and human societies has always been closely linked with technology and, consequently, with the materials which permitted technologies to be effective. This happened since very early times in our evolution and it is not by chance that the longer period, that lasted about 3 million years, was named after stone, one of the most extensively and successfully exploited materials, together with perishable wood, bones and skin. However, we consider here materials not existing in nature, but created by man, whose appearance had revolutionary effects on human societies and civilizations. Many technologies sprung from new materials such a long time ago that we now take them for granted, hardly realizing how disruptive their advent was. Some of them go back to the first technological revolution, some thousand years ago.

1.1.1 Ancient disruptive materials

1.1.1.1 Ceramic

Despite sporadic early appearance and because of its fragility, pottery was unsuitable for the nomadic life of Palaeolithic hunters–gatherers. The sedentary lifestyle of Neolithic farmers opened new possibilities. As early as late eighth millennium Before the Common Era (BCE), farmers living in some Neolithic villages in the Middle East, such as Tell Mureybet in present-day Syria, began moulding, drying and firing clay [1]. In this way, pottery started and ceramic waterproof containers such as vases, basins and jars could be made. Clues exist that pottery was derived from basketry, because early waterproof containers could be obtained by spreading clay on a basket and drying it. When firing was introduced, only clay was used, avoiding the time-consuming basket manufacturing. Regardless of this development, pottery constituted a new material for light and versatile containers, with notable advantages on basketry. It allowed not only a better conservation of a variety of foods, both solid and liquid, such as water and

[1]Department of Industrial Engineering, University of Padua, Padova, Italy

fruit juices, but also new ways of cooking vegetables such as cereals and legumes into of soups and broths, which became an important part of the diet of the Neolithic man. Alcoholic beverages were also a consequence: the earliest evidence of wine production by fermentation of grape juice was found in present-day Georgia and dates back to 6000 BCE. Pottery therefore changed the eating habits and induced an enrichment of the diet. It also had other social consequences: it gave birth to new artisans, potters and thus to a social diversification (Figure 1.1).

The potter's wheel, dating back to 4500 BCE, constituted the first use of rotary motion in productive work and turned out to be a revolutionary and fruitful concept over the time. Pottery had another important impact. The need for good firing promoted the construction of closed kilns, capable of the high temperatures needed to transform the chemical structure of clay into ceramics. Those kilns were instrumental in the development of furnaces and major technologies such as brick firing, metal smelting and glassmaking, mortar production. Archaeologists qualify those new cultures with a specific name, as Pottery Neolithic. Under these points of view, we can hardly disregard ceramics as a disruptive material. And it is

Figure 1.1 Kamares-style pottery of the old palatial period (2100–1700 BCE) at the Archaeological Museum of Heraklion (Wolfgang Sauber in Wikimedia Commons). Impermeable containers of any form allowed the safe conservation of grains and beverages, with a deep impact on dietary habits (https://en.wikipedia.org/wiki/File:AMI_-_Kamaresvase_1.jpg)

interesting to note that still now advanced ceramic materials play a crucial role in the development of futuristic technologies, e.g. thermal shields for space shuttles.

1.1.1.2 Brick

Bricks also were made from clay, shaped in simple forms, dried and eventually fired. They constituted a kind of artificial stone that was easy to obtain and that was actually the first use. It occurred with unfired bricks (called abodes in America) in the Levant around 7500 BCE while fired ones appeared in present-day Iran around 3500 BCE. It hardly needs to be mentioned how many ancient and classical civilizations made a vast use of bricks in their buildings and monuments. And it is really amazing that such a simple and old artefact still plays a paramount role in modern architecture and civil engineering.

1.1.1.3 Metals

Jeroen van der Veer, the former CEO of Royal Dutch Shell, claimed that 'the Stone Age didn't end because they ran out of stones – but as a result of competition from the bronze tools which better met people's needs'. In other words, the Neolithic came to an end when man started controlling metal smelting and the importance of metals was such that the new phases in human history are named after them.

1.1.1.4 Gold and silver: early uses

Although small pieces of gold were used by Palaeolithic men in Spain around 40 thousand years ago, early gold artefacts were cold-formed from rare native nuggets *ca* 5,000 years before BCE. The discovery of gold nuggets is not surprising, because this metal is chemically unassailable from atmospheric agents and therefore does not turn into oxidized compounds (gold reacts only with mercury, cyanide and aqua regia). Thus, gold immediately revealed itself in its fascinating colour and lustre, which led to its first uses for ornamental purposes and soon its history intertwined with wealth and power [2]. Over time, a second form of finding was exploited, powder dispersed in alluvial deposits and in the sands of rivers from which it was filtered, initially with rudimentary sieves, such as a sheepskin, as told by one of the most ancient Greek myths, the journey of Jason to Colchis (a Caucasian region facing the Black Sea, in present-day Georgia) to take possession of the golden fleece (the golden skin of a goat). Indeed, Colchis is where the most ancient gold mine was found [3] and in that area the metal was worked since the fourth millennium BCE [4].

Unlike gold, silver combines with other elements into oxidized compounds, so that initially the pure native form was rarer than gold and silver was regarded as more precious. But in the fourth millennium BCE, people of the Middle East learned to extract silver from galena minerals (PbS), in which it is bound with lead. Once available in greater quantities, silver became less valuable.

Due to their low structural strength, gold and silver were unsuitable for technical uses in antiquity and did not promote the growth of metallurgy but were widely used ornamentally and as money. By the early fourth millennium BCE, gold and silver were smelted in good kilns (melting points are 1,063 and 960 °C, respectively) and their use spread. In the early third millennium BCE, the Sumerians of Mesopotamia systematically melted gold, silver and their alloy electrum into precious objects and fine techniques were developed, such as filigree, inlay, granulation (tiny grains applied on a sheet) and cloisonné (pieces of enamel edged with thin gold edges). The skills of Sumerians goldsmiths are testified by splendid artefacts, dating back to the mid-third millennium BCE. In the same period, similar refined goldsmith techniques were used in the Caucasian Colchis and Svaneti.

1.1.1.5 Gold and silver: money

Barter, the first form of trade, was prone to constraints, which were overcome with money. Its earliest known form consisted of fixed amounts of barley, which began to be used by the Sumerians in the thirtieth century BC, at the time when writing was invented. While the latter allowed administrative activities to develop, the former promoted economic exchanges. In the beginning, it maintained an intrinsic value: a well-defined quantity of goods stored in the granaries.

A major turning point occurred in Mesopotamia in the middle of the third millennium BCE, when money began to be used which had no intrinsic functional value, but was easier to store, transport and exchange. It consisted of a predetermined quantity of silver: the 8.33-g shekel. Pieces of gold, copper and lead were then also used, always of a predetermined weight. The use of money marked the transition to a monetary economy in which each object assumed the value corresponding to a certain quantity of gold or silver. Money could be used to sell and acquire goods and services and to defer over time these two aspects of a commercial exchange, unlike bartering. Money allowed giving an objective and comparable value to any goods. It made possible any kind of commercial mediation, facilitated the provision of services and allowed accumulating wealth in a lasting way.

The next step came when weight and value of standardized pieces of precious metal were guaranteed by some higher authority by means of an imprinted effigy (Figure 1.2). Coinage was invented towards the end of the seventh century BCE by the Lydians, a people of western Anatolia who based their prosperity on artisanal production and trade [5]. Coins allowed more effective trade and gave new importance to gold and silver. Early Lydian coins (made of electrum) and shortly after Greek (silver) and Persian (gold) were minted in pieces of large denomination and high value. But later, smaller coins were introduced, which promoted the small trade of agricultural products and handicrafts, allowing farmers and artisans to free themselves from a subsistence economy. On the other hand, coinage favoured usury and debt slavery.

However, the very true value of money is not so much the intrinsic value of the pieces of precious metal, but rather the value formally recognized by those who trade and can therefore take different forms. After all, money is anything having a shared value and it was invented many times in different places throughout history and other cultures have used different objects as money: beads, shells, written notes and now also intangible information recorded on the Internet, like bitcoin. Its

Figure 1.2 Early Lydian coins, in electro, late seventh century BC (Wikimedia Commons). Coins, consisting of standardized pieces of precious metals with an identifying effigy, had a profound impact on societies, greatly facilitating trades (https://en.wikipedia.org/wiki/File:KINGS_of_LYDIA._Alyattes._Circa_620-10-564-53_BC.jpg)

introduction did not require the invention by a technological genius, but rather the acceptance of a social convention, namely a mental revolution. The more people accept money, the more different goods one can buy and this produces a general acceptance. Thanks to this flexibility, money conquered merchants and producers, who very willingly became its faithful servants whatever their religious or political faiths. Indeed, 'money is the most universal and most efficient system of mutual trust ever devised' [6]. Even today, peoples of the Earth speak different languages, obey different laws and worship different beings, but they all believe in the dollar and the euro, because money is absolutely tolerant, free from linguistic, legal, cultural, religious, race, age, gender and sexual discrimination. It is a trust system that can overcome any cultural barrier.

1.1.1.6 Copper

Early copper artefacts were cold formed from nuggets starting 8700 BCE, earlier than similar gold objects. However, also these nuggets were rare and used to make voluptuary objects with no socio-economic impact. Copper came to a larger and socially important use when man learned to smelt ores such as malachite $Cu_2CO_3(OH)_2$ around 4500 BCE, starting the Copper Age (also called Chalcolithic, Eneolithic, Aeneolithic), that was a transitional period of coexistence of stone and copper manufacturing. Earliest evidence of copper smelting was found in Serbia dating back to 5000 BCE [7], followed by Anatolia (4200 BCE) and then Armenia, Mesopotamia (Sumer), Elam and Egypt (Sinai). Pure copper melts at 1,083 °C, but malachite, the first smelted mineral, often contains impurities that lower the melting point, facilitating the process. The initial production may have been serendipitous, resulting from the accidental presence of copper ores in pottery kilns capable of reaching high temperatures, also thanks to beneficial ventilation. It is likely that the production of copper objects took place in two phases: first the ores were smelted to obtain pure metal and then this was remelted and poured into moulds. The artefacts were then brought to their final shape by beating and annealing. Early smelting and casting were tricky and resulted in rare and costly artefacts for elite uses, while traditional materials, whose processing techniques were widely known, continued to be used commonly for a long time, such as stone tools, both chipped and polished. Such coexistence of an old technology and a new alternative appeared repeatedly throughout history. But even within such conditions, copper smelting marked the beginning of the metal age and brought to an end the Neolithic and, more generally, the Stone Age.

Compared to stone, copper tools were less strong, less sharp and prone to losing their edge. However, casting of molten copper into carved stone forms, once established, provided tools more quickly and in greater quantities, so as to evade a growing demand, fed by agriculture and the demographic boom. Such needs were unknown to the Palaeolithic or Mesolithic man, who for hundreds of thousands of years had obtained from stone technology the tools suited to his needs. Over time, hardening by beating was perfected, which produced an increase in the yield point (Brinell hardness could reach 130 HB compared to 85 HB of pure annealed copper) and the use of stone gradually reduced. By the beginning of the fourth millennium BCE, copper artefacts became more available and their use gradually diversified, bringing Middle Eastern civilizations into the Age of Metals [8]. The systematic production of metal artefacts required a number of mutually interdependent activities: mining, initially in the open air and later in tunnels, ore raise from deeper and deeper levels; ore transportation and smelting in high temperature furnaces; casting in moulds and forging, i.e. metal processing of metal softened by heat. These activities promoted a complementary technical evolution. Furnaces capable of smelting temperatures were needed which were fuelled with wood and charcoal from the third millennium (50 kg of charcoal were needed to smelt a kilogram of copper in those kilns). The kilns had to be equipped with bellows, consisting of simple bags made with animal skins, to blow oxygen-rich air so as to raise the fire temperature. Tools for working the artefacts were needed, such as pliers and pincers. The systematic practice of metallurgy led not only to a wider work specialization and therefore wider social diversification but also to the need for new activities such as provisioning and subsistence of workers engaged in mining and metallurgical activities, which might be sited in isolated or remote areas. In this way metallurgy resulted in major socio-economic consequences. The development of civilizations became deeply intertwined with metallurgy, which was pursued even when mining imposed extreme danger and fatigue. Metals gradually became instrumental for the survival and supremacy of societies and their technology became indispensable: waiving metals would have implied regressing to the Stone Age.

1.1.1.7 Bronze

The transition from the Copper Age to Bronze Age induced the systematic and extensive use of the latter alloy and posed an end to Neolithic cultures. The oldest bronze was arsenical. Artefacts were found in the Iranian plateau and dated back to the fifth millennium BCE. Arsenical bronze alloy (typically 90% copper and 10% arsenic) was obtained accidentally by melting ores containing, in addition to copper, arsenic impurities. Around 3600 BCE, the Sumerians cast this bronze deliberately by using proper proportions of different ores smelted together in closed kilns. They were followed by cultures of Anatolia (Turkey) around 3200 BCE. Compared to pure copper, arsenical bronze presented a lower melting point, better melt fluidity and greater hardness when solidified in artefacts. On the other hand, arsenic is a slow-action toxic element and, even without knowing the causes, ancient peoples were aware of that danger, as their mythologies attest: the Greeks and Romans attributed physical defects to their forging deities, Hephaestus and Vulcan [9].

Throughout a long period of coexistence, tin gradually replaced arsenic in copper alloys. Apart from being not toxic, its advantages were the greater fluidity when melted – which facilitated pouring – and the greater hardness once solidified: the Brinell hardness could reach 120–150 HB (much higher than the 85 HB of pure annealed copper) [8]. Tin bronze allowed new civilizations to establish, which were named after it. Apart from jewels, it could be shaped into weapons, namely strategic artefacts for overcoming enemies, as well as agricultural and craft tools, namely artefacts ensuring economic prosperity. Long blades, very difficult to chip in stone, could be cast (Figure 1.3). Akkadian warriors led by Sargon the Great wielded such swords when they built the first empire in the twenty-fourth century BCE. As bronze objects took hold, the technology consolidated and costs reduced, making available a greater and greater variety of tools. Over time, the metal trade

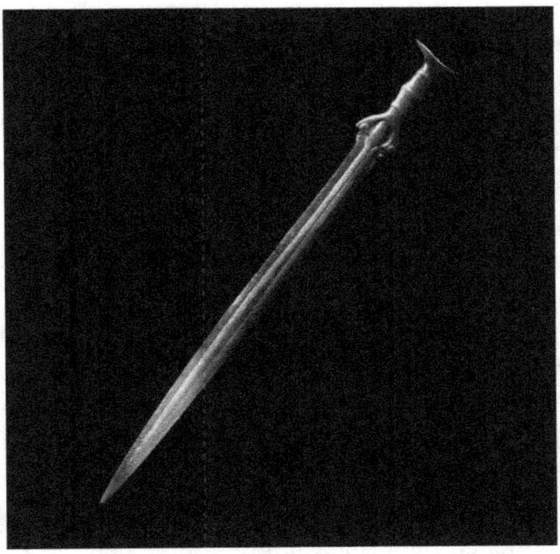

Figure 1.3 Bronze sword from Lugdunum at the Gallo-Roman Museum of Lyon-Fourvière, Lyon, France (Rama in Wikimedia Commons). Weapons like this could only be manufactured with metals and shaped warfare until the invention of firearms (https://en.wikipedia.org/wiki/File:Bronze_swords-MGR_Lyon-IMG_9732-1.jpg)

developed over longer and longer routes. Armies, farming and artisanship of many ancient civilizations and empires stood on bronze: Sumerian, Indus Valley, Yellow River – Chinese, Early European, Akkad, Minoan, Mycenaean, Amorites, Middle Kingdom Egyptian, Chimú, Inca, etc.

1.1.1.8 Iron

The very first iron artefacts were produced in Egypt and Sumer around 3500 BCE by cold forming with Neolithic techniques from rare meteoric monoliths (proven by the presence of nickel). Iron ores (siderite $FeCO_3$, hematite Fe_2O_3, limonite $FeO(OH){\cdot}nH_2O$, etc.) are abundant in the Earth's crust but their smelting remained long elusive, because the high melting point of *ca* 1,550 °C remained inaccessible to ancient kilns, which were capable of 1,100–1,200 °C even if fed with charcoal. When intensive exploitation for bronze production led many tin mines in Anatolia to run out around 1500 BCE, people living in southern Caucasus started obtaining a partial melting of iron ores [10]. To those first smiths, iron appeared as an anomalous metal, given that, unlike the other known metals, it appeared unable to melt completely. By filling their kilns with alternating layers of ores and charcoal and burning the latter, they obtained a spongy and pasty mass called bloom. This had to be forged, with a long hammering process alternated with red-hot heating to keep it soft, and slag was gradually expelled in the process and the bloom was packed progressively into wrought iron (with a small carbon content less, less than 0.8%). Iron forging was slow and tiring but the longer it was, the better the iron obtained. It was also expensive because it required large quantities of wood. The good quality of the first Anatolian iron depended on the presence of titanium in the ores mined there. It was a fortuitous fact that favoured the success of the new technology.

The first wrought iron had mechanical characteristics lower than the best bronzes of the time and was impossible to be worked in sharp weapons. But around 1350 BCE, cementation was invented. It consisted in steeping forging iron into the charcoal embers, thus causing the absorption of carbon in the outer layer, so that the melting point was locally lowered to 1,300–1,400 °C and a pasty iron was obtained that was easier to work, although not totally melted. Most importantly, cementation transformed the surface layer into a steel alloy (iron–carbon) of increased hardness (Brinell hardness up to 200 HB), while the bulk remained elastic and not brittle. A good cemented iron was harder than the best bronzes.

Initially considered a precious metal for ornaments, Homer gives such a description in the Iliad, iron slowly became the metal of choice for superior swords and daggers and the strength of the stroke from such iron weapons was limited only by the force of the arm. Forging and cementing techniques were first developed by the Hittites, who dominated the Caucasian region. When the Hittites of Muwatalli faced the Egyptians of Ramses II in the battle of Kadesh in 1286 BCE, they had iron weapons. When iron spread to other cultures after the tenth century BCE, it was the castes of blacksmiths who kept its working secrets. In fact, iron metallurgy was for a long time kept an esoteric art of divine origin: in the Greek and Roman pantheons Hephaestus and Vulcan were forging gods.

Iron became the strategic material of new powerful empires: Assyrians, Chaldeans, Greeks, Macedonians, Romans, Chinese, Maurya, Byzantines, Arabs, etc. not to mention Phoenicians, Etruscans, Celts and others. From the third century BCE, high-quality steel, ideal for weapons, was obtained in the Western World by means of a time-consuming process in which layered sheets with alternating high and low carbon content were pounded and folded on themselves several times. A similar and even more sophisticated layered steel was obtained by Japanese smiths starting in the sixth century CE to manufacture samurai Katanas and Tachis and a couple of centuries later by the Arab smiths in the Damascus steel. The Chinese were the first to use coal in advanced furnaces to fully melt iron ores into pig iron, which could be poured into moulds in the fifth century BCE. Less than two centuries later, they started decarburizing pig iron to obtain steel [11]. Around the third century BCE, smiths in Mauryan India were also able to melt steel, the so-called Wootz steel, in crucibles. Despite the great importance of iron and steel in the Western World, Europeans could melt pig iron for the first time only around the twelfth century CE. The production of iron soared when firearms spread, in the Far East from the fourteenth century and in Europe to a much larger extent from the fifteenth century, reaching an estimated 100,000 t in 1540 and starting a new phase in the Age of Iron. It was on those iron firearms that the European nations founded their colonial empires which extended all over the Earth in the following centuries.

1.1.1.9 Concrete

Very early hardened granular materials were used by the Assyrians, Egyptians and Greeks. But it was the Romans who developed a superior concrete and mastered its use, starting in the early second century BCE. It consisted of a mixture of crushed stone and liquid mortar made of lime (obtained by cooking limestone) and pozzolan (a sand rich in silica). These latter two substances, once mixed with water, reacted allowing the mortar to harden quickly even in the absence of air, resulting in a high-strength binder. Since Roman concrete was capable of solidifying in water, it made possible the construction of artificial piers, greatly increasing the possibilities offered by the relatively few safe natural dockings, which promoted Rome's maritime expansion. By pouring liquid concrete into properly shaped shuttering, it was possible to obtain quickly and easily an artificial stone of any form. Arches, vaults and domes made of concrete widespread into unpreceded buildings which gave life to the Roman Architectural Revolution. Buildings like the Colosseum (80 CE) and the Pantheon (135 CE) were made of concrete [12] (Figure 1.4). The latter's dome, spanning over 43 m in diameter, was built in concrete with differentiated compositions, stronger at the bottom and lighter at the top, where crushed pumice was used. Its size remained unparalleled for 13 centuries and is still the world's largest unreinforced concrete dome. The recipe of concrete went lost after the fall of the western Roman Empire and another 13 centuries passed before John Smeaton reinvented a similar material, in 1756, which was perfected with the Portland cement of Joseph Aspdin in 1824.

Figure 1.4 Interior of the Pantheon dome spanning 43 m in diameter, with the oculus, the central hole having both a technical function (avoided closing the dome at the top) and an aesthetic purpose, 135 CE (Arnold Dekker at Wikimedia Commons). Built in differentiated concrete, the dome remained unparalleled for more than 13 centuries (https://commons.wikimedia.org/wiki/File:PantheonOculus.jpg)

1.1.1.10 Blown glass

Early glass was likely produced in Egypt, around the mid-third millennium BCE. It is supposed that this occurred by chance with the accidental fusion of Natron (hydrated sodium carbonate, available in the beds of dried lakes) and silica (silicon oxide, forming sand) at the high temperature reached in metallurgic or ceramic kilns. That early glass was opaque and coloured and, due to the difficult production, it was for a long time considered a kind of artificial precious stone, and small glass beads were set in jewels [13]. Only a millennium later glass-coated vases and entire glass vases were produced by forming molten glass on clay casts. Two further millennia later, around 20 BCE, glass blowing was developed in Phenicia, which had been conquered by Roman general Pompey in 64 BCE. The technique relied on improved ovens, capable of reaching temperatures of complete melting, which made the glass blowable and transparent. Blowing simplified and greatly speeded up the production of artefacts, causing the cost to drop. Glassmakers learned to select silicates in order to control the colour of the glass and it was possible to produce colourless glass, by using manganese oxide. Glass became in fact a new material, in terms of cost, appearance, workability and final uses. When glassmakers migrated to Rome, attracted by the rich city markets, glass started spreading in the West. The first stained-glass windows were used in public buildings and palaces in the first

Figure 1.5 Blue blown glass vase and a glass terrine depicted in a fresco both from Pompeii, 79 CE (Wikimedia Commons). Blowing transformed glass into a cheap material, well suited for countless uses (https://commons.wikimedia.org/wiki/File:Brocca_in_vetro_blu,_da_pompei.JPG, https://commons.wikimedia.org/wiki/File:Glasschale,_Villa_Boscoreale.jpg)

century CE and by the end of the century blown glass spread throughout the empire. Glass objects became common for everyday objects, alternatively to pottery, at such a breadth that its diffusion is indicated as Glass Revolution in Ancient Rome [14] (Figure 1.5).

Some centuries later blown transparent glass could be shaped in the form of lenses. At the turn of the first millennium CE, Arab scholars such as Abū al-Haytham (Latinized as Alhazen) used them to explore refraction, setting the basis of modern optics. Around 1290, eyeglasses were invented in northern Italy and soon produced in Venice, with a major social impact because they allowed the extension of the productive life of elder people. At the turn of the sixteenth century, such lenses were combined into telescopes and microscopes, which opened the door to modern optical astronomy, starting with the seminal observations by Galileo Galilei of 1609, and to microbiology, two pillars of the Scientific Revolution [15].

1.1.1.11 Silk and porcelain

Two other ancient materials which were devised in China are worth citing, although they did not have major technological applications. The earliest silk finds were produced by the Yangshao culture of Henan, China, around 3630 BCE. The production of excellent quality silk was made possible by the domestication of the *Bombyx mori* worm, so that intact cocoons and long and regular burrs were obtained. The spinning and weaving of the almost invisible filament in such ancient times constituted an extraordinary technical achievement by itself. The value of Chinese silk was such as to motivate very lucrative long-distance trades to the Near East beginning in the eleventh century BCE.

The network of routes which started being established in the first century BCE, later extended towards the Roman Empire, became famous as the Silk Road [16]. In the first to second century CE, under the Han dynasty, the first true porcelain was produced from kaolin and fired at 1,400 °C, resulting in a translucent ceramic with an amazing vitrified enamel finish. Those Chinese handicrafts, highly valued on the international markets, contributed to expanding the Silk Road. And those century-long trades brought an extraordinary exchange of ideas, culture and knowledge between civilizations far from each other, deeply influencing the history of man. Only in the sixth century CE silk manufacturing was started in the Byzantine Empire, therefore expanding to Western Europe. And only in 1710 European porcelain could be manufactured in Meissen, Saxony.

1.1.2 Materials of the industrial revolution

1.1.2.1 Cast iron and steam engine

Iron and steel entered a new era during the industrial revolution. The first European high-quality melted steel was obtained by clockmaker Benjamin Huntsman in England in 1740. He used a calibrated mixture of soft iron and cast iron that he placed in a clay crucible to avert any contact with the burning coal. The furnace reached 1,600 °C, at which the mixture melted completely and the two components mixed. The crucible size contained about 15 kg of iron and the process lasted about 3 h, so that the steel obtained was very expensive, although of excellent quality, castable or forgeable.

The production of large quantities of iron and steel was achieved along a different path. In 1709, Abraham Darby began using coke, a pretreated coal with reduced sulphur content, in his iron furnaces to obtain cast iron of better quality. In addition, the coke resilience to crashing allowed larger quantities to be stacked in higher and bigger blast furnaces, where larger ore loads could be smelt at reduced costs [17]. In those blast furnaces that were cheap and of good cast iron reached the melting point and could therefore be cast into moulds. It was of paramount importance for Newcomen's steam engines. Coke was used systematically from 1735, allowing English iron industry to soar. But that iron remained too fragile for hammering, due to residual sulphur and phosphorus content. In 1748 Abraham Darby II succeeded in producing forgeable cast iron by a careful selection of ores with low phosphorus content and the production of iron by means of coke achieved full success. Although higher quality steel still had to be bought in Sweden and Russia, British imports dropped and English iron began to be exported. Much better guns could be cast with Darby's iron, which constituted one of the strategic oversea advantages, together with the navy, in establishing the British Empire (Figure 1.6).

Figure 1.6 *Cast iron bridge on the Severn at Ironbridge, England, erected by iron industrialists Abraham Darby III, John Wilkinson and others, 1781 (Wikimedia Commons). It was the first large iron structure ever built and demonstrated the potentials of cast iron obtained in large quantities and at relatively low cost with coke and puddling (https://commons.wikimedia.org/wiki/File:Ironbridge002.JPG)*

Steel puddling consisted of mixing pasty cast iron, to allow carbon in the melt iron to oxidize at the air and ultimately to reduce the carbon content, resulting in wrought iron or even fair quality steel. The procedure was expensive, because slow and only small charges could be worked at a time. Puddling gained momentum in England about 1768 thanks to Henry Cort, who also adopted the rolling mill for the direct production of steel ingots from wrought iron in 1783, a process 15 times faster than hammering, that had already been introduced in Sweden by Christopher Polhem in 1745 [18]. In 1784, Cort introduced puddling in reverberation furnaces, which allowed one to obtain fairly cheap steel. This was another achievement of strategic importance for Great Britain because it allowed it to satisfy the high demand for steel coming from soaring productive sectors, notably steam engines, which until then were made with costly steel imported from Sweden and Russia. Thanks to Cort's iron making technology and to the growing industrial demand, English steel production soared from 200 t in 1740 to 80,000 in 1840. James Watt relied on that kind of iron in building his steam engines. In the same period, Sweden and Russia gradually lost their leadership in iron and steel making in Europe.

The Bessemer converter, which was introduced in 1855 and perfected 1875, and the Martin-Siemens blast furnace of 1864 constituted new iron breakthroughs. In the former process, the melted cast iron poured into the converter was blown with air from the bottom so that atmospheric oxygen reacted with the carbon contained in the iron removing it, while the exothermic reaction heated the melt up to 1,600 °C and burned other impurities. In this way, carbon content was reduced from above 4% to below 1.5% and cast iron loads up to 25 t were converted into good steel without additional fuel in just 20 min [19]. It was a quick and inexpensive process that made the cost of steel production fall by seven times. In the latter process, fuel gas produced from low-quality coke in an external generator was fired in the furnace that heated up to 1,650 °C, reducing the carbon content and other impurities. The process was suitable for all types of ferrous ores and also for scrap. It was therefore inexpensive both in fuel and raw materials. Bessemer and Siemens-Martin steels fed a new phase in the industrial revolution. From 1862 Bessemer steel was used in railway tracks 15 times more durable than wrought iron. Good steel was instrumental to steam engines, locomotives, steamers, electrical machines, bridges, skyscrapers and monuments, and to more powerful heavy artillery. Wrought iron had a major part, too, notably in the structure of the Statue of Liberty of 1886 and the Eifel Tower of 1889, both were designed by Gustave Eiffel.

The increase in production was impressive. British steel, which in 1870 was 220,000 t of which 5,000 by Bessemer converter, in 1900 had risen to 4.9 million tons, of which 1.75 were Bessemer and the rest Martin-Siemens. The most sensational progress was achieved in the United States and Germany. In 1900, compared with a world production of 28 million tons, US production was 10 million tons, two-thirds of which were produced with the Bessemer converter, while German production was 8 million tons. World production of Bessemer steel peaked in 1913 with 10.5 million tons (56%), compared to 7.3 (39%) for Martin-Siemens steel. A new Iron Age was spreading into the twentieth century. Or maybe it was a new phase in a single Iron Age lasting over millennia and changing its face century after century, while supporting man in his achievements and social evolution.

1.1.2.2 Special steels

Special steels are alloys of steel with metals such as chromium, nickel, tungsten, molybdenum, manganese and vanadium, which had been discovered mostly during the eighteenth century and became more and more available in the following. Those additions yield better mechanical, thermal, electromagnetic and/or chemical properties, at a point that special steels had a deep impact in the development of advanced products and processes and on the overall iron industry.

Chromium and nickel steels were first obtained in the laboratory by Michael Faraday in 1819. Nickel steel was then made in 1824 by the Swiss metallurgist Johann Conrad Fischer in his steel mill, but industrial production begun only after 1888, particularly by the Schneider steel mills, in France, in different composition: high resistance (1%–3.5% nickel) and self-hardening (4%–5% nickel). Franz Köller of the Reichraming steel mills in Austria began working tungsten steel in 1855, becoming the first special steel to be industrially produced. The American Julius Baer patented a method of producing chromium steel in 1865, which was brought to commercialization also in France in 1877. The British metallurgist Robert Forester Mushet mastered special steel. He obtained tungsten–manganese steel, also called Mushet special steel (or RMS, 7% tungsten, 2.5% manganese and 2% carbon) in 1868, which was capable of self-hardening in air. This was the first steel for tools, which could cut common steel at an amazing speed, specifically it was twice faster and five to six times more durable than carbon steel. Mushet ceded the rights to a Sheffield steel mill, which spread it around the world. In 1889, by varying the tungsten contents and replacing manganese with chromium (18% tungsten, 4% chromium), high-speed steel (HSS) was obtained. To exploit these special alloys to the best of their possibilities, new operating machines were developed capable of producing cutting speeds up to about 7.5 m/min, under operator control.

Steel with a manganese content of few per cent was known as a brittle and useless alloy, but in 1882 the English metallurgist Sir Robert Hadfield increased the content to an unusual 12%–14% and treated the alloy with 1,000 °C heating and rapid cooling, obtaining a manganese steel that exhibited high hardness and wear resistance. It was used to manufacture objects such as grinders, safes, engine components (shafts, rods, gears, etc.) and railway parts. It was later partially replaced by nickel–chromium steel (typically 3.7% nickel, 0.8% chromium and 0.2% carbon) but is still widely used. In 1886, Hadfield patented silicon steel that was intended for mechanical uses, e.g. springs. However, after 1900, it was discovered that silicon contents up to 6% increased the electrical resistivity (therefore reducing eddy electric currents and related Joule losses) and also narrowed the hysteresis loops (therefore reducing hysteresis losses and related losses), which made it ideal for iron cores and armatures of transformers, motors and generators. Silicon steel sheets at 3%–4% began to be used in rotating and static electrical machines in 1902, initially in Great Britain and Germany. Scientific investigations on the metallographic properties of this alloy, e.g. anisotropy, were started in the 1930s.

Mechanical engineer Frederick Winslow Taylor and entrepreneur Maunsel White of the Bethlehem Iron Company, Pennsylvania, achieved a major development in 1898, when they perfected a high temperature quenching treatment that made steel alloys (in particular, tungsten–chromium) capable of maintaining their hardness at higher temperatures. Tools made of such HSS could cut standard steel at speeds of 18 m/min, which grew to 30 m/min by 1910, and at much greater cutting depths than previous tool steels. In the early years of the twentieth century, vanadium steel was developed which presented a high ultimate tensile strength and fatigue limit and found wide use in the rising automotive industry (gears, crankshafts, axles, etc.), but also in tools and in other critical components. Molybdenum in content of 3%–4% was added to steel for tool steel, permanent magnets and also for heavy artillery and tank armour during World War I.

In 1912, Brenno Strauss and Edward Maurer, two engineers of the German Krupp company, patented austenitic stainless steel, an iron alloy with chromium and nickel. A year later, a martensitic stainless steel was obtained by Harry Brearley in Great Britain, who was working on new special steels for firearms in Sheffield. It had an unusually high chromium content (>10%). In the same period, stainless steels were also developed by Max Mauermann in Austria, by Léon Guillet in France and by Elwood Haynes in the United States. These steels proved able to resist erosion better than any other iron alloy and with other additional elements acquired different and often unprecedented properties, further expanding the use of iron.

Alnico, an iron alloy with aluminium (Al), nickel (Ni) and cobalt (Co) that was developed by T. Mishima in Japan in 1931 exhibited excellent magnet properties, having double coercivity than other magnetic alloys of the time. Nickel–chromium steels were developed for high-temperature uses. One of them is Nimonic 80 alloy which was developed in Great Britain in 1941 and was soon used in aeronautical gas turbine blades, which operated at 800 °C.

1.1.2.3 Electrolytic copper, communications and power transmission

In the nineteenth century, the traditional methods of refining copper with fire, improved over the millennia, were still suitable for producing good copper for domestic tools, burners of steam engines and for wrapping ship hulls, but not for the electrical conductors of dynamos, electric rotary machines and electric lines. Refining with fire provided copper containing small quantities of arsenic, nickel and iron, which altogether amounted to not more than 0.1% but were enough to reduce the electrical conductivity to 50%–70% with respect to pure copper. Electrolytic refining was introduced shortly after the mid-century, in particular with a patent of James Balleny Elkington of 1865, who first industrialized it in Wales around 1885. The first American electrolytic copper refinery was established near New York in 1892. Copper electrolysis was produced by passing an electric current between two copper electrodes placed in a copper sulphate bath. Most impurities deposited on the bottom of the electrolytic bath as a mud from which precious metals often contained in copper ores, namely gold, silver, platinum, rhodium and iridium, were recovered. The process needed cheap electricity that was generated by magnetoelectric machines capable of high electrical powers at low costs, which had come available after 1871. In this way, a synergy was established: powerful electric generators allowed electrolytic refining of copper of unpreceded purity and conductivity and this copper was instrumental in the manufacturing of more powerful and efficient electrical machines, as well as in the construction of long electric lines for telegraph, telephone and power. At the turn of the century, copper with impurities of a few fractions per thousand was produced. In 1914 the International Annealed Copper Standard was established that stated a conductivity of 58 MS/m for pure copper: it was obtained by leaving a very small content of oxygen (0.03%–0.05%) to capture residual traces of impurities, preventing these from affecting the conductivity of the copper.

American Telephone and Telegraph Company was founded in 1885 to develop long-distance lines. In 1899, it absorbed the mother company Bell Telephone Company and soon assumed a monopolistic position in the United States, remaining for many years the largest telephone company in the world. By using also electrolytic copper conductors of large cross-section, by 1887–1889 it had laid several telephone lines that extended up to 700 km. A 2,100-km line was also built between Boston and Milwaukee, but it performed very poorly because efficient amplifiers were still missing. In the following decades, electrolytic copper was pivotal in the expansion of telegraph and telephone lines, submarine cables, as well as power lines, which extended to longer and longer distances, from the 11-km 10-kV 4-MW of the Deptford-London line of 1889 to the 180-km 110-kV of the Stanislaus River–San Francisco line of 1908, to the 1,000-km 1,200-kV 6-GW of the Kasachstan–Tambov line of 1976 [20].

1.1.2.4 Electrolytic aluminium and airplanes

Aluminium has a much more recent story than copper and iron. It was first isolated in 1825 by Danish physicist Hans Christian Ørsted and 2 years later by German chemist Friedrich Wöhler. The first industrial production method was developed in 1854 by French chemist Henri Étienne Sainte-Claire Deville and consisted in obtaining alumina (aluminium oxide) from bauxite using caustic soda and then in obtaining aluminium through reduction by liquid sodium. The process was very expensive, so much so that the final aluminium cost more than gold and was only used to produce precious objects. It remained like that even when American chemist Hamilton Young Castner found a cheaper way of obtaining sodium, in 1888, and caustic soda, in 1890 [21].

When electromechanical generators capable of producing cheap electricity became available, after 1871, the method of producing aluminium by electrolytic reduction of alumina became practicable. The process was identified in 1886 by two men independently, Charles Martin Hall in America and Paul-Louis-Toussaint Héroult in France, who were united by a singular twist of fate, in fact not only did they achieved their remarkable invention in the same year but also were born and died in the same years, 1863 and 1914 [22]. Their method, called Hall–Héroult, made use of cryolite and alumina. It required high electrical currents at low voltage and 18–20 MWh of electrical energy allowed producing 1 t of aluminium, which was enormously cheaper than the previous one, provided cheap electricity was supplied. Suddenly aluminium became an economic metal available for large-scale industrial exploitation, which promoted the great electrochemical industry. The Héroult cells began to be used in Neuhausen, Switzerland, in 1887 exploiting electrical hydropower produced at the local Rhine waterfall and the country remained the world's leading producer for several years. Hall formed the Pittsburgh Reduction Company in America in 1888, which exploited the low-cost hydropower of Niagara Falls. In 1907 this company became Aluminum Company of America (ALCOA), still an industrial giant. These companies made widely available a metal of incomparable lightness (its density is 2,700 kg/m^3 compared to 7,874 kg/m^3 of iron and 8,920 kg/m^3 of copper), high electrical conductivity and high corrosion resistance. Aluminium and its improved alloys (duralumin, Y alloy, Aldrey, etc.) developed in the following century found important industrial uses and, notably, had a paramount importance in the expansion of the aeronautical industry, since early aluminium plane appeared during World War One, e.g. the Junkers J.1 of 1918.

Starting from aluminium, the German chemist and industrialist Hans Goldschmidt invented the aluminothermic reduction process in 1895, which exploited the heat of the exothermic oxidation reaction of metallic aluminium with ferric oxide to produce cheaply metals such as chromium, manganese, molybdenum and tungsten (melting point 3,422 °C) which were needed in high-quality steel alloys and otherwise were hardly accessible. After a few years, thermite welding, which exploits the great heat of the aluminothermic reduction, was developed and alumina began to be used as an abrasive material for industrial grinding, thanks to its hardness, and as a refractory, thanks to the high melting point.

1.1.2.5 Tungsten, ribbon glass and the incandescent lamps

An attractive use for tungsten was the filament of incandescent lamps, because its high fusion point (3,422 °C) promised much higher working temperatures and ultimately much more efficient lighting than carbon filament bulbs. But this use was elusive, because tungsten is brittle and non-malleable so that it refused drawing in very thin filaments [14]. A process for manufacturing tungsten filaments was first patented by Hungarian Sándor Just and Croatian Franjo Hanaman in 1904, who succeeded in drawing the metal. Tungsten bulbs were marketed that same year by a Hungarian company. William Coolidge, a researcher at the General Electric (GE) Research Laboratory, independently achieved a similar result in 1907. The high operational temperature allowed tungsten bulbs to reach a luminous efficacy of 8 lm/W, compared to 3.5 lm/W of carbon types. By exploiting the Coolidge process, GE produced bulbs which ensured an important energy saving and marked the definitive sunset of gas lighting. In 1913, Coolidge used his invention to perfect the tungsten hot cathode X-ray tube, which could be powered at lower voltages than cold cathode type and found wide use in medical diagnostics. A further major improvement in light bulbs was introduced in 1913 by Irving Langmuir, another research fellow at GE Research Laboratory, who devised the gas-filled incandescent lamp, using nitrogen first and argon later, which reached 12 lm/W and lasted longer than vacuum bulbs* [23]. Another major advancement of incandescent bulbs consisted in the ribbon machine developed by the American Corning Glass Works around 1926 to produce glass bulbs without welding very cheaply and on a large scale, at a rate of 300 pieces per minute (later tripled). The success of the technology was such that by 1970 all bulbs produced in the world came out of only 15 machines of this type. It was again GE to patent a practical tungsten lamp filled with halogen gas (iodine) in 1959, capable of 19–35 lm/W. Thanks to these technological advancements, incandescent bulbs achieved a huge success all throughout the twentieth century. Their decline started in the new century, when LED

*Langmuir (1881–1957) had earned his doctorate in Göttingen, Germany, working with Nobel laureate Walther Nernst. Perfecting in Europe was quite common among talented American researchers of the time. After providing huge profits to the company with the gas-filled bulb, he was left free to develop his own research interests at GE, which was a happy choice, because his scientific creativity became prodigious. He pioneered the chemistry of surfaces, for which he was awarded the Nobel Prize in Chemistry in 1932, the first to an industrial chemist. He defined the chemical valence, introduced the concepts of covalent and ionic bonds, pioneered studies on ionized gases and introduced the term 'plasma', now a central topic in nuclear fusion research. He invented the atomic hydrogen welding (AHW, capable of temperatures up to 4,000 °C) and the thermionic probe (Langmuir probe). He studied methods of removing ice from aircraft wings and more … He registered 63 patents.

*Figure 1.7 SEM image of a tungsten filament in an incandescent light bulb (*Дагесян Сакис Аменакови *in Wikimedia Commons). Electric lighting the first mass use of electricity, a revolutionary technology spreading in the twentieth century (https://en.wikipedia.org/wiki/File: Tungsten_filament.JPG)*

lamps became available, capable of up to 170 lm/W. The EU and several other countries began phasing-out them in 2009 (Figure 1.7).

1.1.2.6 Pure semiconductors and solid-state electronics

In the late 1940s, an event occurred which had paramount consequences in the following decades, arriving to revolutionize information and electricity technologies. The electronics of thermionic valves, born in 1904–1906, had allowed sensational developments in communications with radio, long-distance telephony, radar and television in the timespan of four decades and very recently had paved the way to early fast digital computers, whose full potential was still substantially unexplored [24]. But thermionic valves were fraught with severe limitations: they were bulky, expensive, suffered relatively short lifetime and limited reliability and consumed a lot of power to heat the emitting filaments. Alternative solutions had long been explored, but with no practical success (Figure 1.8).

In December 1947, two research fellows at Bell Labs, theoretical physicist John Bardeen and experimental physicist Walter H. Brattain, created a small device consisting of a semiconductor (germanium) base equipped with two gold-tipped electrical contacts, the emitter and the collector, which was capable of amplifying an electrical signal and which they called *transistor* (from transfer resistor) [25]. A month later the group leader William B. Shockley built the first doped germanium junction transistor, building on previous research by others. Due to the strong antagonism between Shockley on the one side and Bardeen and Brattain on the other, the two devices were developed independently and all three inventors were

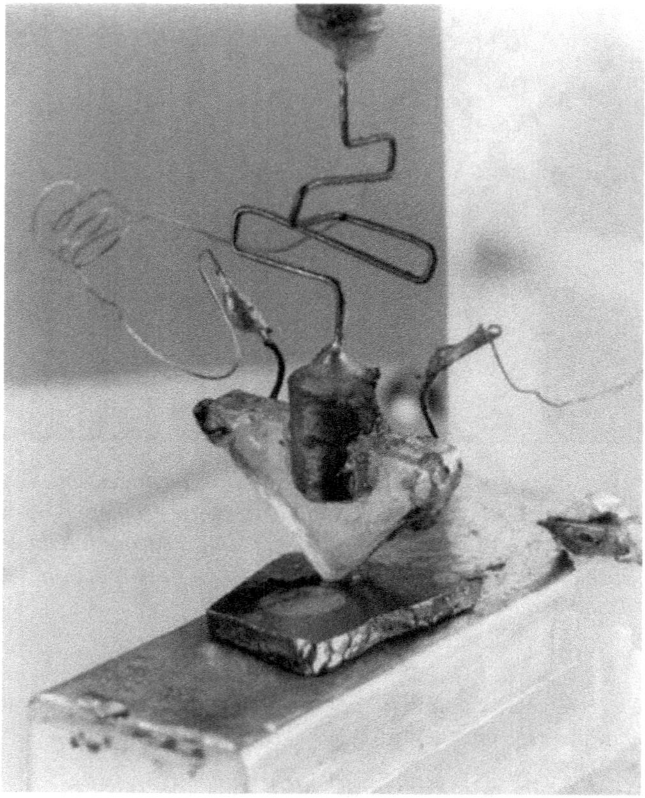

Figure 1.8 *A replica of the first working transistor, of the point-contact type, built by John Bardeen and Walter H. Brattain in December 1947 (Radio Marconi). It stated the age of pervasive solid-state electronics (http://www.radiomarconi.com/marconi/transistor/index.html)*

awarded the 1956 Nobel Prize in Physics.[†] They anticipated by 2 months Herbert Mataré and Heinrich Welker, two German physicists who had been engaged in radar research during the war and who were working at a French company controlled by Westinghouse. These two physicists independently invented the contact tip transistor on a germanium crystal, which they called *transistron*. Western Electric put into production the germanium contact tip transistor in 1951 and Mataré founded Intermetall in Düsseldorf to market it in 1952.

[†]Bardeen received a second Nobel Prize in Physics in 1972, with Cooper and Schrieffer, for the CBS theory of superconductivity. He remains the only person to have received two Nobel prizes in physics. Besides him, only three people have obtained two Nobel Prizes: Maria Skłodowska Curie, biochemist Frederick Sanger and chemist Linus Pauling.

Obtaining very pure semiconductor was vital for the transistor reproducibly and for taking it to a commercial success. This hard challenge was won by Shockley, Gordon Teal and Morgan Sparks at Bell Labs in 1950, resorting to a process for producing single crystals that Polish chemist Jan Czochralski had invented by chance in 1916. Two laboratories independently developed the method for mass production of Shockley's junction transistors in 1952. In 1954, Morris Tanenbaum, also at Bell Labs, made the first silicon transistor and the same year Tael, who had passed to Texas Instruments, produced a silicon transistor that was immediately marketed by that company. Thanks to these developments, the transistor began gaining success from the mid-1950s, particularly in the United States, where it could rely on significant financial support from the US Department of Defense, which aimed at its potential military applications. As Nathan Rosenberg wrote: 'The semiconductor industry represents perhaps the most relevant success story of the entire postwar period for the United States, in terms of government policies aimed at stimulating technical progress, as well as the growth of production and occupation' [26]. By the mid-1905s, the era of solid-state electronics had begun. The transistor would become in a few decades the workhorse of almost all modern electronics, for signal conditioning, information processing and power applications, being capable of both signal modulation and of two-state switching. Its importance derives from the capability to be mass produced in highly automated processes at extremely low cost, which, together with its flexibility and reliability, have made it as ubiquitous as few other products.

The idea of a single device containing multiple electronic components was conceived and patented in 1949 by Werner Jacobi at Siemens & Halske, although with no practical developments. In the summer of 1958, Jack S. Kilby, a new employee at Texas Instruments, found himself working alone in the laboratory while his colleagues were on vacation. Taking advantage of the temporary freedom, he thought of building a transistor, three resistors and a capacitor onto a single germanium substrate of a few square centimetres. This was the first integrated circuit (IC) [27]. The following year, Robert Noyce at Fairchild Semiconductor created the planar version, easier to make and therefore suitable for low-cost large-scale production. The IC was another epochal breakthrough with a series of following fundamental benefits:

- avoided circuit assembly and related working time, cost and error risk;
- drastically reduced the circuit sizes;
- drastically reduced supplying powers;
- paved the way for the circuit miniaturization;
- paved the way for new economic empires, of companies which could master the technology quicker.

The road to miniaturization had been envisaged by Richard Feynman in his prophetic lecture 'There's Plenty of Room at the Bottom', held at Caltech (Pasadena, California) in December 1959. In the late 1960s computers had taken the path and were in full expansion, thanks to IC technology and powerful machines were produced by big companies with International Business Machine (IBM) above all.

But some talented young people were at work to explore new frontiers. One of them was Federico Faggin, who had joined Fairchild Semiconductors in 1968 where he had designed the first digital IC based on metal–oxide–semiconductor silicon gate technology. It allowed producing higher density ICs (the so-called large-scale of integration), thus making possible semiconductor memories, in place of previous magnetic core ones [28]. In the same year, Robert Noyce and Gordon Moore left Fairchild Semiconductor and, supported by two venture capitalists, founded Intel Corporation (from Integrated Electronics), which had to become the world's largest manufacturer of ICs in a few years. Faggin soon joined them to contribute in a more ambitious project, a new IC with a high number of miniaturized components that had been designed by Marcian Edward Ted Hoff Jr, with contributions from Stanley Mazor (another past-Fairchild) and Masatoshi Shima of the Japanese company Busicom that had commissioned the device to use it in a new electronic calculator (Figure 1.9).

Faggin devised the technology to implement the circuit and in less than a year made *Intel 4004*, the first microprocessor, or chip. It consisted of a silicon substrate of 13-mm^2 housing a miniaturized digital programmable circuit of 2,300 transistors that operated as a 4-bit logic unit with a clock speed of 108 kHz. It was a real

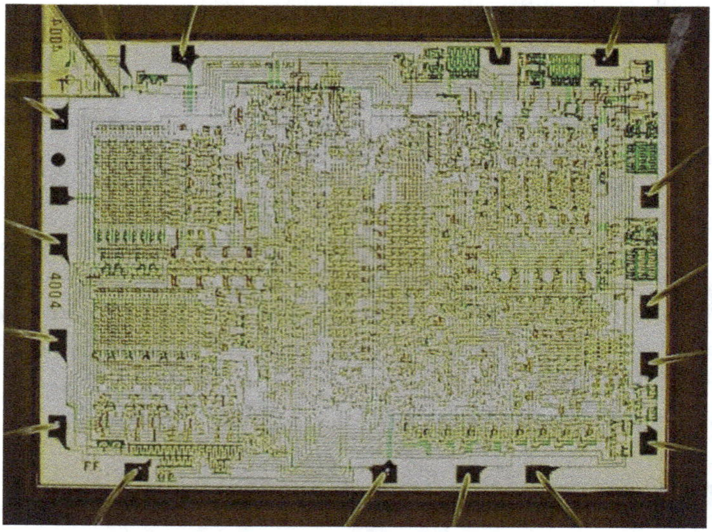

Figure 1.9 A microscope view of the Intel 4004 with the signature F.F. of inventor Federico Faggin in the bottom left corner, 1971 (Intel Free Press in Wikimedia Commons). It was a first chip, starting the revolution of microprocessor the astonishing rise of ICT in a matter of 30 years (https://en.wikipedia.org/wiki/File:Legendary_Chip_Designer_Betting_on_Human_Mind.jpg)

central processing unit (CPU) enormously cheaper, smaller, more reliable and also more powerful than early electronic computers (e.g. ENIAC, 1945). Faggin showed that 4004 could be used for different applications, convincing Noyce to renegotiate the exclusivity with Busicom so as to freely market the chip in 1971, paving the way for a new era in computer science. He designed the 8008 at Intel, the first 8-bit CPU and then, the 4040 and 8080 models in 1974, the latter resulting in a huge success. Faggin left Intel in 1974 and co-founded Zilog, where he created Z80 in 1976, one of the most successful microprocessors ever and a direct competitor of the Intel 8080. After leaving Zilog in 1980, Faggin co-founded Cygnet Technology in 1982. In 1986 he was one of the founders of Synaptic, with Carver Mead, which he headed until 1999. Synaptic developed neural networks, which combine the capabilities of biological brains and electronic brains, starting with I-1000, the first neuronal chip capable of recognizing handwriting, and created cutting-edge solutions in human–machine interfaces (touch-pad, touch-screen, etc.).

On a close inspection, we can recognize that semiconductor technology, with its enormous impact, is largely due to a few people, who worked in close connection with each other, a condition that had already happened in the past in other innovative technological sectors like modern metallic mechanics in England at the turn of the eighteenth century. In this case, the progenitor is William Shockley, who initiated this revolution when he created the research group at Bell Labs that built the first transistors in 1947–1948. After bringing semiconductor technology to the San Francisco Peninsula, Shockley hired Moore and Noyce in his Shockley Laboratory in 1956. Both left 1 year later, as part of the 'traitorous eight', to found Fairchild Semiconductors, where Noyce developed the first IC, independently of Kilby. After founding Intel in 1968, Moore and Noyce hired Hoff and Faggin, who created the microprocessor. All the latter contributed to creating innovative start-up companies, from which a part of the innovation continues to spring today. In this case also, those researchers formed a highly innovative school of excellence aggregated in a spontaneous way where knowledge spread apart from formal education. They formed a network of high-tech semiconductor companies, all located in the same limited area that began to be called Silicon Valley in 1971. Today it hosts many other companies, large and small, which operate also in different sectors, but always with a strong vocation for technological innovation. Silicon Valley is considered the Olympus of microelectronics and information technology, and of other engineering sectors. Several billion transistors are manufactured every year as single devices but especially inside ICs (ICs, microchips, VLSI, etc.) with an estimated *ca* 60 million per year per inhabitant of the Earth. The transistor has conquered areas traditionally dominated by other technologies: it is often easier and more convenient to use a microcontroller (a type of IC) and write a program into it to perform a control task, rather than to perform the same function with a mechanical controller.

1.1.2.7 Plastics

Only few lines are reserved here to the very large family of artificial materials based on polymers. Cellulose was the first raw material used for this aim, with increasing success along the nineteenth century. The process for pulping wood fibres was invented in 1844, to produce cheap cellulose paper that was instrumental in popularizing press and books, and ultimately in mass alphabetization. Nitrocellulose, the first modern explosive, was invented by chance in 1846 starting a new era in these materials which had many major effects in peaceful activities and warfare. Celluloid was perfected in 1870, resulting in the first successful synthetic material. Viscose was also obtained from cellulose in 1884 and 1892 and was the first synthetic fibre, also marketed as artificial silk and Rayon. Cellophane, a very innovative wrapping material also suitable for insulation in electrical devices, was developed in 1908. Bakelite was an artificial material obtained instead from phenol and formaldehyde in 1907. Notably, it was the first one thermosetting (allowing easy shaping before hardening) and exhibiting extremely high electrical resistivity, good mechanical strength and heat resilience, for which it encountered a very wide use in electrical devices, but also in common-use objects.

However, the vast majority of industrial synthetic materials introduced in the twentieth century were developed from coal first and hydrocarbons then, typically after very expensive research programmes developed by huge companies. The most important among them were polyvinyl chloride (PVC, 1926), polyester (PS, 1930), neoprene (1931), polyethylene (PE, 1933), polyamide (PA, 1934), polymethyl methacrylate (PMMA, 1936), polyurethane (PUR, 1937–1942), polytetrafluoroethylene (PTFE, 1938), polycarbonate (PC, 1953), polypropylene (PP, 1955). These plastics often present exceptional properties and low costs resulting in huge industrial success and production amounting at tens of millions of tons per year, with 68 Mt/year of PP at the top. Undoubtedly, synthetic materials have a paramount importance in our everyday life, as we can realize just looking at an endless number of objects around us. Just to mention one, the Lego bricks, which stimulated our creativity and filled our playtime when we were children, are made of acrylonitrile butadiene styrene. The reverse of the coin is that, due to their durability, these materials have become most responsible for global pollution, and plastic waste disposal only recently has emerged in all its dramatic importance, calling for wiser uses and quick global recycling policies. Ocean accumulations fed by rivers are an awful piece of evidence: the Great Pacific Garbage Patch, discovered in 1988, is estimated to extend over 1.6 million square kilometres.

1.1.2.8 The emergence of composite materials

Composite materials have a long history, and bricks made from mud and straws are possibly the most ancient, having been documented in Egyptian frescos *ca* 3000 BCE. Concrete is another ancient composite that, after being used by other civilizations, was mastered by the Romans, as we have already seen. Composite bows made of horn and wood were in use in Eurasia during the Bronze Age and were later adopted by other civilizations, including the Roman Empire. The light and powerful short composite bow made with wood and organic fibres was developed in Central Asia in the fifth century CE and centuries later became the weapon of choice of Mongol mounted conquerors. Although thin layers of wood orthogonally glued were used by ancient Egyptians and Greeks, modern plywood was described only in a 1797 patent and was produced some 50 years later. However, few composites materials appeared overall before the industrial revolution, due to the limited number of raw materials which were available to be combined together. Reinforced concrete, which was introduced in France between 1850 and 1892, combined the tensile strength of cheap steel then available with the formability into complex shapes of concrete, opening new frontiers to architecture, civil engineering and art of constructions which spread during the following century. In recent years, several types of engineering wood have been developed, which are manufactured by binding or fixing strands, particles, fibres, veneers or boards of wood with advanced adhesives or other methods of fixation. They are used in a variety of technical and architectural applications, including buildings, covers, watercraft and furniture. Laminates are another family of composite materials, which include Mallite, Formica, plastic coated paper, Arborite, Micarta, and are used in a vast number of applications (Figure 1.10).

As regards mechanical applications, only in the twentieth century, a major leap forward occurred starting with fibre reinforced plastic made with fibreglass and Bakelite at Owens Corning Company in 1935. A more suitable impregnating resin was used by du Pont in 1936. Progress in fibre-reinforced plastics, made of a polymer matrix, occurred in the 1940s and they have evolved later into advanced composites. Fibre-reinforced plastic made with glass fibre (often called fibreglass) is relatively cheap, stronger by weight than many metals, chemically inert and electrically insulating, which make it suitable for a wide number of applications. Carbon fibre reinforced polymers are more expensive, but extremely light, stiff and strong, resulting in a material of choice in applications like aerospace, watercraft, automotive, civil engineering and sport equipment. In more recent years a number

Figure 1.10 *The Airbus A350, released in 2013, makes large use of composite materials in its structure and fuselage (Eric Salard in Wikimedia Commons) (https://en.wikipedia.org/wiki/File:F-WWCF_A350_LBG_SIAE_2015_(18953559366).jpg)*

of different composites have been developed, including shape-memory polymer, high strain composites, metal matrix composites, ceramics matrix composites, organic matrix/ceramic aggregates, sandwich-structured composites. In this evolution, the concept of metamaterial has emerged, namely a material that has a combination of properties not present in natural materials. To do so, metamaterials consist of assemblies of several components made of composite materials. Typically, these materials are shaped in repeating patterns, having dimensions smaller than the wavelengths of the events they are designed to manage. It is this structure that provides a metamaterial with its smart properties, rather than the materials (metals, plastics, etc.) it is made from. Chapter 3 of this book provides a presentation of the possibilities offered by 3D printing in producing numerically identified optimal shapes.

The optimized design of composite materials and metamaterials has attracted computational studies to which most of this book is devoted [29]. In this way, composite materials are becoming more and more performing compared to bulk materials. And even smarter, being capable of combining different functions, from sensing to actuation, from computation to communication [30].

1.2 Computing machines and computers

1.2.1 Ancient mechanical computing

The previous section on silicon and microprocessors introduces us to electronic computers. It is the case now to consider when computing tools were first used. This occurred with the abacus, far into antiquity and by different civilizations. The oldest known was used in the Akkadian Empire *ca* 2200 BCE to compute sums and differences. The Egyptian used it in the seventh century BCE, the Chinese about the same time and the Greek a little later. Some remained in use until recent times in some cultures, allowing skilled users to execute complex computation at a sensational speed, e.g. the soroban, the Japanese evolution of the Chinese *suanpan*, introduced in the fourteenth century CE. The *yupana* was a system of knots on strings used in the pre-Colombian Inca Empire (Figure 1.11).

The Antikythera mechanism is an extraordinary device dated *ca* 80 BCE that was recovered from a Roman shipwreck in 1902 [31]. It was a three-quadrant astronomical calculator capable of determining the position of the five known planets, the phases of the moon, the equinoxes, the months and days of the week, thanks to a complex kinematic system consisting of at least 30 gears, including an epicycloidal one and a differential gear with ratio 235:19, so as to reproduce Metonic cycle for which the moon that makes 235 sidereal revolutions every 19 tropical years. It witnesses the level reached by Roman–Greek mechanics that later went forgotten.

1.2.2 Mechanical calculators

In modern Europe, progress in computing machines was brought by the Scientific Revolution, starting with the sector, or proportional compass, that was shaped as a

Figure 1.11 *The Antikythera mechanism, dated ca 80 BCE. Fragment A – front; the largest gear, ca 13 centimetres in diameter, is visible in the front (Wikimedia Commons). It was an ancient complex time-computing machine with very sophisticated gears (https://en.wikipedia.org/wiki/File:NAMA_Machine_d%27Anticyth%C3%A8re_1.jpg)*

graduated calliper and was built in many models by several mathematicians, including Fabrizio Mordente, James Hood and Galileo Galilei. It anticipated the slide ruler, which was developed by William Oughtred in 1622 building on other's achievements, notably Napier's logarithms and bones. By resorting to clock technology, German Wilhelm Schickard built the first mechanical calculator of the modern world in 1624. It could perform the four arithmetical operations and was aimed at assisting Schickard's friend Johannes Kepler in his astronomical calculations but went destroyed before delivery, in a fire during the Thirty Years War, and went forgotten for centuries.

Blaise Pascal was one of the most influential mathematicians, physicists and philosophers of the century, a follower of Galileo's theories and a great supporter of the existence of the void. At the aim of facilitating his father in the work as a tax collector, Pascal conceived a mechanical calculator at the age of nineteen in 1642. This device, which was based on cogwheels and came known as the Pascaline, could perform additions and subtractions [32]. It took Pascal 3 years of work to build it, and it was then replicated in some twenty units. Although not very efficient due to mechanical shortcomings, it gained popularity, spreading the idea of mechanical calculation and motivating other people to build similar machines, e.g. Tito Livio Burattini. Gottfried Wilhelm Leibniz was a prominent diplomat, philosopher, jurist and mathematician. From 1675 he developed infinitesimal calculus independently of Newton and in 1679 he studied binary arithmetic as pure theoretical speculation. Inspired by Pascal's machine, he undertook the construction of the Stepped Reckoner at the age of 25 in 1671. This was a calculator capable of performing all four arithmetical operations, thanks to the Leibniz wheel, a drum with stepped gears that he had invented. The machine was completed only in 1694, 23 years later. Again, Leibniz's invention motivated the construction of similar calculators, e.g. by Giovanni Poleni in 1709, who used an alternative mechanism to the Leibniz wheel. These replications must not surprise, because they were the effect of the spread of scientific information nurtured and promoted by the Scientific Revolution (Figure 1.12).

Building on Leibniz's device, the Frenchman Charles Xavier Thomas de Colmar made and patented a mechanical calculator called the *arithmometer* in 1820. De Colmar presented it at the Great Exhibition of Crystal Palace in 1851 and then successfully industrialized it. It was improved repeatedly and was produced in several thousand units by 1915. It was the first successful and widely used calculator.

The American inventor Frank Stephen Baldwin devised an alternative mechanism to the Leibniz wheel, based on two coaxial wheels for each digit, and used it in a mechanical calculator built in 1873. A similar device was brought to success by the Swedish instrument maker Willgodt Theophil Odhner in 1875 which was superior to de Colmar's calculator and was also called an *arithmometer*. By 1918, when production ceased, Odhner's calculator had been made in about 23,000 units. After the patent expired in 1907, it was replicated in several clones. Other successful mechanical calculators of the time were the *Brunsviga*, built in Braunschweig, Germany, from 1892, and the *Saxonia*, built from 1895. In 1887 the Swiss Otto Steiger (1858–1923) built a different type of mechanical calculator,

Figure 1.12 Replica of Leibniz's stepped reckoner at the Deutsches Museum (Eremeev at Wikimedia Commons). The original calculator was built between 1671 and 1694 and was capable of performing the 4 operations, thanks to the Leibniz wheel (https://commons.wikimedia.org/wiki/File:Rechenmaschine_von_Leibniz_(Nachbau)_05.jpg)

called *Millionaire*, which performed multiplications by automating the method of Napier's bones and could divide and extract the square root.

Other mechanical calculators were the hand-cranked Burroughs *adder*, built in the United States in 1886 by William Seward Burroughs I, and the *comptometer*, a keyboard adder built in the United States by Dorr Eugene Felt at the age 20, which was put into production in 1887. The keyboard drive was easily motorized and the first model powered by an electric motor appeared in 1906. Mechanical calculators were widely used in the twentieth century, until the advent of electronic calculators and programmable computers.

1.2.3 Mechanical and electromechanical computers

Charles Babbage was the visionary mathematician who first conceived a programmable computer. In the nineteenth century, logarithm tables were needed in several scientific and technical computations and were hand-calculated based on polynomial expansions. But these calculations, performed manually by people called computers, were prone to frequent errors. Together with friend astronomer John Herschel, Babbage envisaged a mechanical alternative in 1822. Babbage knew that a polynomial of a suitable degree n can properly approximate any differentiable function, according to Taylor's theorem, and that a polynomial $p^n(x)$ of degree n presents equal n-order differences at constant increments of x. He devised a mechanical calculator that was capable of calculating differences up to the sixth order of 20-digit numbers, to exploit that principle. He called it the *Difference Engine*. Supported by a substantial government funding of 1,700 pounds, Babbage undertook the construction of the 25,000-part machine, initially assisted by Joseph Clement, one of the finest British mechanical engineers of the day. For 14 years, with a stop of five, Babbage worked on the grandiose project, without bringing it to completion, despite the government funding had risen to £17,000 in the meantime, a colossal amount, as much as the cost of two warships. Babbage resumed the project in 1847–1848 with the *Difference Engine No. 2*, which also was never completed, but which was built in 1989–1991 at the Science Museum, London, based on the original design and resorting to the mechanical technology of the time, thus demonstrating that Babbage's idea was feasible. The machine can provide results to an accuracy of 31 digits. Starting from 1832, Babbage conceived a much more advanced machine, called the *Analytical Engine*, which could be programmed to perform any desired algorithm. Babbage devoted himself to this project from 1837 and, although working for the rest of his life, he never completed it. It had to be a programmable computer based on mechanical technology. The program was provided through punched cards borrowed from the Jacquard loom of 1804 and was placed in the store (today we call it memory) capable of 1,000 numbers of 50 digits. The mill (now we say CPU) performed the four operations, the square root, comparison operations, sequential control, loops and conditional branching. Lady Ada Byron, the daughter of the poet and countess of Lovelace, and a fine mathematician, wrote for the *Analytical Engine* an algorithm capable of calculating Bernoulli numbers. This is now considered the first computer program and Lady Lovelace the first programmer [33].

The Swedish lawyer and inventor Per Georg Scheutz and his son Edvard took up Babbage's idea of the Difference Engine in 1843 and arrived to build an operating model that was presented at the 1855 Paris World Exhibition and was sold to British Government in 1859. A second model, built in 1860, was sold to the United States Government for use in the Albany Astronomical Observatory (New York). The Scheutzs published the first mathematical tables elaborated by means of a machine.

There are a couple of reasons for citing the tabulating machine built by Herman Hollerith in 1889, which successfully automatized the US Census operations of 1890. One is that it was the first device that used electricity for data processing. In fact, it used electric contacts through the holes of Jacquard-like punched cards, which closed an electric circuit to increment a counter, recording information. The other reason is that the Tabulating Machine Company that Hollerith founded in 1896 to exploit his invention eventually evolved into the IBM in 1924.

The idea of a computer capable of executing complex calculations re-emerged in the 1930s in two different concepts, the analog computer and the digital computer. Both were initially pursued with mechanical and then electromechanical technology. The first includes the *differential analyser* built by Vannevar Bush and co-workers starting in 1928, which consisted of wheel-and-disc mechanisms and could solve systems of differential equations with 18 unknowns, a mathematical problem inaccessible by hand [34]. It gained good success, being replicated in several units. The electromechanical digital computer was first conceived in America by George Stibitz, who, after a first simple prototype made at home with spare parts (the Model K, from kitchen) of 1937, built the *Complex Number Computer* (later renamed *Model I Relay Calculator*) with Samuel Williams at Bell Labs in 1940. This was a programmable binary electromechanical computer made of 3,000 relays and 800 km of connections. It could compute complex algebra algorithms with conditional branching. On the occasion of a demonstrative test, it was used in the first telematic link, over a distance of 360 km. Howard Hathaway Aiken, a reputed mathematician at Harvard University inspired by Babbage's Analytical Engine, built ASCC (automatic sequence controlled calculator) between 1937 and 1944, with financial and technical support by IBM. This remarkable electromechanical digital computer, later renamed *Harvard Mark I*, was conceptually similar to Babbage's machine, though less ambitious, being able to memorize 72 numbers of 23 digits plus the sign. And it was programmable, indeed the first American memorized-program machine. This 16-m long 4.5-t calculator consisted of 765,000 electrical and mechanical components and 800 km of electrical connections. Data input resorted to IBM punched cards and the program was loaded onto a punched paper ribbon. It could tackle any mathematical-scientific problem, at the speed of a multiplication in 6 s and a division in 15.3. During the war it was used by the US Navy to perform cumbersome ballistic calculations and by John von Neumann within the Manhattan project.

Unaware of American projects, Konrad Zuse took the same path in Germany, to cope with the calculations he had to face in aeronautical designs. He constructed a programmable binary mechanical computer, later known as *Z1*, which was completed in 1938 [35]. It had keyboard data input, a processing unit, a memory unit and a punched-card control unit. In 1940, he built a second machine, *Z2*, of the electromechanical type, because it was based on relays, thus being more efficient than *Z1*. Zuse completed a third machine in 1941. *Z3* was a more advanced model that worked in floating point and was Turing-complete, although Zuse ignored Alan Turing's computational theories. Indeed, *Z3* was the first fully operational programmable computer. After that these machines went destroyed by bombing during the War, Zuse started building a more advanced model in 1942, *Z4*, which was completed after the war and sold to ETH, the Zurich Polytechnic, in 1950, becoming the first operative computer to be marketed. Zuse's machines were conceptually superior to contemporary American electromechanical machines and much more compact.

1.2.4 Wartime electronic computers

Electromechanical computers built on two-state relays were naturally suited for handling digital logics and Boolean algebra, as Claude Shannon had theoretically proved in his famous MIT master thesis *A Symbolic Analysis of Relay and Switching Circuits* published in 1938 [36]. At that time, most specialists did not believe that electronic valves were a viable option. But John Vincent Atanasoff, a young physics professor at Iowa University, toyed with the idea in 1936. He started building a binary computer assisted by fresh engineer Clifford Berry, and the machine was partially completed when Atanasoff was called into active military service in 1942. Later called ABC (Atanasoff–Berry Computer), it was the first computer in the world resorting to electronics, since it used 300 vacuum valves as fast switches and a memory of 1,600 capacitors, capable of thirty 50-bit numbers, but had also electromechanical parts and rotating drums hosting the capacitors. Data entry resorted to punched cards, and the machine operated in binary digital logics. It was designed to compute systems of linear partial differential equations (PDEs), up to 29 and the solution took 15 days. Consequently, it was not general-purpose programmable.

Coding and code breaking had a major part in World War Two. In this framework, Alan Turing and co-workers at the Government Code and Cypher School at Bletchley Park, United Kingdom, had built a complex electromechanical deciphering machine, called *Bombe*. When Nazi codes became more complex, code breaking with such machines resulted too long. Tommy Flowers, an electrical engineer at Bletchley Park, thought of a computer made only of electronic switches, understanding that they were much faster than relays and sufficiently reliable in that use. A design team led by Flowers and mathematicians Alan Turing and Max Newman was formed, who completed *Colossus* in December 1943. Specifically designed for codebreaking, it was the first fully electronic computer, consisting of 1,500 valves. Consequently, it was much faster than electromechanical machines

[37]. In June 1944, a more powerful version was completed, *Colossus Mark 2*. Given their strategic importance, these machines were kept under military secrecy for many years after the war. They were declassified only in the 1970s, and their designers having been long deprived of public recognition.

Aware of Atanasoff's work on ABC, Physicist John William Mauchly and engineer John Presper Eckert of the University of Pennsylvania's Moore School of Electrical Engineering conceived a fully electronic programmable computer in 1941. In view of its potential to quickly compute complex ballistic calculations for the US Army, it was funded and classified under military secrecy. The construction of ENIAC (electronic numerical integrator and computer) began in 1943 and ended in 1946. This 30-t machine occupied 200 m^2, consisted of 18,000 vacuum valves and 70,000 resistors and consumed 250 kW of electrical power. It operated in decimal digital logic and had a memory of only 20 numbers of 10 digits, much less than Aiken's *Harvard Mark I*. But it was the first fully electronic general-purpose programmable computer and, performing an algebraic operation in 10^{-4} s, it was 100–1,000 times faster than electromechanical machines. In 30 s, it could perform a ballistic calculation which took 20 h with traditional methods. But it was not very rational: programming was done by manually acting on connections and switches, which took a long time and was prone to errors. It found large use in postwar activities for over 10 years, particularly in science and statistics and within the hydrogen bomb project (Figure 1.13).

After developing a deep analysis of ENIAC shortcomings with Mauchly and Eckert, John von Neumann wrote the *First draft of a report on the EDVAC* in 1945,

Figure 1.13 ENIAC, the first fully programmable general-purpose electronic computer, completed by Mauchly and Eckert at the University of Pennsylvania in 1946 (Wikimedia Commons) (https://en.wikipedia.org/wiki/File:Eniac.jpg)

where he proposed a conceptual scheme consisting in a single memory for data and program, to be used in EDVAC, the next project of Mauchly and Eckert [38]. This concept that had actually already been used by Babbage, Zuse and Aiken constitutes what became known as the von Neumann architecture and is now adopted in all computers. Incidentally, the premature distribution of the *First Draft* later nullified the patent claims of *EDVAC* by Mauchly and Eckert.

1.2.5 Generations of electronic computers

In the early postwar period, existing technical background merged with von Neumann architecture to give life to the first generation of programmable digital electronic computers, both in Great Britain and in the United States.

Building on Aiken's Harvard Mark I experience, IBM started the SSEC (selective sequence electronic calculator) project in 1944 and completed it in February 1948. This was a gigantic machine that conformed the principles of Babbage's analytical engine and von Neumann architecture, then a consultant to the company. Being both electromechanical and electronic, it was not very fast, but it was the first computer that performed both accounting-commercial and scientific calculations. It was used only for demonstration and advertizing purposes: it remained on display in Manhattan, to exhibit the technological capabilities of IBM, which thus gave a clear signal of its interest in the rising information technology.

The first stored program digital (binary) electronic computer was SSEM (small-scale experimental machine), also called *Baby*. This was a small demonstrative machine completed in 1948 at Victoria University in Manchester by a group led by Frederic C. Williams and Tom Kilburn. It had an original memory device, the Williams tube, which was a special cathode ray tube. *Baby* was not intended for operational use but served to demonstrate the feasibility of the concept, after that the *Manchester Mark 1* was built, which was operational in June 1949. Alan Turing worked on its software. The Ferranti company built its commercial version, *Ferranti Mark 1*, and put it on sale in February 1951, thus becoming the first commercial electronic computer.[‡] The Williams tube enjoyed some success for a few years, being adopted in computers built also in California, Sweden and the Soviet Union.

The first operational stored program (i.e. with von Neumann architecture) digital electronic computer was EDSAC (Electronic Delay Storage Automatic Calculator), which was built by the group of Maurice Wilkes at the University of Cambridge and started operation in May 1949, before *Manchester Mark 1*. EDSAC had a memory of 512 words (used for data and instructions) of 34 bits each, made with a delay line technology, just devised by Eckert. It consisted of tubes filled with mercury and with a transducer at one end that converted an electrical impulse (representing a bit) into a sound. This took some time to travel

[‡]After that Zuse had sold *Z4* to the ETH, Zurich, in 1950.

along the tube at the speed of sound and at the other end it was converted into an electrical impulse by a second transducer. By continuously entering the lines, data remained stored. EDSAC used groups of subroutines and routines which were assembled to make programming work easier. The idea proved fruitful and the use of assemblers spread. The same technology was used for the 384-word 32-bit memory of Pilot Model ACE (automatic calculating engine), which was built at the British National Physical Laboratory under the lead of Alan Turing and became operational in May 1950. It had a very rational design, which used 1,000 vacuum tubes only.

After completing *ENIAC*, Mauchly and Eckert left the *EDVAC* project at the University of Pennsylvania and founded the Eckert–Mauchly Computer Corporation (EMCC) in 1947, to commercially exploit their know-how. They built BINAC (Binary Automatic Computer) for Northrop Aircraft Company and delivered it in 1949, but this machine never became operational. Due to financial difficulties, the company was absorbed in 1950 by Remington Rand, then a producer of office machines. In 1951, EMCC built *UNIVAC I*, an electronic computer with von Neumann architecture, for which Eckert created the delay line memory (first adopted in EDSAC). For data input and output *UNIVAC I* used magnetic tapes, borrowed from sound recording technology, which ensured much shorter access times than punched cards, a bottleneck 2,000–3,000 times slower than electronic processing. The first *UNIVAC I* was placed on the market in March 1951 and entered service in June (after *Ferranti Mark 1*). The fifth *UNIVAC I* was used by CBS in the exit polls of the 1952 presidential election and predicted Eisenhower's victory with sensational accuracy. In 1955, Remington Rand merged with Sperry Corporation to become Sperry Rand, which in the 1960s was a major main competitor of IBM.

Jay Forrester, a professor at MIT, envisaged providing electronic computing capabilities with video output, in order to obtain a real-time virtual simulator for training US Air Force pilots, a valuable device in the growing Cold War feeling. *Whirlwind* was operative in 1951, used a cathode ray tube as a monitor and 6,000 thermionic valves. It had a 1,000-word 16-bit RAM (random access memory) of the dielectric type to ensure very fast access times, but this memory was not satisfactory so that the more performing magnetic memory was developed in 1953. It used matrices of 1,024 small toroidal magnetic cores placed at each intersection of a double grid of 32×32 wires capable of magnetizing the cores in two opposite directions, thus storing a bit, and to read the state of magnetization. With these cutting-edge solutions, *Whirlwind* became the progenitor of the computers used by the SAGE defence system of the US Air Force and of the commercial computers of the 1960s.

These machines, based on valves and different types of memories, constituted the first generation of electronic digital computers which met the demand of a new rising market. The second generation used transistors, ferrite core memories and high-level languages. The first of them was a demonstrative machine made by Tom Kilburn at Victoria University of Manchester in 1953 using germanium point-contact transistors, but also vacuum valves. TRADIC (transistorized airborne digital computer), the first American transistorized computer, was made at Bell Labs in 1954, using 800 transistors and vacuum valves. It was used aboard military aircrafts. The first fully transistorized models, in 1955–1958, were *CADET* of Atomic Energy Research Establishment of Harwell, *TX-0* built at MIT, *Mailüfterl*, created at the Vienna Polytechnic with the support of Philips company, Japanese *ETL Mark III* and Canadian *Defense Research Telecommunications Establishment Computer*.

The first fully transistorized commercial computer, *Metrovick 950*, was produced in 1956 by Metropolitan-Vickers in the United Kingdom. Shortly thereafter, other companies came out with their transistorized models: IBM with model *7070* sold at US$813,000 in 1958 and model *7090* sold at US$2.9 million in 1959; Sperry Rand with *UNIVAC Solid State* in 1958; Siemens & Halske with model *2002* in 1959; Digital Equipment Corporation (DEC)[§] with *PDP-1* that priced 'only' US $120,000 in 1959; Olivetti with *Elea 9003* in 1959. Generally speaking, these machines were 10 times cheaper, faster, more powerful and more reliable, and 100 times smaller and less consuming than vacuum valve models. The computer market grew very rapidly and 26 companies in the United States, 7 in Great Britain, 3 in Germany, 2 in Holland, 1 in France and 1 in Italy were competing by 1960. The technological and commercial policies of some US companies were so effective that they achieved a dominant position in the international market. IBM, although not the most technologically advanced, was able to increase its market share more than others, reaching 70% in 1960.

The third generation appeared around 1964 featuring ICs, semiconductor memories (which were a kind of ICs) and operating systems. Multitasking and time sharing became common in these computers, which prevented long waiting queues and allowed real-time operation at video terminals. The most significant products were American. IBM announced the *360* system, a family of differently sized machines, in 1964. A model with a computing speed of 500 KIPS (kilo instructions per second) and a power of 256 kB (kilo bytes) priced US$111,000. DEC marketed *PDP-8* in 1965, with a computing speed of 300 KIPS and a power of 6 kB, which cost US$16,000. It was the first successful minicomputer, sold in 50,000 units. Typically, these machines were 100 times faster and cheaper, 1,000 times smaller, more powerful and more reliable and 1,000 times less consuming than vacuum valve models of the first generation.

1.2.6 Supercomputers

Supercomputers are extremely powerful machines aiming at extreme performance. The first was Atlas, built at the University of Manchester in 1962 and capable of 1,000 KIPS. The guru of these machines was Seymour Cray, who built *CDC 6600* at Control Data Corporation (CDC) in 1964, which performed 2-MB RAM and 10-MHz clock rate and priced US$7 million. Cray founded Cray Research Inc. in

[§]DEC was an MIT spinoff created in 1957 to exploit the TX-0 know-how.

Figure 1.14 The Summit supercomputer, in operation at the Oak Ridge NL, United States, was the fastest supercomputer in the world as of November 2018 (Carlos Jones/ORNL at Wikimedia Commons). With a measured power efficiency of 14.668 Gflops/W it was also the third most energy efficient in the world. Since 2020 it has been surpassed by Japanese Fujitsu Fugaku, capable of 415.5 PLFOPS (https://en.wikipedia.org/wiki/File:Summit_(supercomputer).jpg)

1972 and produced *Cray-1* in 1976, which used 200,000 ICs and an innovative architecture to achieve performance then fantastic: 80-MHz clock rate, 8-MB RAM, 80 MFLOPS (millions of floating-point operations per second) speed and 303-MB secondary storage. It was sold at US$8.8 million and was very successful (Figure 1.14).

Microprocessors, starting with *Intel 4004*, revolutionized the computer world, allowing personal computers to be built and extreme powers to be increasingly achieved. Compared to the *Cray 1* of 1976, an iPhone 11 of 2019 is capable of 2.66-GHz, 4-GB RAM, 512-GB secondary storage and prices some hundred dollars. On the other hand, supercomputers have established new records. Today, *Fujitsu Fugaku*, which started operation at Kobe, Japan, in 2020, boasts a speed of 415.5 PLFOPS (millions of billions of FOPS). It outperformed IBM *Summit* of Oak Ridge National Laboratory that is capable of 148.6 PLFOPS and was put into service in 2018. The growth in computing power has revolutionized the exploitation of accurate mathematical models in science and engineering, and subtle numerical methods which allow the implementation of very detailed models of the real world. In this way, supercomputers have a disrupting impact in scientific research and technological development, allowing extremely intensive computations in various fields, including quantum mechanics, weather forecasting, climate research, oil and gas exploration, molecular material modelling, physical and engineering simulations and cryptanalysis.

1.2.7 Software

The first high-level programming language, *Plankalkül*, was developed by Konrad Zuse for *Z4* in 1943–1945. With the emergence of von Neumann's architecture, the need for similar programming languages quickly emerged also in Great Britain and the United States. The first British was *Autocode*, developed for *Manchester Mark 1* in 1952. In America, the first person to deal extensively with high-level programming languages was Grace Hopper, a former programmer of Aiken's *Harvard Mark I* [39]. Hopper joined EMCC in 1949 to work on *UNIVAC I*, developing *A-0*, the first American compiler, in 1951. Building on *A-0*, she directed the development of *MATH-MATIC* in 1955–1958 and *FLOW-MATIC* in 1955–1959, which were the first compiler-based languages intended for scientific and business computing, respectively.

In 1954–1957, John Backus of IBM developed *FORTRAN* (Formula Translator) for scientific computing. Continuously improved, it is still widely used more than 60 years after its birth. A similar success has *LISP*, a family of languages created by John McCarthy at MIT in 1958, which has been and is used preferentially in applications of Artificial Intelligence, a term coined by McCarthy.

ALGOL (ALGOrithmic Language), the first version of which was written by the Dutch Edsger W. Dijkstra in 1958, was a family of languages written on rules established and shared by American and European researchers. It was an important step towards the creation of an international computer language. In 1959, Hopper promoted the creation of a working group formed by companies and research institutions to create COBOL (COmmon Business-Oriented Language), based largely on *FLOW-MATIC*. It was conceived specifically for accounting and business calculation and was also the first capable of running on different computers, albeit in different versions, called dialects.

Operating systems, capable of managing hardware and of providing basic functions, appeared in the 1960s, particularly with the *IBM 360* series machines, and allowed computers to perform multiple functions and instructions simultaneously. In 1972, Kenneth Lane Thompson of Bell Labs developed the first version of the *UNIX* cross-platform operating system, which quickly became an industry standard, suitable for heterogeneous applications such as computer-aided design, numerical control machine systems and numerical simulations. In 1991, Finnish Linus Torvalds released Linux, the kernel of open-source Unix-like operating systems, which are now provided by the GNU Project. In 1994, a group of researchers led by Canadian James Arthur Gosling produced *JAVA*, the programming language for multi-platform networked applications. While progress in software remains parallel to that of hardware, programming languages and operating systems have greatly simplified the writing of computer programs, reducing development time and errors and simplifying debugging.

1.3 Numerical methods

1.3.1 Numerical methods in antiquity

Numerical approximation is as old as mathematics. Babylonian mathematicians of the eighteenth–seventeenth centuries BCE could compute the square root of two with accuracy at the sixth digit. An iterative method for computing square roots was conceived by Archytas of Tarentum, a Greek statesman, mathematician and philosopher living in Magna Graecia in the fourth century BCE and the great Archimedes of Syracuse, also in Magna Graecia, computed π at the third digit in third century BCE. Liu Hui, a Chinese mathematician living in the same century, explored iterative processes and Chinese mathematicians of the second century CE used the elimination method to solve systems of three linear equations (what became much later known as Gauss elimination) (Figure 1.15).

Zu Chongzhi, another Chinese mathematician living in the fifth century CE, computed π at the seventh digit, a precision that remained long unparalleled. After than early forms of linear interpolation were used by Babylon astronomers and by Hipparchus of Rhodes, both in the Hellenistic period, second-order interpolation was first used in China by astronomer Liù Zhuó around 600 CE and third-order interpolation was used again in China around 1,280 CE. In the same period, Chinese mathematician Zhū Shìjié explored early finite-difference computation. In India, second-order interpolation was proposed *ca* 625 CE by Brahmagupta, who

Figure 1.15 A Babylonian clay tablet with cuneiform writings of the approximation of the square root of 2, corresponding to 1.41421296, with an accuracy at the sixth digit, ca 1800–1600 BCE (Bill Casselman at Wikimedia Commons) (https://en.wikipedia.org/wiki/File:Ybc7289-bw.jpg)

also introduced zero, negative numbers and their algebra in the decimal positional numbers, thus giving a fundamental contribution to the Hindu–Arabic numeral system. Parabolic interpolation schemes were used by Persian mathematicians al-Bīrūnī in the eleventh century [40].

1.3.2 Numerical methods in the early modern period

Persian mathematician al-Kashī was able to go further, computing π at the sixteenth digit in the fifteenth century. Many others improved this computation later, notably with Austrian astronomer Christoph Grienberger reaching 38 digits in 1630 by means of a polygonal approximation, that still remains a record for manual computation. The Euler number *e* was discovered by Jacob Bernoulli in 1683 and was later computed with increasing precision, starting with Roger Cotes who reached 13 digits in 1714, Leonhard Euler with 23 digits in 1748, John von Neumann with 2010 digits on the ENIAC in 1949 and Steve Wozniak with 116,000 digits on Apple II in 1978, to cite a few.

Isaac Newton described the eponymous iterative method for finding the root of a real-valued function, $f(x) = 0$, in 1669 and proposed an early form of polynomial interpolation and the minimal resistance problem, namely an early approach to the calculus of variations, in his *Principia Mathematica* of 1687. In Europe, an early polynomial interpolation formula was proposed also by British Edward Waring in 1779 and the same result was published by Italian–French Joseph-Louis Lagrange in 1795 to whom it is now usually attributed. It is also a corollary of a formula published by Leonhard Euler in 1783, who also proposed the eponymous single step method in 1768–1780. This is a first-order numerical procedure for solving ordinary differential equations (ODEs) with a given initial value occurring in problems of celestial mechanics [41] and was later encountered in modelling chemical reaction rate, electrical circuits, etc. In addition, Euler first formalized and named the calculus of variations in 1756, building on previous results by Lagrange [42].

1.3.3 Numerical methods in the modern period

In the nineteenth century, advancements in the interpolation theory were proposed mainly by Augustin-Louis Cauchy and Charles Hermite in France and by Carl W. Borchardt and Leopold Kronecker in Germany. Late in the century and in the beginning of the next, approximation theory was developed, with major contributions by Germans Karl Weierstrass, Carl Runge and George Faber, French Charles Méray and Hungarian Lipót Fejér. In the first part of the nineteenth century, several physical phenomena developing in continuum media were formulated in mathematical models consisting of PDEs of elliptic form (e.g. Poisson equation), parabolic form (e.g. diffusion equations, Fourier heat conduction equation, Navier–Stokes equations for fluid-dynamics) and hyperbolic form (e.g. wave equations, Maxwell electromagnetic wave equation). The solution of PDE systems called for new numerical tools.

Carl Friedrich Gauss was likely the first to propose an iterative method for a solving system of linear equations, now known as Gauss elimination, between 1798 and 1810. It consists in a sequence of operations to modify the matrix of coefficients until a lower triangular matrix is obtained. The theory of system of linear equations progressed greatly in the nineteenth century, starting with German mathematician Hermann Grassmann, and then progressing with the work by British and French scholars such as James Joseph Sylvester, who introduced the term matrix, Arthur Cayley, Augustin-Louis Cauchy and William Rowan Hamilton, who took the matrix and determinant concepts to maturity. The Gauss–Jordan elimination is an improvement with contribution by German geodesist Wilhelm Jordan. The Gauss–Seidel method and the Jacobi methods are also iterative schemes for solving systems of linear equations, the former being due to improvements by Philipp Gustav von Seidel (and inappropriately attributed also to Gauss) and the latter was proposed by Carl Gustav Jacobi within celestial mechanics investigation [43]. All these German mathematicians were active in the nineteenth century. In the mid of that century, Ada Lovelace, inspired by Babbage's Analytical Engine, conceived the algorithm for computing Bernoulli's numbers, as already mentioned. During the nineteenth century, important progress occurred also in the calculus of variations for the search of maxima and minima of functionals [44].

Building on the Euler single-step method, British astronomer John Couch Adams and co-workers developed multi-step methods for solving ODEs in the nineteenth century. Similar work was done by German mathematicians Carl Runge and Martin Kutta, who developed the eponymous family of implicit and explicit iterative schemes for the approximate solution of ODEs encountered in studying aerodynamics around 1901.

1.3.4 Numerical methods in the twentieth century

The early decades of the twentieth century saw new steps ahead towards the establishment of numerical analysis. Important progresses were made on series convergence. In 1909, the Swiss physicist Walther Ritz developed the variational method for solving PDEs related to acoustic and elasticity that consisted in converting the differential problem into the solution of a matrix equation and anticipating the finite-element method (FEM). The method is now credited (improperly) also to John W. Strutt (Lord Rayleigh). In 1915, Russian engineer Boris Galërkin published an article in which he proposed an approximate method for converting a differential equation to a discrete problem by using weak formulations of the original problem, which later became seminal to the development of discretization approaches such as FEMs and spectral element methods.

Applied mathematics boomed in America from the 1930s, mainly due to two events. On the one hand, from the late 1930s analog computers and then digital ones became operative which provided much more computing power than previous tools. On the other hand, refugees escaping Nazi Germany and controlled countries stated arriving from the 1930s. One of them was Richard Courant, who gave a major contribution to wartime mathematical investigations. Generally speaking, a revolution occurred in numerical analysis in the 1940s, as a consequence of the participation of large numbers of mathematicians in war efforts (especially in the

USA, Germany and the former USSR), such as the Manhattan Project that aimed at constructing the atomic bomb and the development of the electronic computers, with their superior computing speed, as we have already noted.

An early development in linear programming (i.e. optimization in linear systems) was proposed by Soviet economist Leonid Kantorovich in 1939 for reducing costs in the army and increasing losses imposed on the enemy. Similar work was developed by Dutch–American economist Tjalling Koopmans, for which the two shared the 1975 Nobel prize in economics. In 1947, American mathematicians George Dantzig and John von Neumann proposed the simplex method to tackle linear programming problems, particularly in finding the best assignment of 70 people to 70 jobs. The Monte Carlo method resorts to randomness to solve complex deterministic problems and Italian–American physicist Enrico Fermi first experimented it in studying neutron diffusion, without publishing [45]. Polish–American physicist and mathematician Stanislaw Ulam used the Monte Carlo method in the late 1940s to solve hydrodynamic problems encountered while working on the hydrogen bomb. John von Neumann contributed further developing the method and implementing it in ENIAC.

American Claude Shannon published the sampling theorem in his celebrated paper on information theory of 1948 [46], which actually had been previously presented in the Soviet Union by Vladimir Kotelnikov and outlined earlier by Herbert Raabe in 1939. Someya introduced the theorem independently in Japan in 1949. Spline interpolation was introduced by Romanian–American Isaac J. Schoenberg in 1946 and consists in fitting low-degree polynomials to small subsets of the values instead of fitting a single, high-degree polynomial to all of the values at once. While Shannon sampling theorem has major applications in communication and signal engineering, splines are largely used in approximation, interpolation, numerical analysis and other branches of mathematics. In addition, splines are largely used in computer-aided geometrical modelling and geometric design, computer graphics, medical imaging and other computer applications.

John Crank and Phyllis Nicolson proposed the eponymous method in 1947 [47], which is a finite difference method (FDM) for solving parabolic PDEs such as the heat equation. It is of the second order and implicit in time, which can be written as an implicit Runge–Kutta method, resulting in numerically stable and often unconditionally stable, e.g. in the case of diffusion equations. A procedure for checking the stability of finite difference schemes was then applied to PDEs, resorting to Fourier decomposition, was developed at Los Alamos National Laboratory and then rigorously formalized by John von Neumann and co-authors [48].

A milestone in modern numerical linear algebra consisted of the study by John von Neumann and American mathematician Herman H. Goldstine on round-off error analysis of matrix factorization and inversion methods, where the concept of condition number $k(A)$ of a symmetric positive definite (SPD) matrix with respect to inversion was introduced [49]. Alan Turing developed a similar analysis for Gaussian elimination in 1948.

The Institute for Numerical Analysis, established on the UCLA campus in 1947, had a pivotal role in the development of the discipline, because INA attracted a stream of internationally recognized applied mathematicians, promoting their fruitful cooperation [50]. They included George Forsythe, Isaac Schoenberg, Olga Taussky-Todd, John Todd, Magnus Hestenes and Eduard Stiefel. Among them was also Hungarian mathematician Cornelius Lanczos, who conceived the eponymous algorithm in 1950 to find the most useful eigenvalues and eigenvectors of a Hermitian matrix [51].

1.3.5 Computerized numerical methods

Numerical methods soared after 1950, when electronic computers started providing full benefits of their power. American mathematician David M. Young pioneered the theory of stationary iterative methods in the 1950s, now used in the numerical solution of large sparse linear systems of PDEs, notably leading to the development of the methods of successive over-relaxation (SOR) in computer applications, a variant of the Gauss–Seidel method, and later of the symmetric SOR method by Young and Stanley Phillips Frankel.

Cornelius Lanczos and, independently, American Magnus Hestenes and Swiss Eduard Stiefel conceived the conjugate gradient (CG) method around 1950–1952. Stiefel implemented it in Zuse's *Z4*, at the Institute of Applied Mathematics of ETH, Zurich [52]. The method demonstrated suitable for systems of linear equations with SPD matrices, working well in the case of sparse systems. It was also suitable for unconstrained optimization problems. In the 1970s, it became well known that conjugacy-based methods work very well for PDEs, especially of the elliptic type, such as the Laplace equation $\Delta u = 0$ and the Poisson equation $\Delta u = f$.

In 1953, Greek–American mathematician Nicholas Metropolis and others proposed a Monte Carlo method named Metropolis–Hastings algorithm to generate sample states of a thermodynamic system [53]. From this algorithm, the simulated annealing (SA) method was eventually derived, which is a stochastic approach to search the global optimum of an optimization problem.

In the 1950s and 1960s, stable and efficient numerical methods for linear systems and eigenvalue calculations were developed by several authors in the United States (e.g. Wallace Givens at Argonne NL and Alston Householder at Oak Ridge NL), in the United Kingdom (e.g. James Wilkinson at National Physical Laboratory and John Francis at National Research Development Corporation), in Switzerland (Heinz Rutishauser at ETH, Alexander Ostrowski at University of Basel) and in the USSR (Vera Kublanovskaya, at Steklov Mathematical Institute).

Francis and Kublanovskaya independently proposed the QR algorithm for computing eigenvalues and eigenvectors in 1961. In the United States, David Young, Richard Varga and Gene Golub provided contributions to iterative methods, while Golub and William (Velvel) Kahan, and others, developed stable algorithms for singular value decomposition-based least-squares and pseudoinverse calculations [54].

Several authors proposed non-iterative methods for solving the Poisson equation on regular grids in the 1960s and 1970s. In particular, these investigations were carried out by Oscar Buneman and Golub at Stanford University while working on plasma physics. They are related to cyclic reduction and to the fast Fourier transform, which gained consensus after its generalization by Americans James W. Cooley and John W. Tukey in 1965 [55]. Non-iterative methods found application also in petroleum engineering.

In more recent years, namely in the 1970s, 1980s and 1990s, other sectors of numerical analysis gained increasing attention from several researchers, including nonsymmetric linear systems, preconditioning, multilevel algorithms, and large-scale eigenvalues solvers gained increasing interest from several researchers. The emergence of parallel computing promoted investigations on domain decomposition.

Among many books on numerical analysis, the canonical text remains the comprehensive treatise edited by Milton Abramowitz and Irene Stegun for the US National Bureau of Standards in 1964 [56]. Much more is available today in digital form.

1.4 Numerical optimization

1.4.1 Deterministic optimization

We have seen that algorithms for finding a minimum of a function were first formulated by Newton and Gauss and that linear programming for solving optimization problems was introduced by Leonid Kantorovich in the Soviet Union in 1939 [57]. Advancements came in the United States in 1947, with the simplex method by George Bernard Dantzig [58] and the theory of duality by John von Neumann. Several other deterministic techniques for dealing with optimization problems were developed in the following years, e.g. the Levenberg–Marquardt algorithm (LMA, 1963), convex programming (CP, 1970), nonlinear optimization (NLO, 1979), gradient methods and support vector machines (SVM, 1995). All these methods are very effective and fast for specific problems, but suffer from some drawbacks too, because they can deal better with unconstrained problems, are sensitive to noise and prone to remain trapped in local minima.

1.4.2 Stochastic optimization

Stochastic optimization methods, which are the most common of the so-called metaheuristic family of techniques, are convenient options to cope with the previous issues. They are iterative, can deal with non-differentiable problems and typically work with populations of candidate solutions. They use random variables which can be introduced in different positions within the algorithm, resulting in different formulations, and which can be trimmed to achieve the best performance. Stochastic optimization methods are simple and effective tools capable of achieving good performance in almost all cases [59]. They exhibit noise robustness and can escape from local optima/minima to seek the global optimum/minimum, so that very good solutions can be obtained for most problems. On the other hand, stochastic methods are computationally expensive in terms of CPU size and processing time but are also usually intrinsically suitable for parallel computation, which makes them profitable in high performance computers. The first of these methods was the evolutionary algorithms first proposed by Nils Aall Barricelli in the 1950s [60].

Evolution-inspired optimization strategies gained momentum after the mid-1970s, notably with genetic algorithms (GA, 1970), artificial immune system (AIS, 1986), differential evolution (DE, 1995). Different forms of foraging algorithms, another group of biologically inspired methods, appeared later, such as particle swarm optimization (PSO, 1995) [61], artificial bee colony (ABC, 2005), seeker optimization algorithm (SOA, 2006). In the same span of years, methods simulating natural and artificial concepts were also proposed. Just to name a few, we can cite SA (1983), Tabu search (1986), Taguchi method (1998), harmony search (HS, 2001), P system-based optimization (PSBO, 2002). A larger number of derived formulations have germinated from the ingenuity of researchers, who dubbed them with fantasy names taken from the inspiring concepts and applied them in different fields of optimization. An incomplete list is ant colony optimization, bacterial foraging optimization algorithms, cross-entropy (CE) method, evolution strategies, parallel tempering, probability collectives (PC), quantum annealing, random search, reactive search optimization, stochastic hill climbing, stochastic tunnelling. Thus, a large number of stochastic optimization algorithms are now available, among which those best fitted to a specific task can be selected and adapted.

Chapter 2 of this book provides a keen presentation of recent advances in numerical methods for design optimization in which the possibilities offered by 3D printing and metamaterials are exploited.

1.5 Numerical models for continuum models

The computational power of computers, immensely superior to human capacity, opened new doors to the numerical analysis of continuum problems formulated in terms of steady-state and time-variable ODEs or PDEs.

1.5.1 FDMs for ODEs and PDEs (1920s)

FDMs discretize differential equations by approximating derivatives with finite differences, so that they can convert a possibly nonlinear system of ODEs or PDEs into a system of linear equations that can be solved by matrix algebra techniques. An FDM was first proposed by English mathematician Lewis F. Richardson to solve PDEs related to stresses in a masonry dam in 1910 [62]. It was re-proposed by A. Thom in the 1920s under the title 'the method of square' to solve nonlinear hydrodynamic differential equations [63]. It was then used by Richardson to solve PDEs in weather forecasting in the 1920s. German mathematicians Richard

Courant, Kurt Otto Friedrichs and Hans Lewy worked on FDMs for solving PDEs and defined the eponymous convergence condition for explicit time-integration schemes in 1928 [64]. All three moved to America in the 1930s, eventually becoming American citizens and contributing to the rise of numerical analysis in that part of the world.

FDM gained momentum in the 1950s, to convert ODEs and PDEs, possibly non-linear, into systems of linear equations to be solved with matrix algebra methods. Major progress occurred in FDMs for PDEs in the 1950s, particularly with the equivalence theorem of Hungarian–American Peter Lax, the alternating direction implicit (ADI) iterative method for solving elliptic and parabolic PDEs and splitting-up methods. Notable contributions to ADI came from Donald W. Peaceman and H. H. Rachford Jr, Garrett Birkhoff, Richard S. Varga and David Young, and Jim Douglas in the USA and by Nikolai Nikolaevich Yanenko and others in the USSR.

Together with the SOR method, ADIs became the methods of choice for the numerical solution of discretized PDEs during much of the 1950s and 1960s. Only in the 1970s the CG method emerged, especially after the development of incomplete Cholesky factorization, which was used as a preconditioner for CG. Pivotal contributions came from John Reid in the United Kingdom, Koos Meijerink and Henk van der Vorst in the Netherlands and Paul Concus, Gene Golub and Dianne O'Leary in the United States [65]. Thanks to ease of implementation, developed numerical methods and computational power of present-day computers, FDM algorithms are still widely used in the solution of a large variety of physics and engineering problems.

1.5.2 FEM for PDEs (1940s)

The FEM is the most widely used method for solving systems of PDEs modelling continuum physics and engineering problems, including structural analysis, heat transfer, fluid flow, mass transport, electromagnetic field and coupled problems. An FEM subdivides the analysis domain into small parts called finite elements. This is achieved by a space discretization based on the construction of a mesh consisting of a finite number of points. The unknown functions are then approximated over the mesh and the resulting simple approximated equations that hold in the finite elements are assembled into a larger system of equations that models the entire problem. The FEM then uses variational methods, or other reduction methods, to approximate a solution by minimizing an associated error function. A system of linear equations is obtained that can be solved by matrix algebra techniques.

Early work leading to the development of FEM for structural problems was carried out by different authors, notably Greek Johann Argyris, Chinese Feng Kang at Steklov Institute in Moscow and Russian–Canadian Alexander Hrennikoff [66]. A seminal work on variational methods was published by Richard Courant in 1943 [67]. FEM gained major impetus with the spread of electronic computers in the 1960s, mainly thanks to contributions from Argyris at University of Stuttgart,

Polish–British Olgierd Zienkiewicz at Swansea University, French Philippe G. Ciarlet at University of Paris 6 and American Richard Gallagher at Cornell University. The spread of the method was helped by the public release of the NASTRAN (NASA STRucture ANalysis) code in 1968 that was developed for NASA under federal funding to enforce the use of a single generic software program by all research centres involved in aerospace programs. Its success was such that it was inducted into the US Space Foundation's Space Technology Hall of Fame in 1988. The SAP code, developed by the Ray Clough and Edward Wilson group at Berkeley University in 1970, was also made widely available for free. Updated into SAP IV by Jurgen Bethe, it became one of the best performing structural analysis codes of the time. FEM then spread to numerical modelling in several other fields.

1.5.3 Other methods (1960s)

The finite volume method (FVM), first introduced in the early 1960s, is an alternative method for representing PDEs in the form of systems of algebraic equations to be solved with matrix algebra methods. In the FVM, volume integrals containing divergence terms are converted to surface integrals through the surfaces of each finite volume, based on the divergence theorem. Then the condition that the flux entering a given volume is identical to that leaving the adjacent volume is applied. The FVM evaluates exact expressions for the average value of the solution over some volume and uses them to construct approximations of the solution within cells. The method is used in many fluid dynamics and electromagnetics packages.

The boundary element method (BEM) was introduced in the 1970s to solve linear PDEs formulated in integral form. BEMs provide approximate solutions of boundary value problems which are exact solutions of the differential equations inside the domain and are parametrized by a finite set of parameters on the boundary. They are used in fluid mechanics, acoustics and electromagnetics.

Meshfree methods, introduced in 1977, do not require connections between nodes of the simulation domain, i.e. a mesh, and are formulated in terms of interaction of each node with all its neighbours. As a consequence, extensive properties such as mass or kinetic energy are assigned to the single nodes. Meshfree methods enable the simulation of some otherwise difficult types of problems, including Lagrangian simulations, in which the nodes can move according to a velocity field.

For the solutions of PDE-based continuum problems, several other numerical methods have been created, such as the control volume FEM (1983) and the finite volume element method (1990). The extended FEM is an improved version of the FEM, developed by Ted Belytschko in 1999 for applications in computational mechanics. It is suitable to model the propagation of various discontinuities either strong (cracks) or weak (material interfaces), by adopting advantages of meshfree methods while alleviating their negative sides.

1.5.4 Key developments in computational electromagnetics

The beginning of computational methods for electromagnetics can be traced back to a seminal treatise by Richard V. Southwell of 1946 [68]. Early development regarded mainly FDMs applied to static linear problems and eddy current problems, in which the computational domains were discretized with structured meshes and lower order approximation was used [69]. More sophisticated schemes were proposed in the 1960s, notably by Alan Winslow, who used a discretization scheme based on unstructured grids, obtained with an automatic generator, and variational algorithms. At later investigations, this approach demonstrated equivalent to an FEM formulation.

Straightforward applications of FEMs to electromagnetics emerged in the 1970s, starting with works on rotary electrical machines by Peter P. Silvester and Madabushi V. K. Chari [70] and then widespread to deal generalized time-dependent 3D problems. In the same period, integral formulations were developed into the moment method, later generalized to non-linear 3D problems. BEM formulations based on Green's integral theorems were also introduced.

Successful improvements of FEMs consisted in the incomplete Cholesky CG used as a preconditioner for dealing with large sparse matrices and Delaunay triangular meshing that was soon generalized to 3D tetrahedral elements. Kelvin transformation allowed managing open boundary FEM problems, while Whitney forms were demonstrated to be an effective approach for computational electromagnetics. Other advanced approaches were introduced to increase the computation efficiency.

Algorithms for electromagnetic force computation, a major effect in magnetic problems, appeared since the 1980s, when advanced modelling of magnetic hysteresis and anisotropy was introduced, too. More recently material modelling was extended to soft magnetic composite and high-temperature superconductors, which both have the potential of a disruptive impact on electric devices.

A number of methods have been introduced to overcome the low effectiveness of FEMs in high-frequency problems. One of them is the finite difference in time domain, proposed by Yee in 1966, which uses two staggered grids in space and time for each electric and magnetic for the electric and the magnetic fields, can deal with nonlinear materials and is effective in such high-frequency problems. Other effective approaches are the method of moments that has been successfully applied to antenna applications, and the transmission line matrix first published by P. B. Johns and R. L. Beurle in 1971, which has emerged as a powerful method for wave propagation.

As regards alternative formulations, the finite integral method proposed by Thomas Weiland in 1977 replaces field components with their integrals over lines and surfaces belonging to two interlacing meshes, and discretization is applied to the constitutive relations. Notably, it can be used effectively across the frequency spectrum. The cell method results in a very similar approach while maintaining multi-dimensional attributes of physical quantities, which are disregarded in FDM, FEM, BEM and FVM. It is also used in solid mechanics, fluid mechanics and continuum mechanics and is a promising tool for modelling heterogeneous materials.

References

[1] W. Barnett and J. Hoopes, (eds.), *The Emergence of Pottery*, Washington (DC): Smithsonian Institution Press, 1995.

[2] A. W. Cramb *A Short History of Metals,* Department of Materials Science and Engineering, Carnegie Mellon University, 1996.

[3] A. Hauptmann and S. Klein, Bronze Age gold in Southern Georgia, *ArcheoSciences*, 33, (2009) 75–82.

[4] A. Okrostsvaridze and D. Bluashvili, Mythical "Gold Sands" of Svaneti (Greater Caucasus, Georgia): Geological reality and gold mining artefacts, *Bulletin of the Georgian National Academy of Science*, 4(2), (2010) 117–121.

[5] N. Ferguson, *The Ascent of Money: A Financial History of the World*, London: Penguin Books, 2008.

[6] Y. N. Harari, *Sapiens, a Brief History of Humankind*, London: Penguin Random House, 2011.

[7] M. Radivojević, T. Rehren, E. Pernicka, D. Šljivar, M. Brauns, and D. Borić, On the origins of extractive metallurgy: New evidence from Europe, *Journal of Archaeological Science*, 37(11), (2010) 2775–2787.

[8] R. E. Hummel, *Understanding Materials Science*, New York (NY): Springer, 2004.

[9] M. Harper, Possible toxic metal exposure of prehistoric bronze workers, *British Journal of Industrial Medicine*, 44, (1987) 652–656.

[10] J. C. Waldbaum, *From Bronze to Iron: The Transition from the Bronze Age to the Iron Age in Eastern Mediterranean*, London: Coronet Books Inc., 1978.

[11] D. B. Wagner, *Iron and Steel in Ancient China*, Leiden: E. J. Brill, 1993.

[12] L. C. Lancaster, *Concrete Vaulted Construction in Imperial Rome: Innovations in Context*, Cambridge (UK): Cambridge University Press, 2005.

[13] A. Macfarlane and G. Martin, *Glass: A World History*, Chicago (IL): University of Chicago Press, 2002.

[14] M. Guarnieri, An historical survey on light technologies, *IEEE Access*, 6, (2018) 25881–25897.

[15] M. Guarnieri, The rise of light – Discovering its secrets, *Proceedings of the IEEE*, 104(2), (2016) 467–473.

[16] C. I. Beckwith, *Empires of the Silk Road: A History of Central Eurasia from the Bronze Age to the Present*, Princeton (NJ): Princeton University Press, 2009.

[17] J. E. Rehder, Change from charcoal to coke in iron smelting, *Historical Metallurgy*, 21(1), (1987) 37–43.

[18] C. Evans, O. Jackson, and G. Ryden, Baltic iron and the British iron industry in the eighteenth century, *Economic History Review, LV*, 4, (2002) 642–665.

[19] K. Tracy and H. Bessemer, *Making Steel From Iron*, Newark (DE): Mitchell Lane Publishers, 2005.

[20] P. Dunsheath, *A History of Electric Power Engineering*, Cambridge (MA): The M.I.T. Press, 1976.

[21] V. H. Lord, Hamilton Young Castner (1858-99) a forgotten pioneer of the aluminum industry, *Journal of Chemical Education*, 19(8), (1942) 353–356.

[22] N. C. Craig and C. M. Bickert, Historical metallurgy: Hall and Héroult: The men and their invention, *CIM Bulletin*, 79(892), (1986) 98–101.

[23] J. M. Anderson, Irving Langmuir and the origins of electronics, *IEEE Power Engineering Review*, 22(3), (2002) 38–39.

[24] M. Guarnieri, The age of vacuum tubes: Merging with digital computing, *IEEE Industrial Electronics Magazine*, 6(3), (2012) 52–55.

[25] M. Guarnieri, Seventy years of getting transistorized, *IEEE Industrial Electronics Magazine*, 11(4), (2017).

[26] N. Rosenberg, *Inside the Black Box: Technology and Economics*, Cambridge (UK): Cambridge University Press, 1982.

[27] M. Guarnieri, The unreasonable accuracy of Moore's law, *IEEE Industrial Electronics Magazine*, 10(1), (2016) 40–43, # 7128828.

[28] F. Faggin, How we made the microprocessor, *Nature Electronics*, 1(1), (2018) 88.

[29] A. Mortensen, (ed.), *Concise Encyclopedia of Composite Materials*, Amsterdam: Elsevier, 2006.

[30] M. A. McEvoy and N. Correll, Materials that couple sensing, actuation, computation, and communication, *Science*, 347(6228), (1689) 126.

[31] T. Freeth, Y. Bitsakis, X. Moussas, *et al.*, Decoding the ancient Greek astronomical calculator known as the Antikythera Mechanism, *Nature*, 444 (7119), (2006) 587–591.

[32] G. C. Chase, History of mechanical computing machinery, *IEEE Annals of the History of Computing*, 2(3), (1980) 198–226.

[33] J. Fuegi and J. Francis, Lovelace & Babbage and the creation of the 1843 'notes', *IEEE Annals of the History of Computing*, 25, 4, (2003) 16–26.

[34] M. Guarnieri, Trailblazers in electromechanical computing, *IEEE Industrial Electronics Magazine*, 11(2), (2017) 58–62.

[35] K. Zuse, *The Computer – My Life*, Berlin & Heidelberg: Springer, 1993.

[36] C. E. Shannon, A symbolic analysis of relay and switching circuits, *Transactions AIEE*, 57(12), (1938) 713–723.

[37] P. E. Ceruzzi, Inventing the computer, *Proceedings of the IEEE*, 101(6), (2013) 1503–1508.

[38] J. von Neumann. First draft of a report on the EDVAC, *IEEE Annals of the History of Computing*, 15(4), (1993) 27–43.

[39] S. Payette, Hopper and Dijkstra: Crisis, revolution, and the future of programming, *IEEE Annals of the History of Computing*, 36(4), (2014) 64–73.

[40] E. Meijering, A chronology of interpolation: From ancient astronomy to modern signal and image processing, *Proceedings of the IEEE*, 90(3), (2002) 319–342.

[41] J. C. Butcher, *Numerical Methods for Ordinary Differential Equations*, New York (NY): John Wiley & Sons, 2003.

[42] T. Rüdiger, Euler and the calculus of variations, in R. E. Bradley, C. E. Sandifer (eds.), *Leonhard Euler: Life, Work and Legacy*, Amsterdam: Elsevier, 2007, p. 24.

[43] H. H. Goldstine, *A History of Numerical Analysis From the 15th Through the 19th Century*, New York (NY): Springer, 1977.

[44] H. H. Goldstine, *A History of the Calculus of Variations From the 17th Through the 19th Century*, Berlin: Springer Science & Business Media, 1980, p. 110.

[45] N. Metropolis, The beginning of the Monte Carlo method, *Los Alamos Science*, (1987), Special Issue dedicated to Stanislaw Ulam, 125–130.

[46] C. E. Shannon, A mathematical theory of communication, *The Bell System Technical Journal*, 27, (1948) 379–423.

[47] J. Crank and P. Nicolson, A practical method for numerical evaluation of solutions of partial differential equations of the heat conduction type, *Proceedings of the Cambridge Philosophical Society*, 43(1), (1947) 50–67.

[48] J. G. Charney, R. Fjørtoft, and J. von Neumann, Numerical integration of the barotropic vorticity equation, *Tellus*, 2, (1950) 237–254.

[49] J. von Neumann and H. H. Goldstine, Numerical inverting of matrices of high order, *Bulletin of the American Mathematical Society*, 53 (1947) 1021–1099.

[50] M. R. Hestenes and J. Todd, *NBS-INA—The Institute for Numerical Analysis – UCLA 1947-1954*, Washington: National Institute of Standards and Technology, Spec. Publ. 730, 1991.

[51] C. Lanczos, An iteration method for the solution of the eigenvalue problem of linear differential and integral operators, *Journal of Research of the National Bureau of Standards*, 45(4), (1950) 255–282.

[52] M. Hestenes and E. Stiefel, Methods of conjugate gradients for solving linear systems, *Journal of Research of the National Bureau of Standards*, 49(6), (1952) 409–438.

[53] N. Metropolis, A. W. Rosenbluth, M. N. Rosenbluth, N. Marshall, A. H. Teller, and E. Teller, Equation of state calculations by fast computing machines, *The Journal of Chemical Physics*, 21(6), (1953) 1087.

[54] G. H. Golub and W. Kahan, Calculating the singular values and pseudo-inverse of a matrix, *Journal of the Society for Industrial and Applied Mathematics, Series B: Numerical Analysis*, 2(2), (1865) 205–224.

[55] J. W. Cooley and J. W. Tukey, An algorithm for the machine calculation of complex Fourier series, *Mathematics of Computation*, 19(90), (1965) 297–301.

[56] M. Abramowitz and I. Stegun, *Handbook of Mathematical Functions with Formulas, Graphs, and Mathematical Tables*, Gaithersburg: National Bureau of Standards, 1964.

[57] L. V. Kantorovich, Mathematical methods of organizing and planning production, *Management Science*, 6 (4), (1960) 366–422. First published in Russian in 1939.

[58] G. B. Dantzig, Linear programming and extensions, *Princeton Landmarks in Mathematics*, Princeton University Press, 1963.

[59] L. B. Booke, D. E. Goldberg, and J. H. Holland, Classifier systems and genetic algorithms, *Artificial Intelligence*, 40 (1–3), (1989) 235–282.

[60] N. A. Barricelli, Esempi numerici di processi di evoluzione, *Methodos*, 1954, p. 45–68 (Numerical testing of Evolution Theories, Acta Biotheoretica, 16 (1–2), (1962) 69–98).

[61] J. Kennedy and R. Eberhart, Particle swarm optimization, *Proceedings of IEEE International Conference on Neural Networks IV*, 1995, 1942–1948.

[62] L. F. Richardson, The approximate arithmetical solution by finite differences of physical problems involving differential equations with an application to stresses in a masonry dam, *Philosophical Transactions of the Royal Society of London, A*, 210 (1910) 307–357.

[63] A. Thom and C. J. Apelt, *Field Computations in Engineering and Physics*, London: D. Van Nostrand, 1961.

[64] R. Courant, K. O. Friedrichs, and H. Lewy, Über die partiellen Differenzengleichungen der mathematischen Physik, *Mathematische Annalen*, 100 (1928) 32–74. (Translated by Phyllis Fox as: On the partial difference equations of mathematical physics, IBM Journal of Research and Development, 11 (1967) 215–234.).

[65] P. Concus, G. Golub, and D. O'Leary, Numerical solution of nonlinear elliptic partial differential equations by a generalized conjugate gradient method, *Computing*, 19(4), (1978) 321–339.

[66] A. Hrennikoff, Solution of problems of elasticity by the framework method, *Journal of Applied Mechanics*, 8(4), (1941) 169–175.

[67] R. Courant, Variational methods for the solution of problems of equilibrium and vibrations, *Bulletin of the American Mathematical Society*, 69 (1943) 1–23.

[68] R. V. Southwell, *Relaxation Methods in Theoretical Physics*, Oxford: Oxford University Press, 1946.

[69] C. W. Trowbridge and J. K. Sykulski, Some key developments in computational electromagnetics and their attribution, *IEEE Transactions on Magnetics*, 42(4), (2006) 503–508.

[70] P. P. Silvester and M. V. K. Chari, Finite element solution of saturable magnetic field problems, *IEEE Transactions on Power Apparatus and Systems*, PAS-89(7), (1970) 1642–1651.

Chapter 2
Advances and trends in design optimisation

Jan K. Sykulski[1]

A design is often understood as a plan or specification for the construction of an object or system or for the implementation of an activity or process; in engineering, it normally refers to a series of steps to be undertaken to create functional products or processes. The design may also be perceived essentially as a 'fitness for purpose' exercise which provides a guarantee that the components and the finished product are fit for their intended purpose. Much of this book is concerned with various design activities.

Optimisation plays a pivotal role in the design process as inevitably we wish to end up with a product which is better, lighter, cheaper and more reliable than what the competition can offer. This necessitates a specification of an objective function – or a set of often conflicting objectives – which we attempt to minimise or maximise simultaneously; we often find, however, that it is impossible to achieve all optima at the same time and we therefore need to compromise, which may not be an easy decision and we therefore need some sound principle to follow rather than making a random choice. To complicate things further, when manufacturing a product, we are often faced with machining tolerances and variability of material properties – and yet we expect all (or at least the vast majority) of our units to meet the performance specifications, not just some of them.

When predicting the future performance of a device or a system, we typically use various models: some may be simple and rapid but not particularly accurate, others very reliable but computationally expensive. Rather interestingly, all such models are relevant as we will argue straight away when introducing the concept of 'hierarchical design'. Then we need to identify a good (fast and accurate) algorithm to tackle the actual optimisation task – there is no shortage of such algorithms and recent years have produced literally hundreds (or even thousands) of possible candidates, many inspired by nature; if anything we are spoiled for choice. The decision of which approach might be the most appropriate for our problem may not necessarily be straightforward. It is also argued that there is no optimisation scheme which would perform efficiently always: clearly if there were such an algorithm we would all be using it without the need to search for alternatives. But some

[1]Electronics and Computer Science (ECS), University of Southampton, Southampton, UK

approaches have been demonstrated to work well most of the time and are available as toolboxes of popular general software.

This chapter provides a review and critical assessment of recent advances in design optimisation, emphasising the main features and the most promising approaches, finally making some projections regarding possible future developments.

2.1 The hierarchical design paradigm

The term was first introduced in the context of creating designs using a hierarchical approach for electrical and logic circuits. The hierarchy refers here to the creation of higher level cells based on lower lever cells. We can extend the concept to cover any type of design. In the framework of the topic of this book, we observe that in order to perform design optimisation, we sometimes use simple representations relying on far reaching approximations (exploiting equivalent circuits, empirical expressions and idealised models) – resulting in fast computation but not very accurate – but we also need advanced numerical models, often quite accurate and reliable, but computationally very expensive. Traditionally, the 'simple' versus 'advanced' models were often regarded as competitive, whereas in the context of hierarchical design both types are used and needed, they simply play different roles in the process.

The hierarchical design concept can be depicted as an inverted pyramid (Figure 2.1). In the context of the design of electromechanical or electromagnetic

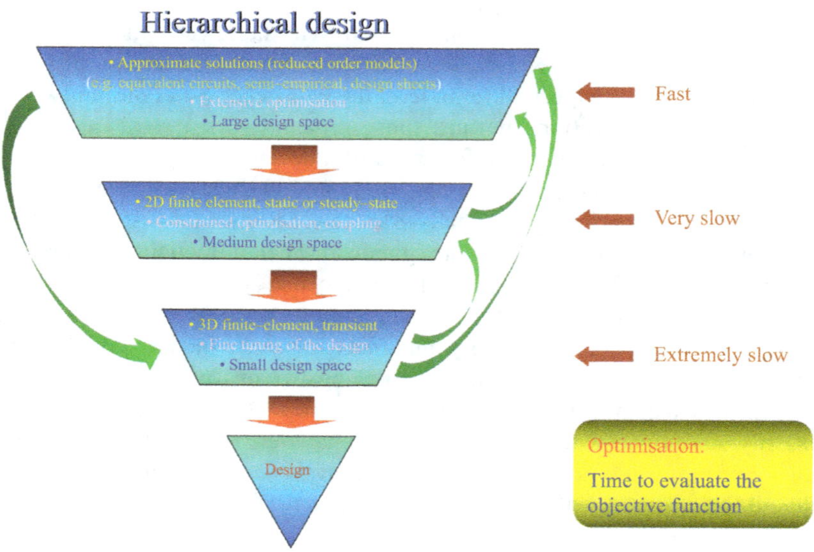

Figure 2.1 The concept of hierarchical design

(EM) devices – which are the focus of this book – we are likely to use something like a full 3D transient finite-element software (more about that in Section 2.6), precise but computationally challenging. But before we reach that stage, we may not need such an accurate tool initially, when all we want is to have a rough estimate of the likely performance, a 'ballpark figure', but we need it fast. Thus, we may be much better off using rapid approximate methods, based on the concept of an equivalent circuit or even empirical, before moving on to the next level with possibly 2D finite-element simulation and finally progressing to the full 3D model. Following the graph, we therefore move 'down' the pyramid, narrowing the design space, improving the accuracy of the prediction but significantly increasing computing times. At each stage we need an appropriate optimisation algorithm, but not necessarily the same one, as the cost of evaluating the objective function is initially small. So, it may be that in the design process, we rely on different algorithms; once we reach the final stage, we are likely to benefit from such approaches which minimise the number of necessary function calls. The times required to perform one iteration in the optimisation process are shown on the right-hand side of the diagram. Moreover, a smooth transition from one model to another is essential, especially as internal loops exist (as indicated by the green arrows) and we need to preserve modelling continuity. We will come back to this issue when talking about future developments.

2.2 The 'no free lunch' theorem

A number of 'no free lunch' (NFL) theorems were proposed in the 1990s and are well summarised in [1] – a graphical representation is suggested in Figure 2.2.

Broadly speaking, NFL theorems ascertain that for any algorithm, an elevated performance over one class of problems is offset by inferior performance over another class. In other words, it is claimed that on average all algorithms perform well; some are better suited for many types of problems, whereas others offer superior performance for a narrow group of selected cases; the former would be good candidates for general-purpose optimisation software, the latter may be very competitive in a particular situation, even if exhibiting poor performance overall. Moreover, NFL theorems demonstrate the danger of comparing algorithms by their performance on a

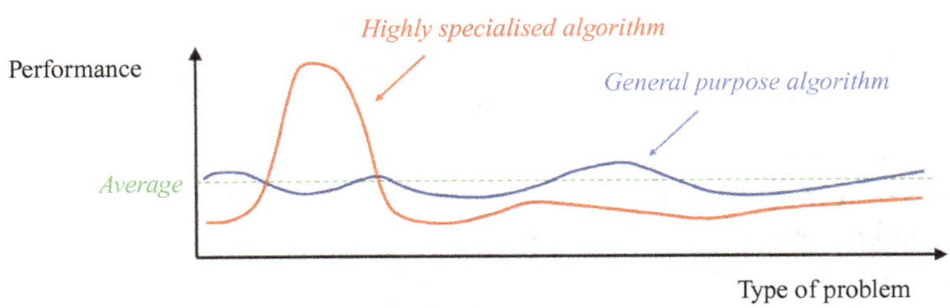

Figure 2.2 A graphical interpretation of the 'no free lunch' theorem

small sample of problems. There are several consequences worth mentioning. For example, it is often possible to show that a particular algorithm outperforms others in some specific application, even if this is not the case elsewhere, so it may not be fair to draw general conclusions based on a limited experience. It is also not surprising to see so many papers published with new algorithms put forward, especially those nature inspired, as nature has so much variety to offer.

From a practical point of view, the NFL theorems not only provide some guidance regarding how to go about selecting an approach which is 'fit for purpose' but also offer a general warning that good experience with a certain class of problems is no guarantee of satisfactory performance for other cases. Some well-established methods – such as genetic algorithms (GAs) and particle swarm optimisation (PSO), to name the two most popular ones – promise a safe choice and decent performance overall but may not necessarily be top best for a particular problem under investigation. Nonetheless, general purpose software needs to satisfy a variety of customers tackling different tasks; it is therefore not surprising that GA and PSO have gained such popularity amongst the designers.

2.3 Surrogate modelling

A surrogate model is essentially an engineering approach applied when an objective cannot be measured or found directly or is computationally too expensive. As mentioned already, for most devices under scrutiny in this book, a 3D finite-element (or equivalent) software is necessary and currently available packages offer powerful tools for that purpose (see Sections 2.5 and 2.6). However, during optimisation process, the objective function needs to be evaluated multiple times – often running into thousands – and it is simply impractical to involve the numerical solution at each iteration. Instead, a surrogate is created that mimics the behaviour of the simulation model as closely as possible while being computationally cheaper. The process consists of the following stages:

- initial sampling to create the simplest possible start-up version of the surrogate (sometimes referred to as active learning);
- construction of the surrogate model;
- finding an optimum relying on the computationally efficient model;
- appraisal of the accuracy of the surrogate: this is normally done by using the original accurate model (e.g. the 3D finite-element software);
- adding the new point to the existing samples and repeating the previous process over and over again until a satisfactory solution has been found.

Popular surrogate modelling approaches include polynomial response surfaces, kriging (see Section 2.10), radial basis functions (RBFs), support vector machines, space mapping, artificial neural networks and Bayesian networks. Further methods recently explored are Fourier surrogate modelling and random decision forests. Depending on the type of surrogate used and the complexity of the problem, the process may converge on a local or global optimum, or perhaps none at all.

2.4 The concept of a robust design

Mathematically we are searching for the optimum (usually minimum) of the specified objective function (or functions in multi-objective optimisation) and once such an optimum has been found (within specified accuracy), we may think that we are 'home and dry'. Unfortunately, real life may be more complicated than that. Practical devices are manufactured typically within specified tolerances and the materials used often exhibit a variation of their characteristics (such as *B–H* curve or value of conductivity) so that every completed product may be somewhat different from its neighbour, even if made using the same batch of supplied components. The question arises to what extent the predicted performance – which has been established for some idealised 'perfect' specimen – is a representative for the whole family of manufactured products. This is addressed by the concept of a 'robust design'.

Consider the shape of the objective function in terms of one particular variable – as depicted by the red curve in Figure 2.3 – and let us assume that we search for the minimum of the specified objective. The global minimum is clearly marked, and it is a good one compared with what is otherwise available. However, the parameter of choice used for illustration may have some 'uncertainty', either because of the machining tolerance or it may be linked to the fact that a particular variable is related to a material property which may have some stochastic distribution of the values within a specified range. Either way, we may be faced with an unavoidable practicality that when the device is actually manufactured the value of the selected parameter may be anywhere within the orange band. It therefore transpires that although some of the samples will indeed exhibit superior performance, many will be nowhere near the expected

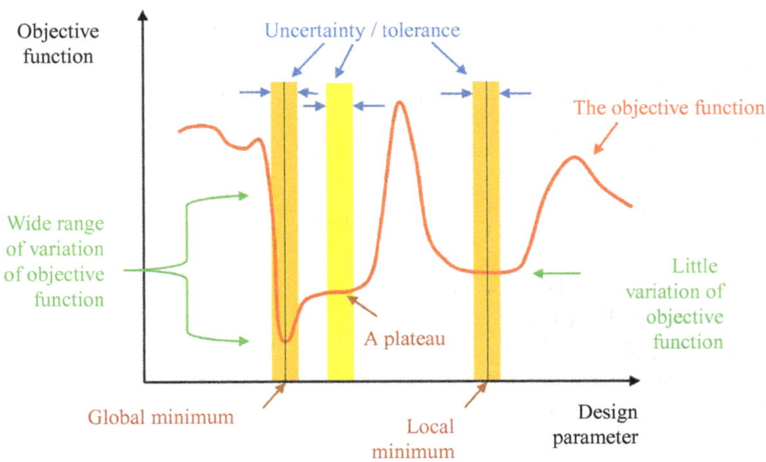

Figure 2.3 The illustration of the concept of the 'robust design'

specification; in other words, the design is not sufficiently 'robust'. As it happens in this example, the local minimum – even if appearing to be 'worse' for the idealised sample in the middle of the band – in reality ensures that all manufactured units perform up to the specified standard. In this particular case therefore the local minimum provides a more robust solution. To complicate things further, notice the plateau in between the two minima (with a yellow band); even if strictly speaking this is not an optimum, if the parameter was chosen to be as in the middle of the plateau, all units would perform well, in fact better than for the local minimum.

The conclusion from this short discussion is that if we apply the concept of a robust design, we not only need the values and locations of different minima, but also their shape in the vicinity of the optimum covering the whole uncertainty area. More information about this aspect may be found in [2].

2.5 The advances in computational electromagnetics

The reader is referred to [3,4] for a comprehensive treatment and extensive literature. This short section will attempt to provide a summary.

Computational electromagnetics (CEM) has come a long way over the past 30 years or so and most commercial packages offer the following functionality:

- *Pre- and post-processing*: fully interactive, a range of supported output devices, automatic and adaptive meshing;
- *Statics*: magneto- and electrostatic solutions for nonlinear, anisotropic and hysteretic materials, permanent magnets, laminated structures;
- *Quasi-statics*: steady-state (single frequency) analysis incorporating complex permeability, background dc fields, voltage-driven formulations;
- *Transient eddy currents*: full transient time simulations, non-linear;
- *Motional eddy currents*: motion-induced eddy-current analysis;
- *Stress and thermal*: mechanical stress or thermal analysis following from EM solutions;
- 2D, 2D axi-symmetric and 3D formulations.

The following is a subjective non-prioritised list of recent advances which have made the greatest impact on the CEM community [4]:

- higher order finite difference time domain (FDTD);
- multiple multipole technique (MMT);
- transmission line matrix (TLM);
- finite-element difference (FED);
- subspace projection extrapolation (SPE);
- finite integration technique (FIT);
- formulations in terms of differential geometry;
- implementation of edge and facet elements;
- improved anisotropy and hysteresis models;

- efficient application of continuum design sensitivity analysis (CDSA);
- multi-objective optimisation.

There are several packages on the market worth further exploring; here is a non-exhaustive list of the most popular software:

- OPERA by SIMULIA [5];
- MagNet by Siemens Digital Industries Software [6];
- Ansys [7];
- Altair Flux [8];
- Integrated Engineering Software [9];
- EFFE by Bathwick Electrical Design [10].

2.6 Computer-aided design

The computer-aided design (CAD) – assisted by CEM – has come of age in electromechanical and EM industry. There are risks, however, arising from sometimes inadequate understanding of relevant underlying theory or inability to appreciate the intricacies of numerical modelling. Thus the importance of *engineering judgement* cannot be overstated. To put it simply, the answer can only be as good as the model created for simulation. The software used cannot be held accountable for the suitability of the model: it is up to the designer to make this judgement. Thus, the designer preserves the central role in the process and takes full responsibility for the final result. But a well-constructed CAD system will assist by offering as much automation as feasible (and practical) to allow the designer to focus on the main design task at hand rather than overburdening himself with the details of how to run the software. Most of the packages available on the market provide such functionality and are effectively already self-contained CAD platforms; moreover, data and information exchange with other systems is often also available.

A typical CAD system for EMs is described in Figure 2.4. Whatever the actual simulation technique used (finite element, finite difference, boundary element, finite integration, etc.), the three familiar stages will be recognised: pre-processor, solver and post-processor. All geometrical data, material properties, sources and boundary conditions are specified either by the user or through transfer from another system (such as AutoCAD). The solver is normally fully automatic (although the user often has the ability to control some of the solution parameters) and inevitably solves a large system of algebraic equations. Finally, the results need to be scrutinised, important global parameters established (such as forces, stored energy) and local and global errors examined. But first the designer needs to create a model, which is then discretised depending on the method used, and after the simulation is finished the results must be assessed and interpreted. There may be a need to revise the model and repeat the process. This system then provides the basic component of the optimisation loop where it works together with a surrogate model (or alternative models) in the search for an optimum.

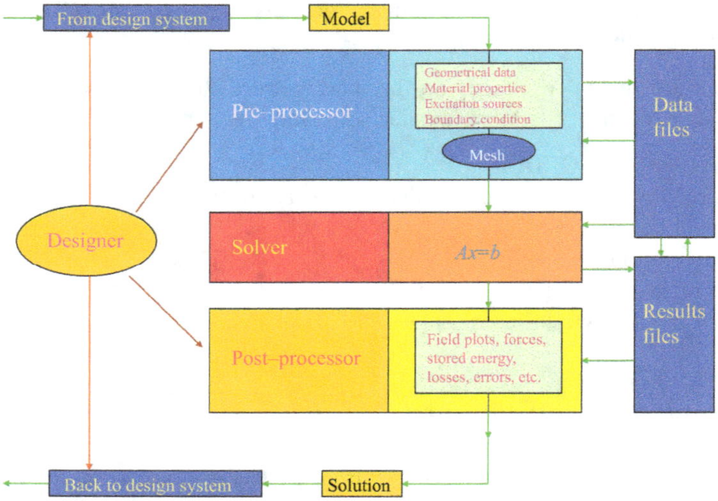

Figure 2.4 A typical CAD system for electromagnetics

2.7 Single- and multi-objective optimisation

A single-objective optimisation problem (SOOP) can be formulated mathematically as

$$\text{Minimise} : f(\mathbf{x}) \tag{2.1}$$

$$\text{Subject to} : g_i(\mathbf{x}) \leq 0, \quad i = 1, m \tag{2.2}$$

$$h_j(\mathbf{x}) = 0, \quad j = 1, p \tag{2.3}$$

$$x_k^l \leq x_k \leq x_k^u, \quad k = 1, n. \tag{2.4}$$

where $f(\mathbf{x})$ is the objective function or goal (usually to be minimised but could be maximised); $g_i(\mathbf{x})$ is the ith inequality constraint; m being the total number of inequality constraints; $h_j(\mathbf{x})$ is the jth equality constraint; p being the total number of equality constraints; $\mathbf{x}$ is the vector of design variables, where $\mathbf{x} = [x_1, x_2, \ldots, x_n]^T$; n being the total number of design variables; finally x_k^l and x_k^u are the lower (l) and upper (u) bounds of the kth design variable x_k, respectively.

Methods to solve SOOP problems are generally classified as deterministic or stochastic. A deterministic algorithm will always follow the same route and return the same answer, whereas a stochastic process will choose a different path every time and is likely to end up with a different outcome on each attempt. Generally, deterministic techniques can only guarantee to find a local optimum (depending on the starting point), whereas stochastic processes promise a better chance of getting a global optimum (but no guarantee). However, the former are usually very fast and converge to the optimum quickly (even if only a local one), whereas the latter may be computationally inefficient (though overall delivering a higher rate of success of finding the global optimum). There are many solvers on offer, some requiring the calculation of the derivates, some not, these may be easily found via Google or as built in procedures (add-on modules) of popular software, such as COMSOL [11].

We will be arguing later (see Section 2.10) that *kriging* is a particularly effective approach to optimisation as it considers the probability (likelihood) of getting closer to the global optimum when the next point for evaluation is selected. So it may be useful to list straight away the most promising criteria. Using *kriging surrogate models* in SOOP, we attempt to evaluate the design vectors to maximise:

- the probability of improvement;
- the expectation value of the improvement (EI);
- the generalised expected improvement;
- the weighted expected improvement (WEI);
- the credibility of a hypothesis about the location of the minimum;
- the 'minimiser entropy' criterion.

Multi-objective optimisation problems (MOOP) are nowadays of particular interest and involve several objective functions to be optimised concurrently. For a *nontrivial* problem, no single solution exists that simultaneously optimises each objective. In that case, the objective functions are said to be conflicting, and there exist a (possibly infinite) number of acceptable solutions. A solution is called *nondominated*, or *Pareto optimal* (see Section 2.8), if none of the objective functions can be improved without degrading some of the other objective values. In mathematical terms, a MOOP can be formulated as

$$\text{Minimise} : \{f_1(\mathbf{x}), f_2(\mathbf{x}), \ldots, f_k(\mathbf{x})\} \tag{2.5}$$

$$\text{Subject to} : \mathbf{x} \in \mathbf{X} \tag{2.6}$$

where $k \geq 2$ is the number of objectives and $\mathbf{X}$ denotes the feasible set of decision vectors, typically defined by some constraint functions similar to SOOP. Two general strategies exist: the *non-scalarising methods* where each objective function is considered individually, and *scalarising methods* where the MOOP is first converted into SOOP and then solved using a SOOP algorithm. There are many methods available for the conversion:

- ε-constraint (ε-C);
- weighting method (W);
- weighted metrics (including the Tchebycheff metric) method;
- achievement scalarising function approach;
- lexicographic ordering approach;
- value function method.

Given the **large number** of possibilities of transforming a MOOP into a SOOP, and a **large number** of selection criteria for solving a SOOP, there exist a **huge number** (large$\times$large) of combinations for solving MOOPs; it will therefore come as no surprise that so many papers are published putting forward new possibilities.

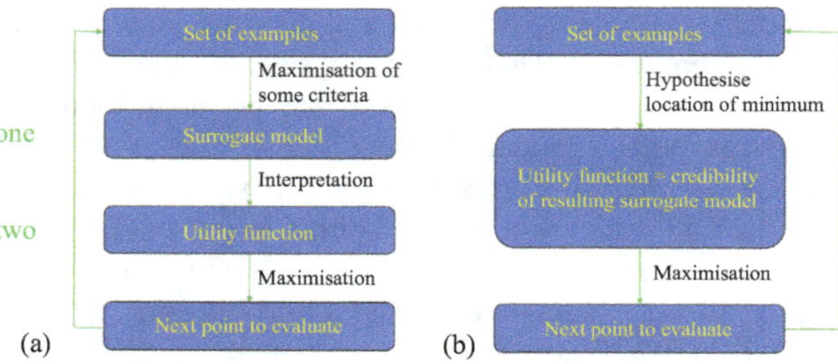

Figure 2.5 (a) Two-stage optimisation and (b) one-stage optimisation

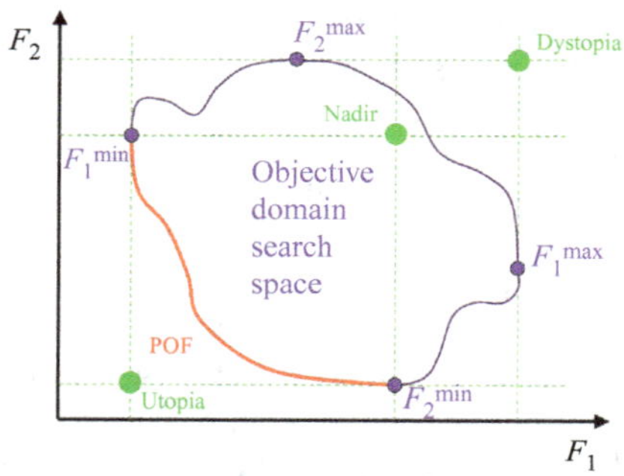

Figure 2.6 Pareto optimisation. POF, Pareto optimal front

We finish this section by providing simple flow charts to illustrate the principles of two- and one-stage optimisation approaches (Figure 2.5). In the former, we create a surrogate model, followed by a selection of an utility function in order to identify a suitable point for evaluation; in the latter we create a hypothesis of where the location of the optimum might be and assess the credibility of our prediction.

2.8 Pareto optimisation

Pareto optimality is named after an Italian engineer, economist and philosopher Vilfredo Pareto (1848–1923), who noticed that many economic solutions helped some people while hurting others. He searched for solutions that helped some people without hurting anyone else; these are now called *Pareto improvements*. Consider a case depicted graphically in Figure 2.6 with two objectives F_1 and F_2, both of which we wish to minimise. Let the *objective domain search space* indicate the area containing all possible solutions to the problem within the imposed constraints, with the minima and maxima of the individual objectives clearly marked.

It transpires that it is not possible to achieve the minima of both objectives simultaneously as this particular solution (marked as the *Utopia* point) lies outside the allowable region. The *Pareto optimal front* (POF) can then be considered representing a family of possible solutions, with the final decision to be made subsequently based on the knowledge of the shape of the front. Without additional subjective preference information, all Pareto optimal solutions are considered equally good. In other words, there is no mathematical 'best' point along the Pareto front. However, as the final decision is ultimately required, it may reflect on the fact that one of the objectives is perhaps more important than the other; if not, the chosen design is often selected as the one on the POF the closest to the Utopia point but may be influenced by other considerations.

2.9 Balancing exploitation and exploration

The stochastic optimisation process may be perceived as a decision-making activity where at each iteration we need to identify the next point for evaluation. We have two conflicting goals: to *explore* the search space hoping (or expecting), there might be a better solution elsewhere, while at the same to *exploit* the knowledge already accumulated in case the optimum is nearby. As the shape of the objective function (often in multidimensional space) is not known *a priori*, there is no definite way of making the judgement and we need to rely on some predictions accepting that there is a trade-off and thus a careful balance has to be maintained between exploitation and exploration. We shall see in the next section how in practice the balance may be controlled in the context of *kriging* techniques, but as achieving a good balance is crucial for attaining efficiency of computation the concept has been emphasised by allocating it a prominent place in this chapter, even if only one short paragraph.

2.10 Kriging techniques

Kriging – as a kind of a regression model – is able to predict response surface of the objective function through exploiting a spatial correlation of data and search for the global optimum using limited information [12–14]. As already mentioned in Section 2.7, kriging considers the probability (likelihood) of getting closer to the global optimum when the next point for evaluation is selected. As we will show later, this leads to very efficient algorithms and considerable savings in computing times.

Kriging is defined as

$$\widehat{y}(x^i) = \sum_{k=1}^{m} \beta_k f_k(x^i) + \varepsilon(x^i) \tag{2.7}$$

where the sum $\sum_{k=1}^{m} \beta_k f_k(x^i)$ – as a linear combination of the values of initial sampled points x^i – may be viewed as a global approximation of the true function, and the coefficients β_k are regression parameters and ε denotes an additive

Gaussian noise, related to uncertainty. Two design vectors x^i and x^j, close to each other in the design space, are assumed to have their errors correlated:

$$R\{\varepsilon(x^i)\cdot\varepsilon(x^j)\} = \prod_{k=1}^{n} e^{-\theta_k|x_k^i - x_k^j|^{Pk}} \tag{2.8}$$

where θ_k controls how fast the correlation between design vectors fades away along the kth coordinate, while p_k defines the smoothness of the function in this direction. The prediction made by the kriging surrogate model can be regarded as a Gaussian process γ. For a single objective to be minimised, the improvement may be assessed by comparing the value accomplished by the objective function with the current minimum of the prediction, which may be written as

$$I(x) = \max\{f_{\min} - \gamma(x), 0\} \tag{2.9}$$

Following [12,15], the expected improvement (EI) may be found by integrating over the likelihood of achieving it. The maximum of EI specifies the position where the new sampling point should be located. The EI and the standard error are defined as

$$EI\{I(x)\} = \begin{cases} (f_{\min} - \widehat{y}(x))\Phi\left(\dfrac{f_{\min} - \widehat{y}(x)}{s(x)}\right) + s(x)\phi\left(\dfrac{f_{\min} - \widehat{y}(x)}{s(x)}\right) & s(x) > 0 \\ 0 & s(x) = 0 \end{cases} \tag{2.10}$$

$$s(x) = \sqrt{\widehat{\sigma}^2\left[1 - r^{\mathrm{T}}R^{-1}r + \frac{(1 - r^{\mathrm{T}}R^{-1}r)}{1^{\mathrm{T}}R^{-1}1}\right]} \tag{2.11}$$

where $\widehat{y}(x)$ is the kriging prediction of the objective function; $f_{\min}$ the minimum of y for the current samples; $s(x)$ the *root-mean-square (rms) error* produced by kriging; R are correlation matrices between x and the existing sampling points; r indicates correlation between x and the other unknown points; finally, Φ and ϕ are the normal density and normal distribution functions, respectively. To be interpolating, the Gaussian distribution $\varepsilon(x^i)$ in (2.7) must be $N(0, \sigma^2)$, with σ^2 to be determined.

Many versions of the kriging approach have been developed. For example, in [16], a modification to (2.10) was put forward by adding *weights* to the two terms on the right hand side, thus

$$WEI\{I(x)\} = \begin{cases} w(f_{\min} - \widehat{y}(x))\Phi\left(\dfrac{f_{\min} - \widehat{y}(x)}{s(x)}\right) + (1-w)s(x)\phi\left(\dfrac{f_{\min} - \widehat{y}(x)}{s(x)}\right) & s(x) > 0 \\ 0 & s(x) = 0 \end{cases} \tag{2.12}$$

where w can be adjusted (either automatically or by the user). It is interesting to note that the first term on the right-hand side of both (2.10) and (2.12) favours searching the promising regions (hence *exploitation*), whereas the second term puts emphasis on searching regions with high uncertainty (*exploration*). By introducing the weight w, we therefore provide some level of control over the all-important balance between exploitation and exploration, as explained in Section 2.9.

2.11 Numerical experiments using test functions

Consider a well-known Schwefel test faction defined as

$$f(x) = \sum_{i=1}^{d} -x_i \sin\left(\sqrt{|x_i|}\right) \tag{2.13}$$

popular for testing optimisation algorithms. The one-dimensional version is graphically illustrated in Figure 2.7(a). It has one global minimum and several local ones. In order to kick off the process we need some initial sampling points which allow the kriging algorithm to make its first prediction, which at this stage is clearly not very good yet. By applying (2.10), we can estimate the EI and then decide to insert a new sampling point where the maximum of EI occurs.

This process can now be applied repetitively; each time a new point is added the kriging prediction improves. After 11 iterations the global minimum is found. Some intermediate stages and the final solution are illustrated by Figure 2.7(b)–(e). Finally, a version using the WEI is also shown in Figure 2.7(f). By a careful selection of the weight (and thus by changing the emphasis on exploration and exploitation), it was possible to reduce the total number of iterations.

The second test example uses a two-dimensional version of a function

$$f(x) = 10 - \sum_{i=1}^{n} \left[\frac{3.5}{1 + (x_i - 5)^2} + \frac{2.2}{1 + (x_i - 15)^2/10} + \frac{1.2}{1 + (x_i - 25)^2/30} \right] \tag{2.14}$$

which is graphically depicted in Figure 2.8. There is a definite global minimum as indicated. A slightly modified version of kriging has been used here called adaptive WEI [16].

Figure 2.9 shows snapshots of the kriging prediction, as well as the distribution of the error, as the iterations progress. The global minimum has been found; moreover, the shape of the function is simulated quite accurately, an important point in relation to our previous discussion about the 'robust optimum'.

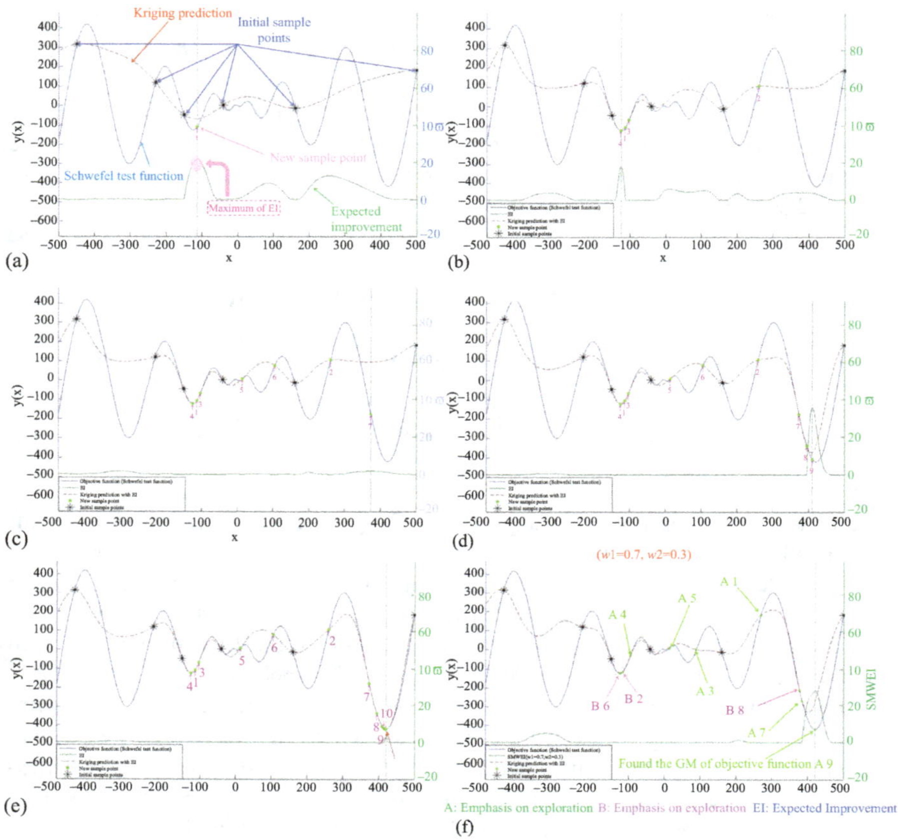

Figure 2.7 The expected improvement kriging applied to a Schwefel test faction: (a) the function and first prediction, (b) after 4 iterations, (c) after 6 iterations, (d) after 9 iterations, (e) after 11 iterations (final solution), (f) the weighted expected improvement kriging with different weights

2.12 An engineering example

The testing electromagnetic analysis method (TEAM) 22 problem [17] describes a superconducting magnetic energy storage (SMES) device, attempted by many researchers in the context of different types of optimisation algorithms. This section summarises the discussion of [18].

There are two versions: the three-parameter and the eight-parameter problems. We initially focus on the smaller version, where all design variables are fixed except the radius, height and thickness of the outer coil (R_2, h_2 and d_2). The design

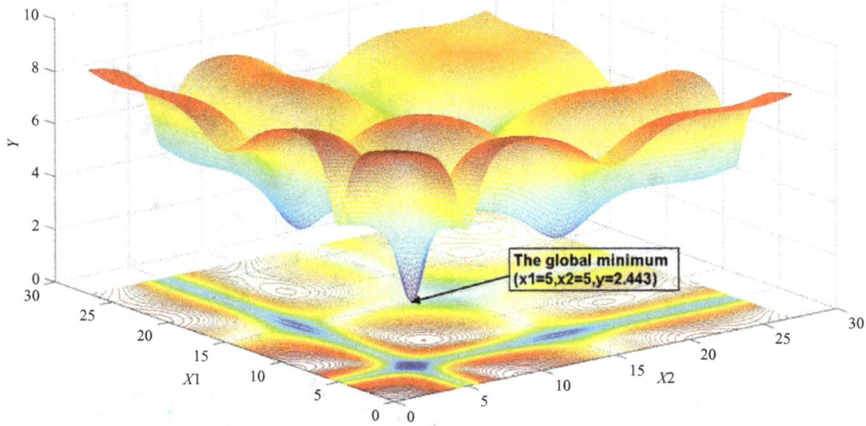

Figure 2.8 The two-dimensional test function

objective is to achieve a minimal stray field while the stored energy E must be as close as possible to 180 MJ. The objective function is specified as

$$OF = \frac{B^2_{stray}}{B^2_{norm}} + \frac{|E - E_{ref}|}{E_{ref}} \tag{2.15}$$

where $E_{ref} = 180$ MJ, $B_{norm} = 3\ \mu$T and B^2_{stray} is defined as

$$B^2_{stray} = \frac{\Sigma_{i=1}^{n=22} |B_{stray,i}|}{22} \tag{2.16}$$

subject to the following inequality constraint, known as the quench condition

$$|J| + 6.4|B| - 54.0 \leq 0 \tag{2.17}$$

where J (in A/mm^2) is the current density and B is the maximum magnetic flux density.

A comprehensive comparison was conducted using several general purpose popular methods (see Table 2.1), as well as three kriging algorithms: EI (see Section 2.10), adaptive WEI (AWEI) (see Section 2.11) and worst case expected improvement (WCEI) [18], a modified version of EI to cater for the robust optimisation version of the problem. A brief description of WCEI now follows.

Early attempts to account for design uncertainties can be attributed to G. Taguchi, the 'father of robust design' [19], who considered performance variations due to noise factors beyond the control of the designer and used the *design of experiments* [20] in order to evaluate different designs. The *worst case* problem

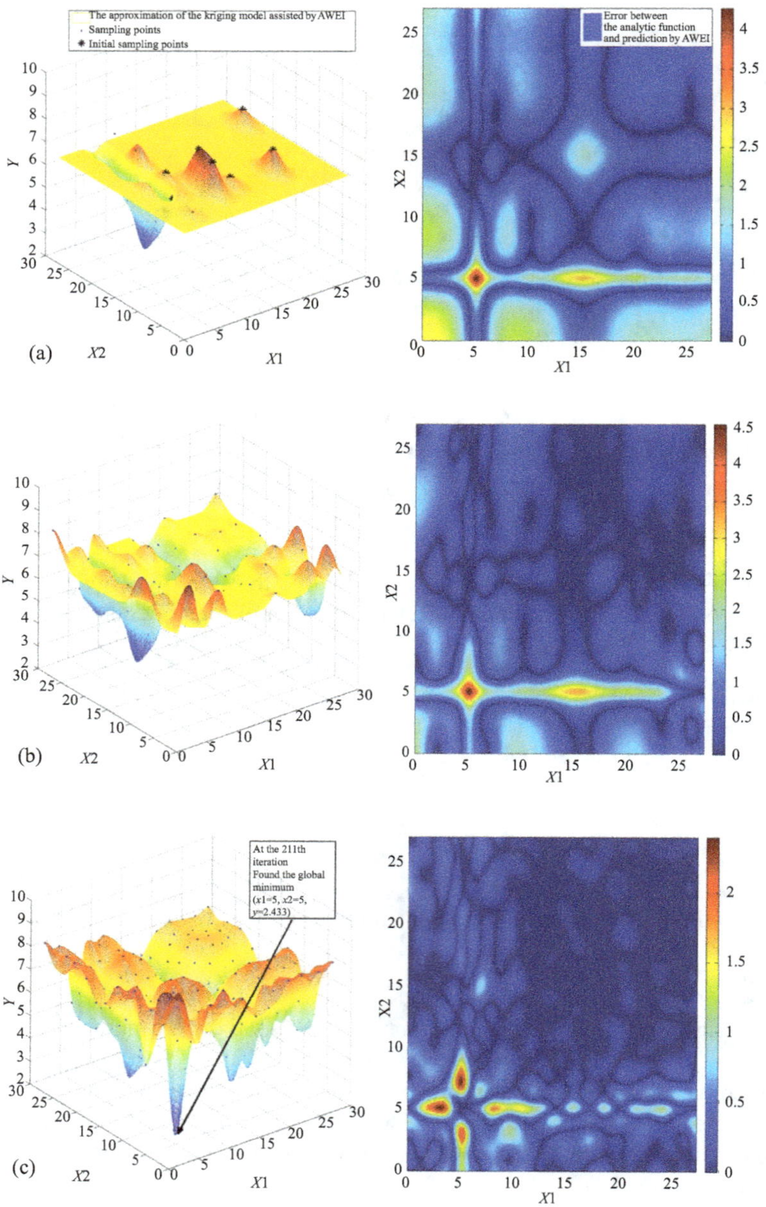

Figure 2.9 Kriging-assisted AWEI: (a) after 20 iterations (b) after 80 iterations, and (c) final solution after 211 iterations

Table 2.1 Performance comparison of various methods for TEAM 22 problem [17]

Algorithm	Optimal value	R_2	d_2	$h_2/2$	Number of function calls
GA	0.134	3.040	0.386	0.240	2,400
SA	0.098	3.078	0.390	0.237	5,025
TS	0.089	3.080	0.380	0.246	3,821
UTS	0.089	3.080	0.370	0.254	1,800
NTS	0.101	3.110	0.421	0.241	3,278
Kriging EI	0.088	3.090	0.394	0.236	211
Kriging AWEI	0.088	3.090	0.400	0.232	323
Kriging WCEI	0.146	3.021	0.391	0.250	277

GA, genetic algorithm; SA, simulated annealing; TU, Tabu search; UTS, universal Tabu search; NTU, new Tabu search; kriging EI (Section 2.10); kriging AWEI (Section 2.11); kriging WCEI (worst case expected improvement).

specification [21] considers the uncertainties so that the output function can be expressed by a simplified equation:

$$f = f(x + \varepsilon) \tag{2.18}$$

where $\varepsilon(-\xi, \xi)$, the distribution of uncertainty ε, is unknown, but the magnitude is bounded by a given range ξ. The algorithm proposed in [18] consists of two stages. First, the kriging model is updated by sequentially adding infill (new sampling) points at each iteration based on the WCEI by taking the minimal EI value within the worst case region of a particular design point:

$$WCEI(x) = \max\{\min\,[EI(x+\varepsilon], 0\} \tag{2.19}$$

$$x + \varepsilon \ \in \ X \tag{2.20}$$

where X is a set of points located within the worst case region of the point x. An extra layer of the problem is embedded within the WCEI: the new infill point will be located where the minimal EI around the target point is the largest. During model updating, the worst case estimation of the objective function is computed simultaneously based on the kriging model constructed using the existing design sites at that iteration. The second stage is triggered when the maximum WCEI within the design space drops below a predefined value or a limit of iterations has exceeded its allowance.

In stage two, the worst case region around the worst case optimum is exploited and validated, but instead of calculating the improvement, an expected *deterioration* is estimated to locate the maximal 'worsening' within the worst case region. Hence a modified process is now governed by

$$E[D(x)] = \begin{cases} (\widehat{y}(x) - y_{wc})\Phi(u(x)) + \widehat{s}\phi(u), & s > 0 \\ 0, & s = 0 \end{cases} \tag{2.21}$$

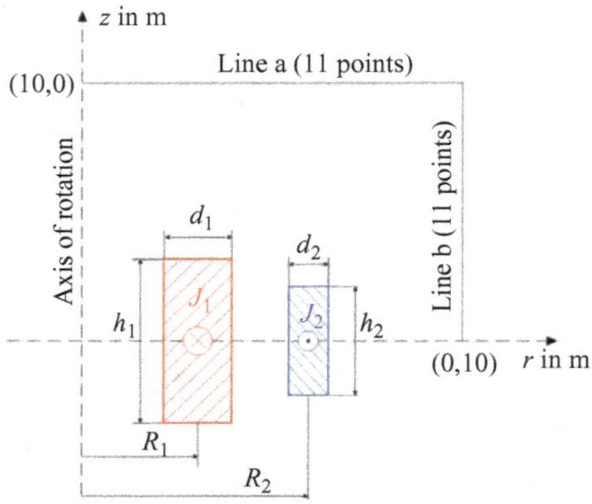

Figure 2.10 The superconducting magnetic energy storage device [17,18]

$$u = \frac{\hat{y}(x) - y_{\min}}{\hat{s}(x)} \tag{2.22}$$

Iterations continue until the value of the expected deterioration is zero or smaller than a predefined value and the location of the worst case estimated optimum is added as the next infill point, while the associated objective function is also evaluated. A more detailed explanation may be found in [18].

The kriging-assisted *WCEI* algorithm has been applied to the SMES device specified at the beginning of the this section (refer to Figure 2.10) while the uncertainty of each design parameter was assumed to be 1% of the respective design range (constraint). In Table 2.1, the performance of three kriging schemes explored in this chapter has been compared with several popular (off the shelf) algorithms, including GA, simulated annealing SA and three versions of Tabu search.

The optimal values obtained using all algorithms are comparable – even if strictly speaking the two kriging-assisted (EI and AWEI) answers are the best (closest to zero) – while the values of the three parameters are also reasonably similar. But the remarkable difference is related to the computational effort needed to locate the global minimum, with the two kriging algorithms requiring an order of magnitude fewer function calls. This is a spectacular saving bearing in mind that each function call requires full finite-element solution, which for this particular problem may not be very time-consuming, but needs to be executed thousands of times!

The bottom row in Table 2.1 (slightly shaded) refers to the robust optimum under the assumption of 1% uncertainty for each of the three design variables, as already mentioned. The optimal value may be somewhat worse (although not significantly), but the design is now robust. Obviously the 1% uncertainty is totally arbitrary and similar calculations could be performed assuming a different value, or even different values for each variable. But clearly the message is that kriging appears to be a very efficient approach to optimisation, including the robust version.

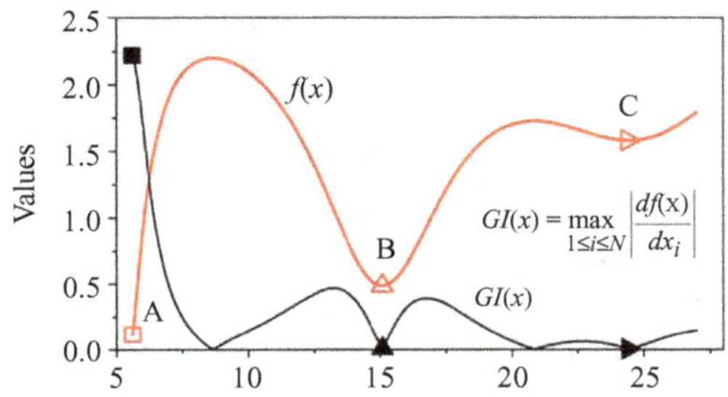

Figure 2.11 The single-variable analytical function and its gradient index

To conclude this section, it might be helpful to mention two other possible methods which could be used instead of the 'worst case' formulation in the context of robust optimisation, namely the gradient index (GI) and six-sigma quality (SSQ) [22]. For the former, the GI may be added as a second objective supplementing the normal minimisation of (2.1)–(2.4), that is

$$\min GI(x) = \max_{1\leq i\leq n}\left|\frac{\partial f(x)}{\partial_{xi}}\right| \tag{2.23}$$

A simple graphical illustration using a single-variable analytical function is shown in Figure 2.11. There are two local minima (B and C) and a global one (A). In the context of the robust design (Section 2.4), when considering uncertainty of the variables, the theoretically best point A may not be a practical optimal solution. The GI may then facilitate making the final choice. The final judgement will be influenced by the margin of the uncertainty and by how much the objective function may change when the design variables assume limiting values. For a minor departure of the parameter from its nominal value, the objective function may still be satisfactory and even a 'sharp' optimum like A may be judged to be acceptable. But under increased uncertainty, the confidence in the final performance may be assured only by a much more 'shallow' minimum, such as B, or even C.

As the size of the uncertainty is relevant, we can define the gradient sensitivity as the difference between the largest and the smallest GI within the uncertainty range. However, the assessment of average performance should also be carried out.

The latter of the mentioned methods, the *SSQ* approach, further improves the quality of the assessment of robustness and reduces computation times. With SSQ, the measure of dispersion is classified into six sigma (standard deviation) levels; the optimisation problem can be reformulated as

$$\min \ \mu_f \text{ and } \sigma_f \tag{2.24}$$

where σ_f is the standard deviation specifying the intensity of variation due to the uncertainty of variables and μ_f the mean value expressing the average performance within the uncertain range. Consider the one-dimensional case of the test function

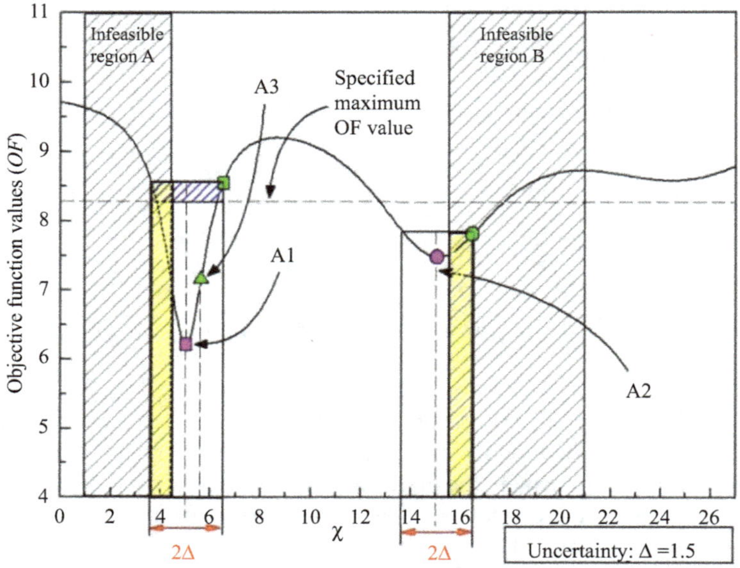

Figure 2.12 The objective function and the standard deviation

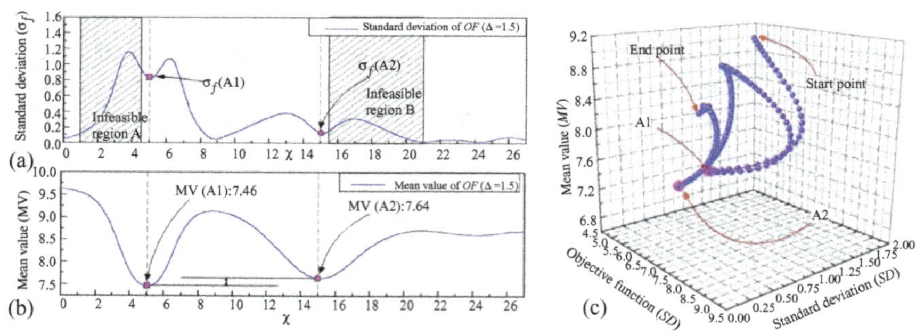

Figure 2.13 (a) The standard deviation, (b) the mean value and (c) the trajectory of the objective function of Figure 2.12

(2.14) from Section 2.11 shown graphically in Figure 2.12 with the corresponding standard deviation for the assumed uncertainty $(x) = 1.5$. In addition, it has been assumed that some constraints exist making certain regions infeasible.

The two parameters, σ_f and μ_f, for the test function (2.14) are shown in Figure 2.13(a) and (b), respectively. The trajectory of the objective function in terms of standard deviation and mean value is also plotted in Figure 2.13(c) with reference to characteristic points from Figure 2.12. The standard deviation of A2 at 0.0302 is less than the value for A1 at 0.1499 implying better robustness.

As reported in [22], linking either GI or SSQ with kriging makes the combination a very powerful and efficient tool.

2.13 A brief review of nature inspired algorithms

Some nature-inspired algorithms are well established and easily available as tools in general purpose software. For example, the original version of the GA was proposed as long ago as 1975. The PSO has been on the market for over 25 years. The last decade, however, has brought literally an explosion of publications reporting on new algorithms inspired by nature. Recent publications [23,24] would be a good starting point to survey what is currently available. The following is a short list of such new algorithms with somewhat 'exotic' titles:

- Artificial swarm intelligence;
- Bee colony optimisation;
- Ant colony optimisation;
- Bumblebees;
- Dragonfly algorithm;
- Firefly algorithm;
- Good lattice swarm optimisation;
- Grey wolf optimiser;
- Monkey search;
- Sheep shepherding algorithm;
- Whale optimisation algorithm;
- Brainstorm optimisation;
- Ecogeography-based optimisation;
- Flower pollination algorithm;
- Japanese tree frogs calling;
- Spotted hyena optimiser;
- Termite colony optimisation;
- Black hole;
- Colliding bodies optimisation;
- Fractal-based algorithm;
- Intelligent water drop;
- Mass and energy balances algorithm;
- Sine cosine algorithm;
- Sonar inspired optimisation;
- Vision correction algorithm;
- Water wave optimisation;
- Cohort intelligence;
- . . . and hundreds more!

Consistent with the *no free lunch theorem* (Section 2.2), every algorithm will perform well for some identified particular problem. It is probably fair to say, however, that on the whole, the general purpose methods – such as GA or

PSO – are likely to be a safe choice, especially that it is easy to find open-source codes for them.

For the benchmarking comparison of different algorithms, conclusions can be affected by the performance metrics used. Typically, comparison studies relate to the accuracy, computational effort, stability and success rate. Moreover, due to the statistical nature of nature-inspired algorithms, results are not really repeatable; hence multiple runs are needed in order to acquire meaningful statistics.

2.14 The challenge of large data sets

As argued in Sections 2.10–2.12, kriging surrogate modelling facilitates effective decision-making concerning selecting the next point for evaluation. This is particularly beneficial in the context of the design of EM devices where computationally time-consuming numerical field modelling simulation is used. The challenge, however, is that correlation matrices (mentioned in Section 2.10 in relation to (2.11)) are essential which, especially for cases with many design variables and multiple objectives, may grow in size slowing down of what otherwise should be a very fast process. Similar problems occur with other methods.

The question therefore arises how to prevent or alleviate this 'combinatorial explosion'. Following the suggestions put forward in [25], let us consider the eight-parameter version of the TEAM 22 problem, the smaller version of which was already discussed in Section 2.12. A large memory would be required to store the response surface produced by kriging, as well as the correlation matrices created during the prediction stage. The size of these matrices would be directly related to the number of variables; hence with eight variables and assuming very modest 10 steps for each variable vector, a 10^8-parameter set would need to be stored. Since the kriging predictor generates a response surface at each iteration – and is of the same size as the task itself (10^8) – a computer with 9-GB available memory would already be challenged. The schemes suggested in [16,25] include matrix partitioning and sub-matrix duplication, as well as a 'zoom in' strategy. The latter has been tried with the three-parameter TEAM 22 SMES problem (Section 2.12 and Figure 2.10) using two versions of kriging, and Figure 2.14 shows some details of the distribution of the sampling points. For kriging with EI (Section 2.10), 79 points were created in the first stage, with further 132 points in the zoomed-in region; in the AWEI case (Section 2.11), 156 sampling points were followed by 167 points in the second stage.

Finally, let us show the final comparison table for the challenging eight-parameter TEAM 22 problem, as detailed in Table 2.2.

Consistent with the previous findings, kriging shows to be extremely efficient. The conjugate gradient method (CGM) also performs cost-effectively and provides an accurate answer, but as it is a direct method it cannot guarantee finding a global optimum. Obviously, no method can provide such a guarantee, but stochastic methods in general, and kriging-assisted methodologies in particular, offer a high

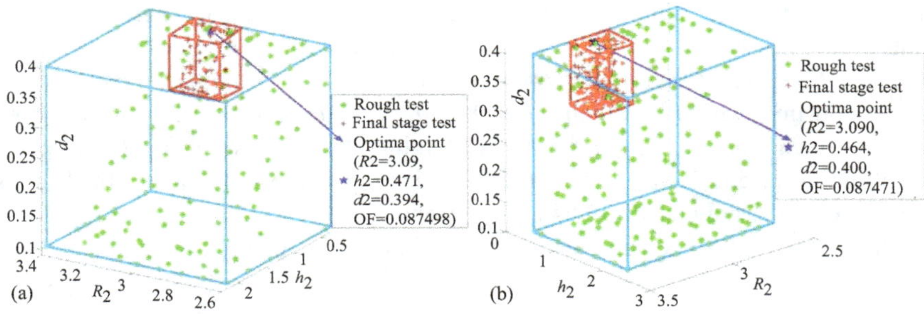

Figure 2.14 Illustration of the 'zoom-in' technique applied to the 3-paramters TEAM 22 problem: (a) EI, (b) AWEI

Table 2.2 Performance comparison for the eight-parameter TEAM 22 problem [25]

Algorithm	R_1	R_2	h_1	h_2	d_1	d_2	J_1	J_2	**Objective function**	**Number of FEM calls**
PSO	1.00	2.26	1.11	1.78	0.5223	0.344	28.18	−5.49	2.3363	~6,000
Q-PSO	1.29	2.61	1.08	2.27	0.397	0.204	30.00	−21.3	0.5735	~6,000
E-QPSO	1.00	1.80	2.06	3.60	0.516	0.285	20.00	−6.36	1.1730	~6,000
GSA	1.94	2.82	1.13	1.10	0.399	0.195	22.50	−22.5	0.0052	17,150
ES	1.99	2.93	1.92	0.94	0.290	0.188	26.60	−26.6	0.0049	4,200
SAA	1.69	2.91	1.61	0.88	0.323	0.207	20.90	−20.9	0.0110	14,000
CGM	1.84	2.76	1.18	1.00	0.395	0.214	22.50	−22.5	0.0248	200
Kriging	1.00	1.80	1.56	1.39	0.400	0.150	30.00	−30.0	0.0536	449
Benchmark	1.20	1.80	2.18	3.03	0.583	0.195	16.96	−18.9	0.0018	–

PSO, particle swarm optimisation; Q-PSO, quantum-behaved particle swarm optimisation; E-QPSO QPSO, using the exponential probability distribution; GSA, global search algorithm; ES, evolution strategy; SAA, simulated annealing algorithm; CGM, conjugate gradient method.

probability of success. In terms of the challenge of the large data sets, the method described in Section 2.15 is of particular interest.

2.15 Points aggregation techniques

In simple words, the points aggregation technique locates the most interesting search area for the next infill (new sample) point, aggregates points outside this area and finally constructs a kriging model for infill point search within the identified area. First, a 'centre' is located, around which a prescribed number of 'interior' points will not be aggregated and inside which the next infill point will be added. Without going into the details, a particular scheme for selecting such a centre and its interior is fully described in [26]. The goal then is to aggregate design points located further from the centre into a smaller number of nodes (the 'knots'), which can be easier stored in the computer memory. The problem of outside points aggregation involves *hierarchical cluster analysis* [27] and a single variable optimisation. Each cluster's centroid may be weighted by a Gaussian function and the weight in this case is obtained from the Gaussian correlation function in the kriging model itself. A graphical illustration is presented in Figure 2.15, where (a) shows the clustering without Gaussian weights, while (b) depicts the clustering with Gaussian weight terms applied. The test case involved 500 observations and assumed that the memory could build a kriging model consisting of up to 250 design points. We specified that 40% of the memory would be used to store the interior points while the remaining 60% used to store nodes related to outside points. The 400 points outside the centre area had been aggregated into 150 nodes.

The point aggregation technique is illustrated using a test function:

$$\begin{aligned} f(x_j) = c + a\prod_{j=1}^{j=2} \cos\{w(x_j - p_j)\}e^{-\{w(x_j-p_j)\}^m} \\ + b\prod_{j=1}^{j=2} \sin\{e^{-\{v(x_j-q_j)\}^k}\} + f_{wave} \end{aligned} \tag{2.25}$$

where $c = 4.5$, $a = 3.5$, $b = 2.8$, $w = 8.4$, $v = 6.4$, $m = 2$, $k = 6$, $p_1 = 0.2$, $p_2 = 0.3$, $q_1 = 6.5$, $q_2 = 8$, and f_{wave} is an interpolation function of a set of randomly generated points.

The kriging model in Figure 2.16(a) was constructed using 100 design points, while the model in (b) contains 60 nodes, including 20 points inside the centre area and 40 nodes outside. In this case, 62 original points have been aggregated. A slight loss of information has resulted – mainly away from the important region – but, as the size of the correlation matrix depends on the square of the number of points, the memory requirements have been reduced by 64%.

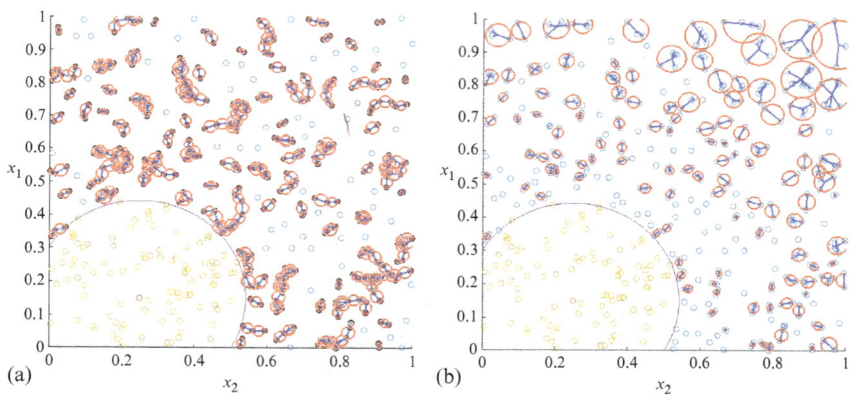

Figure 2.15 Clustering (a) without and (b) with Gaussian weight functions

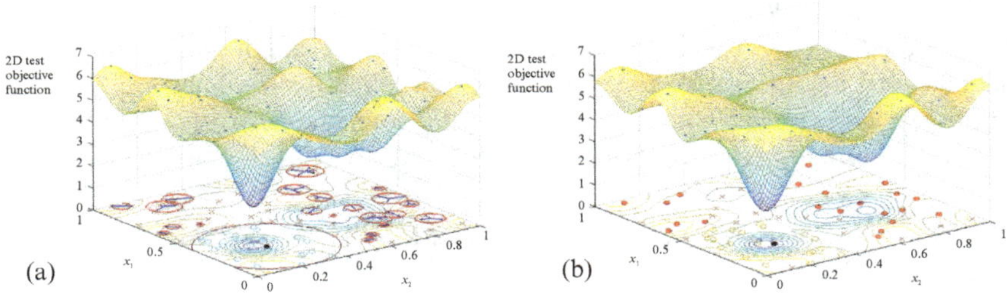

Figure 2.16 Kriging estimate of a 2D test function (a) using original 100 design points and (b) after point aggregation

Similar savings for the eight-parameter version of the TEAM 22 problem have been reported in [26] suggesting that the points aggregation technique can alleviate noticeably some of the memory storage problems with little or no loss of accuracy.

2.16 Design sensitivity aspect

Sensitivity analysis determines how different values of independent variables affect dependent variables (outcomes) under a given set of assumptions. In other words, sensitivity analysis measures how the impact of uncertainties of one or more input variables can lead to uncertainties in the output. There is even a toolbox in MATLAB® called SAFE (sensitivity analysis for everybody) [28]. Global sensitivity analysis – used by SAFE – is a term describing a set of mathematical techniques to examine how the variation in the output of a numerical model can be attributed to variations of its inputs. Freely available global sensitivity analysis tools include the repository of MATLAB and Fortran functions maintained by the Joint Research Centre, the Sensitivity Analysis package for the R environment, the GUI-HDMR MATLAB package, the C++ based PSUADE software and the Python Sensitivity Analysis Library SALib – all relevant links may be found in [28].

In the context of optimisation there are two particular aspects worth mentioning, the first one concerned with the philosophy of the design process. In engineering practice, it is often the case that the designer is not necessarily too bothered if a global (or even local – if sufficiently good) optimum has been found, as long as there is *improvement* to the existing design. But in order to recognise what improvements are possible, if any, a study must be conducted to identify the design parameters worth pursuing to achieve the expected performance advancement. Several partial derivatives may need to be estimated; an interesting example of this approach may be found in [29] with the aim to establish a parametric EM design environment.

Sensitivity analysis may also be incorporated directly into the optimisation process. An example of such an approach may be found in [30] where the concept of CDSA – first introduced in the 1980s – has been implemented in combination with standard EM software. CDSA provides a methodology where the EM analysis is treated as a 'black box', i.e. the internal details of the EM analysis system are irrelevant, as the CDSA method only requires the inputs and outputs of the system. This is a very interesting feature of CDSA providing a simple interface even with commercial software with no need to access the source code.

To derive the sensitivity formula, an objective function F is expressed as

$$\bar{F} = \int_{\Omega} g(\boldsymbol{A}, \text{curl}\,\boldsymbol{A})d\Omega + \int_{\Omega} \boldsymbol{\lambda}[-\text{curl}(\nu\,\text{curl}\,\boldsymbol{A} - \boldsymbol{M}) + \boldsymbol{J}]d\Omega \tag{2.26}$$

where g denotes a scalar function differentiable with respect to the magnetic vector potential, $\boldsymbol{A}$ and curl $\boldsymbol{A}$, which are in turn implicit functions of the design variable vector $\boldsymbol{p}$, ν is the reluctivity, $\boldsymbol{M}$ the permanent magnetisation, $\boldsymbol{J}$ the current density and $\boldsymbol{\lambda}$ means the Lagrange multiplier vector interpreted as the adjoint variable. The process is then applied to topology optimisation (TO) and shape optimum design (SOD) – incidentally involving some rather complicated mathematical expansions – and results in the following expressions for the sensitivities:

$$\frac{d\bar{F}}{d\boldsymbol{p}} = \int_{\gamma} [(\nu_1 - \nu_2)\ \text{curl}\,\boldsymbol{A}_1 \cdot\ \text{curl}\,\boldsymbol{\lambda}_2 + (\boldsymbol{J}_2 - \boldsymbol{J}_1) \cdot \boldsymbol{\lambda}_2 + (\boldsymbol{M}_2 - \boldsymbol{M}_1) \cdot\ \text{curl}\,\boldsymbol{\lambda}_2]\boldsymbol{V}_n\, d\Gamma \tag{2.27}$$

for SOD (involving surface integration along the moveable boundary γ between two different materials), where $\boldsymbol{p}$ is a design variable vector relevant to the geometry of the design model, and $\boldsymbol{V}_n$ is a design velocity vector normal to γ. And

$$\frac{d\bar{F}}{d\boldsymbol{p}} = \int_{\Omega} \left[\left(-\frac{\partial \nu}{\partial \boldsymbol{p}}\right)\ \text{curl}\,\boldsymbol{A} \cdot\ \text{curl}\,\boldsymbol{\lambda} + \frac{\partial \boldsymbol{J}}{\partial \boldsymbol{p}} \cdot \boldsymbol{\lambda} + \frac{\partial \boldsymbol{M}}{\partial \boldsymbol{p}} \cdot\ \text{curl}\,\boldsymbol{\lambda}\right] d\Omega \tag{2.28}$$

for TO (where volume integral needs to be performed). Here, $\boldsymbol{p}$ is composed of permeability μ, current density $\boldsymbol{J}$ and permanent magnetisation $\boldsymbol{M}$ assigned to design regions of interest. Using these analytical formulae, the first-order gradients of an objective function with respect to the design variables can be easily established from the field and potential values obtained in the dual solution system consisting of the primary and adjoint systems; this can be done without any amendments to the finite-element software source code. The iterative process involves creating an FE model of the primary system, solving it, finding the objective function, calculating the adjoint system forcing terms (from the primary system), solving the adjoint system and computing the sensitivities and the objective function. These steps are repeated iteratively until the objective function converges or the solution is adjudged to be satisfactory. A simple flow chart describing the process is shown in Figure 2.17.

A remarkable feature of the CDSA formulation – as reported and illustrated through engineering examples in [30] – is the incredibly fast rate of convergence of the iterations, attributed to the fact that computing times do not depend on the number of design variables. A small price to pay, however, is that two solutions are required (of the primary and of the adjoint system).

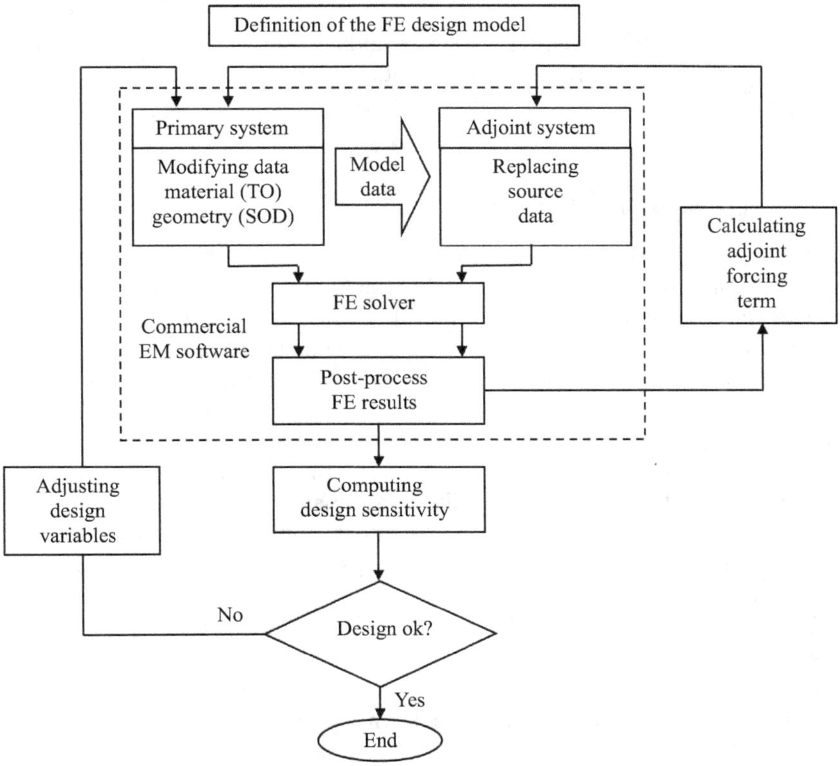

Figure 2.17 Flow chart of the CDSA implementation for SOD and TO problems

2.17 Future trends

We have now reviewed the state of the art – albeit somewhat superficially – and the question arises where do we go next! The methods and techniques reviewed continue to be areas of intensive research and new ideas and improved algorithms emerge all the time. As argued before, there is rich body of literature as it is always possible to find an example where a particular scheme works better than others (refer to the *no free lunch* theorem of Section 2.2), while incremental enhancements are also very welcome. But what is the next possible 'big step' (or steps)? Watching the current developments – and having been involved in the relevant research myself for some years now – I have identified four areas which in my view will influence the art of design optimisation. The following suggestions are subjective, but recent publications appear to suggest that there are new milestones on the horizon.

As an introduction to this 'gazing into future' activity, it might be helpful to recall our discussion from Section 2.1 and the concept of the hierarchical design.

We already mentioned that for this design pyramid to work efficiently, a smooth transition is needed between different stages and models. But some models are very simple (e.g. equivalent circuits) whereas others are complicated (e.g. finite elements). Can they somehow be connected? Here are some suggestions how to proceed.

2.17.1 Model order reduction

Several definitions of the model order reduction (MOR) exist depending on the context. Originally, MOR was developed in the field of control theory with the aim to reduce the complexity of the properties of dynamical systems while preserving – as much as possible – their input–output behaviour. Nowadays, MOR is a flourishing area of research, both in control theory and in numerical analysis.

The MOR attempts to protect the essential features of a structure or a process. Thus, even in a simplified version, the most basic properties of the original model must already be present. Figure 2.18 illustrates the concept in a graphical way, demonstrating that sometimes very little information is needed to capture the essential features. This example of the 'Stanford Bunny' shows that – even with only a few facets – the rabbit can still be recognised. This example was constructed for an entirely different purpose and does not contain any reference to the way MOR is performed mathematically; however, it can still be used to explain in an easy-to-understand way what MOR is about.

Let us suppose that a certain quantity, calculated using a discretised numerical model such as finite elements, varies as shown in Figure 2.19(a). An approximated version of this variation is suggested in (b), which clearly misses some of the details and subtleties of the accurate solution but nevertheless preserves the main features of the function behaviour. Importantly, this has been achieved with an order of magnitude fewer elements of the relevant mesh (or grid), resulting in very significant savings in computational effort required should the reduced model be used. Thus the benefits are clear, although this example still does not explain what steps need to be taken to arrive at a simplified but still acceptable model.

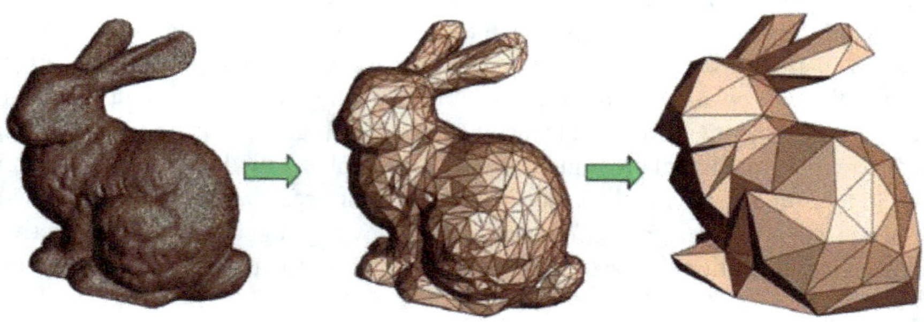

Figure 2.18 Graphical illustration of model order reduction

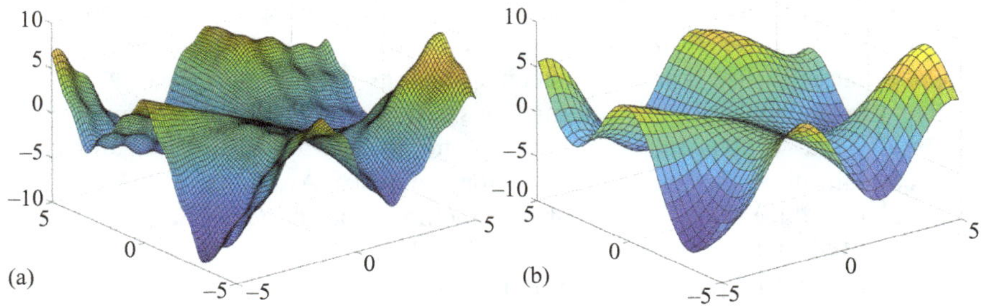

Figure 2.19 Function variation: (a) full model, (b) reduced model

Figure 2.20 An image processing example of the model order reduction; (a) original picture, (b) to (f) stages of simplification

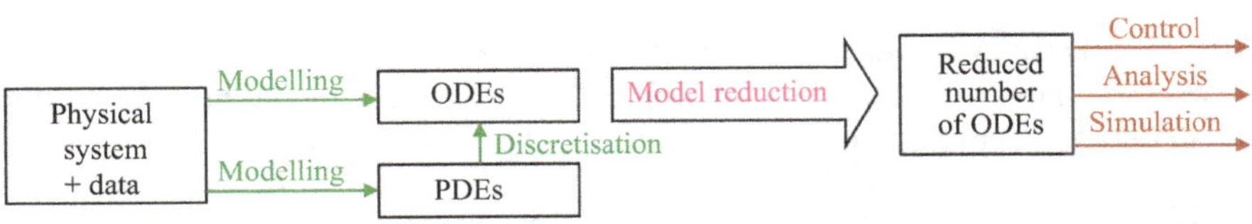

Figure 2.21 Model order reduction concept. ODE, ordinary differential equation; PDE, partial differential equation

A very revealing case from image processing is shown in Figure 2.20. The original picture (a) is gradually simplified, initially with a little loss of the detail, but (d) already shows signs of deterioration, while (e) and (f) are clearly no longer very accurate, although both still offer recognisable images. But the amount of data in (f) is significantly less than in (a).

Finally, the concept of the MOR may be summarised using a simple flow chart of Figure 2.21. The original physical system is typically modelled and discretised (using finite elements, finite differences, boundary elements or related method), resulting in a normally very large number of degrees of freedom (DOFs). The MOR process then creates a smaller and simpler version with a much reduced number of DOFs, which requires significantly less computational effort. The challenge, however, is to preserve the primary features of the solution, while accepting reduced accuracy when it comes to details.

The fundamental mathematical methods underpinning MOR were published in the 1980s and 1990s, including asymptotic waveform evaluation, truncated balanced realisation, Hankel-norm reduction, Laguerre, Arnoldi and PRIMA, Padé via Lanczos, reduced basis, Krylov-based methods and proper orthogonal decomposition (POD). A helpful review may be found in [31].

The MOR techniques have only fairly recently entered the field of CEM, but immediately with great impact. Of particular, relevance and interest will be publications [32–36], although several other authors have also tackled the problem. Although this is still relatively early-stage research, the findings are encouraging.

Following the argument introduced in [32] and retaining the original notation, a finite-element model (FEM) may be formulated for a parameterised linear magnetostatic problem on a domain $D(\mathbf{p})$ resulting in a linear system of equations:

$$\mathbf{S}(\mathbf{p})\mathbf{X}(\mathbf{p}) = \mathbf{F}(\mathbf{p}) \tag{2.29}$$

where $\mathbf{p} = (p_1, \ldots, p_M)$ denotes the set of M input parameters (currents, material characteristics, etc.), $\mathbf{S(p)}$ is the $N\times N$ stiffness matrix, $\mathbf{F(p)}$ is the $N\times 1$ source vector, $\mathbf{X(p)}$ is the $N\times 1$ vector of unknowns and N represents the number of DOF. The solution $\mathbf{X(p)}$ makes it possible to determine the field distribution and also the quantities of interest, which are often quadratic functions of $\mathbf{X(p)}$ (such as energy or force). If we designate G as the quantity of interest (being a quadratic function of $\mathbf{X}$ $(\mathbf{p})$), we can write

$$G = \mathbf{X}(\mathbf{p})^t\mathbf{G}(\mathbf{p})\mathbf{X}(\mathbf{p}) \tag{2.30}$$

where $\mathbf{G(p)}$ is an $N\times N$ matrix. If formulations (2.29) and (2.30) are used in a design process, the number of required solutions of the FE model can be high, especially for a large number of parameters M. To lessen the computational burden, either the number of calls could be limited (but this would lead to unacceptable loss of accuracy), or the size of the linear system (2.29) should be reduced using an MOR method. Of the many options, let us explore the POD methodology.

Consider the Z-parameter set realisations $(\mathbf{p}^1, \ldots, \mathbf{p}^Z)$ and the $N\times Z$ matrix $\mathbf{X}$ of the related solutions $(\mathbf{X}(\mathbf{p}^1), \ldots, \mathbf{X}(\mathbf{p}^Z))$, often referred to as 'snapshots'. A linear space K can be defined spanned by the vectors $(\mathbf{X}(\mathbf{p}^1), \ldots, \mathbf{X}(\mathbf{p}^Z))$. The $N\times R$ matrix $\boldsymbol{\Psi}$ can be introduced (where $R \leq Z$) of the vectors $\left(\boldsymbol{\Psi}^1, \ldots, \boldsymbol{\Psi}^R\right)$, referred to as the orthogonal basis of the space K. The matrix Ψ can be obtained by a singular value decomposition of the matrix $\mathbf{X}$. The basic idea of the POD method is to search for an approximation of the solution of (2.29) in the space K, so that $\mathbf{X(p)}$ will be approximated by the linear combination:

$$\mathbf{X}(\mathbf{p}) \approx \boldsymbol{\Psi}\mathbf{X}_r(\mathbf{p})\boldsymbol{\Sigma}_{i=1}^{R} x_{ri}(\mathbf{p})\boldsymbol{\Psi}^i \tag{2.31}$$

After replacing $\mathbf{X}(\mathbf{p})$ by the approximation (2.31) in (2.29), the equation system becomes overdetermined. By applying the Galerkin method, the vector $\mathbf{X}_r(\mathbf{p})$ has to satisfy an $R \times R$ linear system of equations

$$\boldsymbol{\Psi}^t \mathbf{S}(\mathbf{p}) \boldsymbol{\Psi} \mathbf{X}_r(\mathbf{p}) = \boldsymbol{\Psi}^t \mathbf{F}(\mathbf{p}) \tag{2.32}$$

The size of the system (2.32) is then equal to R, which is much less than N, the size of the full system (2.29). Even if the matrix $\Psi^t\mathbf{S}(\mathbf{p})\Psi$ is dense, the solution is significantly faster. From the solution $\mathbf{X}_r(\mathbf{p})$, an approximation of the FE solution can then be derived as $\mathbf{X}(\mathbf{p}) = \boldsymbol{\Psi}\mathbf{X}_r(\mathbf{p})$. The accuracy of the method depends crucially on the choice of the parameters $(\mathbf{p}^1, \ldots, \mathbf{p}^Z)$. It is worth noting that the POD method can be seen as a solver to an FE equation system, as it leads to an approximate solution $\boldsymbol{\Psi}\mathbf{X}_r(\mathbf{p})$ of the same format as the FE solution of the full problem (2.29).

There are further modifications and improvements depending on the actual implementation of the method. The mentioned references are therefore worth pursuing for all the details and application examples. These include adaptive procedures [32], combination with the discrete empirical interpolation method [33], incorporation of the RBFs [34], as well as an attempt to create an equivalent circuit from an FE model using MOR and a general application to rotating electrical machines [35].

To complete this section, let us show an example of the MOR technique applied to an electrical machine. The flux density distribution obtained from a full FEM (the 'original') is shown in Figure 2.22(a). By applying the MOR – with a significant reduction of the number of simultaneous equations to be solved – a reduced model yields the flux density map of Figure 2.22 (b), which to a naked eye looks identical to (a), although some small discrepancies do exist, albeit almost negligible, as shown in (c).

As argued before, the MOR approach looks very promising in addressing the challenge of providing a smooth transition between different level models within

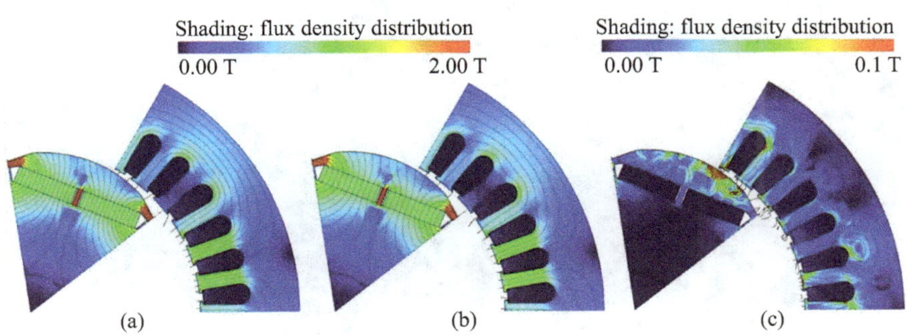

Figure 2.22 Flux density distribution in an electrical machine: (a) original FE model, (b) POD approximated model and (c) their difference

the hierarchical design paradigm. It is, however, interesting to note that this approach starts from the 'bottom' of the inverted pyramid, that is, a full model is first established and then reduced to make it computationally more attractive and efficient.

2.17.2 Deep learning

Let us approach the problem from a completely different perspective. *Deep learning* is an artificial intelligence (AI) activity that emulates the functions of the human brain in processing data and constructing patterns to facilitate decision-making. Deep learning – also known as deep neural learning or deep neural network – is a subset of machine learning in AI that has networks capable of unsupervised learning from unstructured or unlabelled data, as depicted by Figure 2.23.

Convolutional neural networks (ConvNets/CNNs) have gained a significant reputation (see some details about their structure in Figure 2.24, but also refer for example to [37]). It is a class of deep neural networks most commonly applied to image visualisation, analysis and processing, an example of which is demonstrated in Figure 2.25. They are also known as shift invariant or space invariant artificial neural networks, based on their shared-weights architecture and translation invariance characteristics.

The feasibility of applying deep learning techniques to predict the solution of Maxwell's equations for low-frequency EM devices has recently been explored in [38]. A deep CNN was trained – in a supervised manner – to learn a mapping for magnetic field distribution for different topologies of variable complexity,

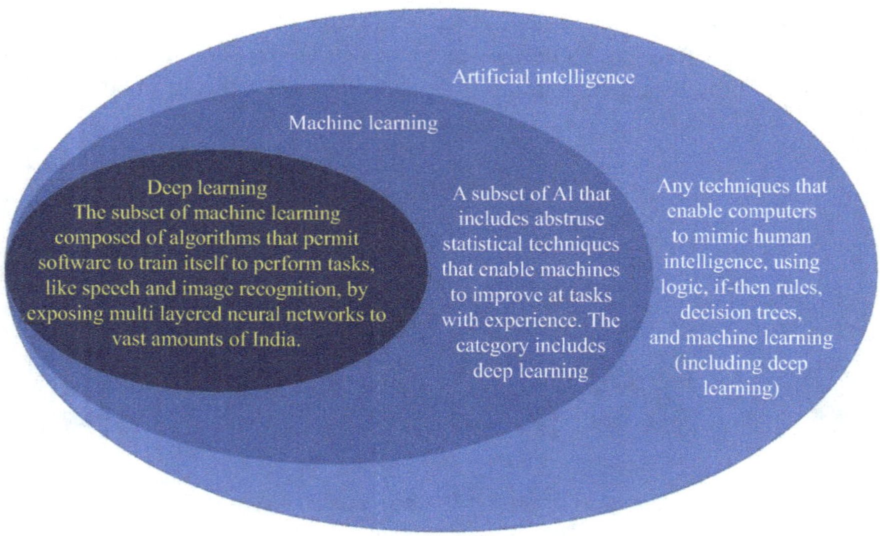

Figure 2.23 The concept of deep learning

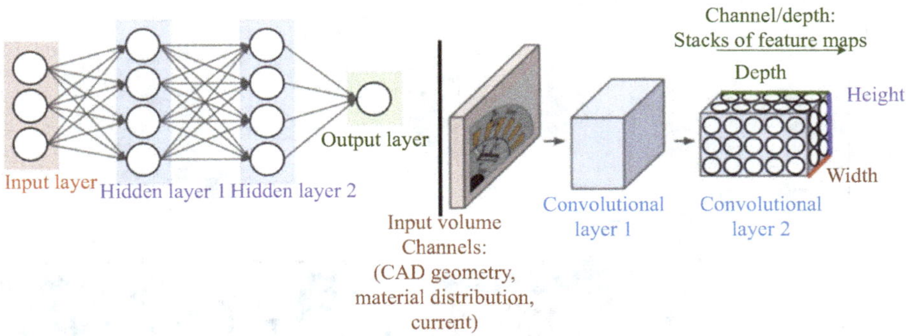

Figure 2.24 A comparison of network architecture between ANN and CNN

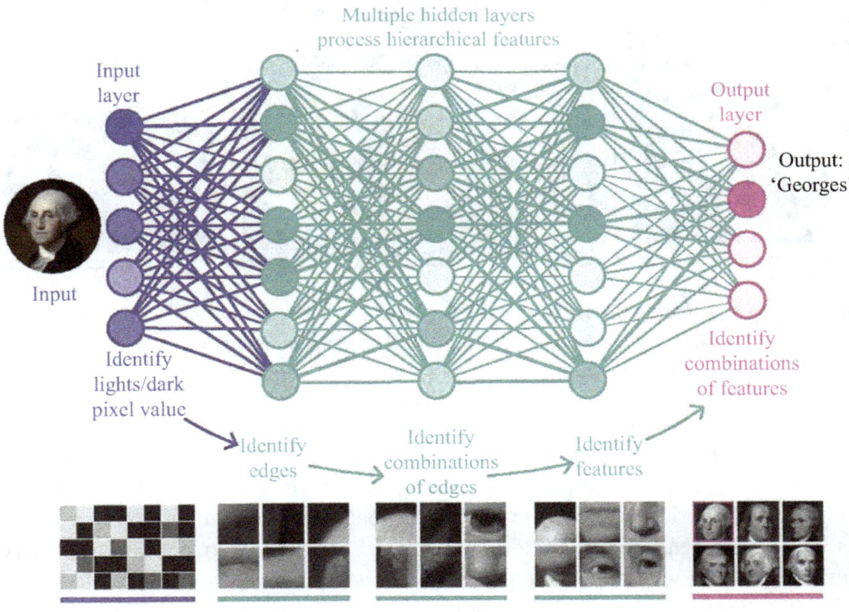

Figure 2.25 A deep learning neural network for image processing

including a simple coil in the air, a transformer and an interior permanent magnet (IPM) machine.

A set of inputs and outputs was generated first for training the networks. The input of each network was formatted as a uniform structured grid containing data about the geometry, materials and excitation at each node of the input. Each of the three 'devices' was parameterised and simulated using a commercial FEM software package. Figure 2.26(a) shows the CAD models and (b) the results (field

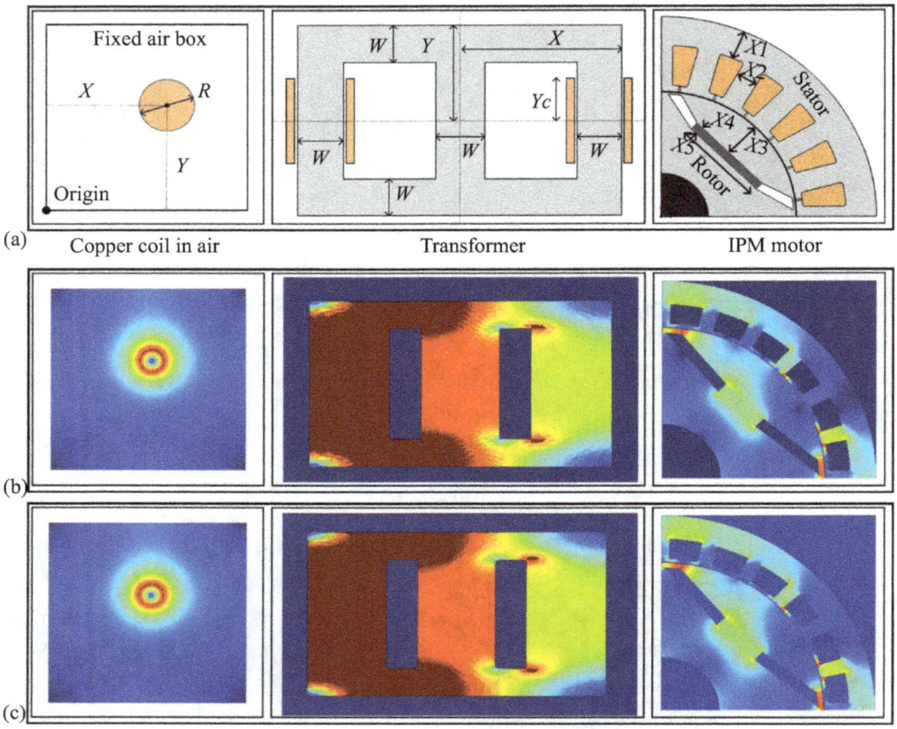

Figure 2.26 A deep learning based magnetic field predictor: (a) CAD models, (b) FEM results and (c) deep learning predictions

distributions) obtained for particular sets of parameters from the finite-element simulations. The deep learning model consisted of a total of 32 layers with 16 trainable convolutional layers, eight pooling/up-sampling layers and eight dropout layers. The network was trained using backpropagation with an aim to minimise the rms error between the prediction and the FEM field distribution. Supervised training was used where the prediction accuracy depends on how close the input is to the training data. The field distributions predicted by the trained deep learning networks are shown in Figure 2.26(c), just underneath the distributions obtained from the finite-element simulations. A visual inspection shows negligible differences; the actual errors were found to be within the range of 0.1%–1%.

There are two bottlenecks in this approach: first is the sheer amount of simulated (big) data required for training a deep network; the second is related to the fact that if geometry is significantly different from the ones in the training data set, substantial errors may be expected. Notwithstanding, the initial findings are very encouraging.

The second example attempts to find directly an efficiency map of IPM machines, such as used in variable-speed drives for hybrid and electric vehicles.

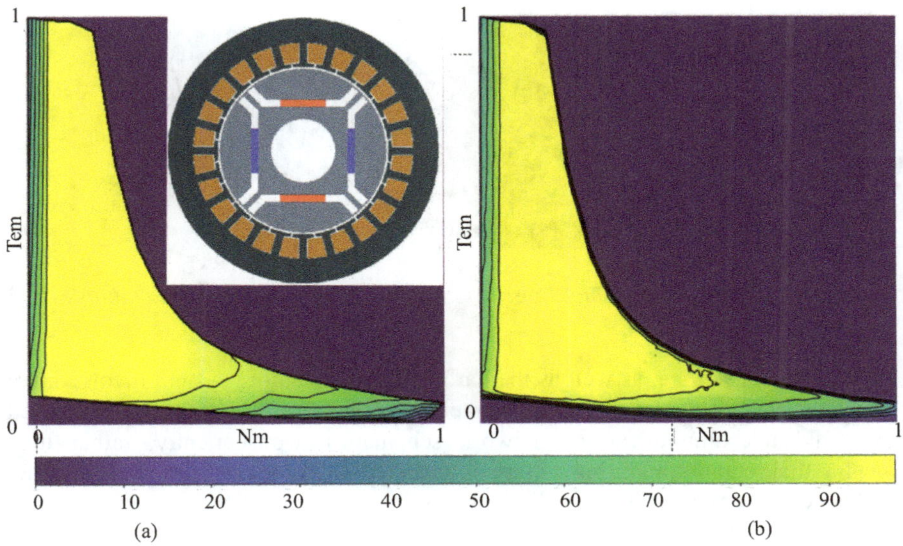

Figure 2.27 The 'end-to-end' deep learning model results: (a) target (from FEM) and (b) prediction (TEM = electromagnetic torque; Nm = maximum speed, both normalised to 1 per sample)

When projected on a 2D plane of speed and torque, an efficiency map consists of varying levels in different regions, with the peak torque constrained by the thermal limits and the maximum speed dependant on the rotor's structure. To predict the efficiency map for a given motor geometry, a large data set of design variations, motor geometric images and efficiency maps is required.

In [39], two deep learning network architectures were employed. In the first one, the two important stages of efficiency map calculations (the flux linkage maps and the torque–speed envelopes) were replaced by a combination of recurrent and feedforward neural networks to account for geometric and operating point variations. For the second architecture, an 'end-to-end' model was trained to predict the same efficiency maps.

The 'target' and predicted results for the end-to-end model are shown in Figure 2.27. The normalised mean-squared error between the FEM generated results and the prediction lies in the range of 1%–1.5% per node/point for a sample. As with the first example, a significant bottleneck was the computational resources required to generate the training data, but the results are nevertheless very promising.

2.17.3 Cloud computing

Cloud computing is usually understood as the delivery of on-demand computing services – from applications to processing power and/or storage – typically over the Internet and on a pay-as-you-go basis. The idea stems from the notion that the

(a) 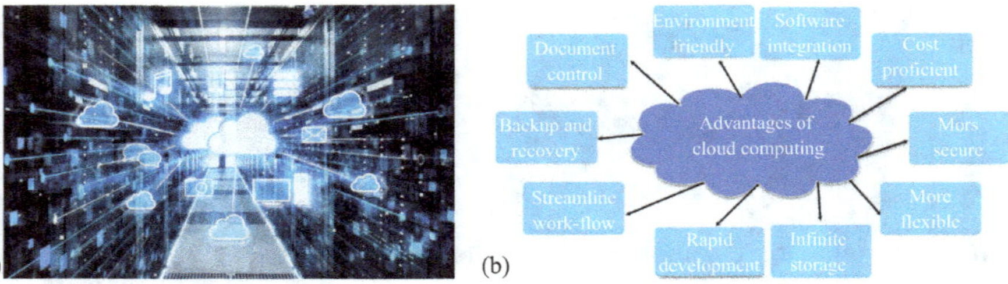 (b)

Figure 2.28 Cloud computing: (a) common graphics and (b) advantages

location of the service, as well as the hardware and operating system involved, are largely irrelevant to the user. The metaphor of the 'cloud' was in fact borrowed from old telecommunications network schematics. Consequently, rather than owning their computing infrastructure or data centres, companies or individuals can rent access to anything from applications to storage from a cloud service provider.

A typical graphical representation of cloud computing is shown in Figure 2.28 (a), while the claimed advantages of the system are depicted in (b). The cost of migrating to and then using cloud computing is not a straightforward calculation, especially as there is much competition between service providers, but also it depends on what services are actually requested. But there is no shortage of offers.

For our discussion the more interesting question, however, is in what way access to cloud computing and availability of different services could be beneficial to design optimisation. The two particular services of relevance would be access to potentially enormous computing power for computationally demanding simulations and huge available storage space for data. Indeed, in many sections of this chapter, even as recently as the last sub-section on deep learning, we argued that significant computational resources may be required to perform various tasks, such as completing the training of the relevant neural network. Earlier on we were concerned about the impact of 'big data' encountered in some formulations and the need for reducing the size of some data sets. Moreover – irrespective of the actual technique used for design optimisation – the more we already know about our device or system the better, as this knowledge can be harnessed in speeding up the optimisation process. With cloud computing and 'reasonable' prices for access to computing power and storage, many mentioned bottlenecks can be alleviated, and many tasks made easier. For example, training of the neural networks could be taken off-line in anticipation of future needs, so that when the design task actually arrives the data may be already there.

Cloud computing has rarely been mentioned in the context of CEM or design optimisation, but clearly there are potential benefits.

2.17.4 Multi-physics

One final important aspect of the design optimisation of most electromechanical or EM devices is the fact that there is much more than just EM problem to be solved.

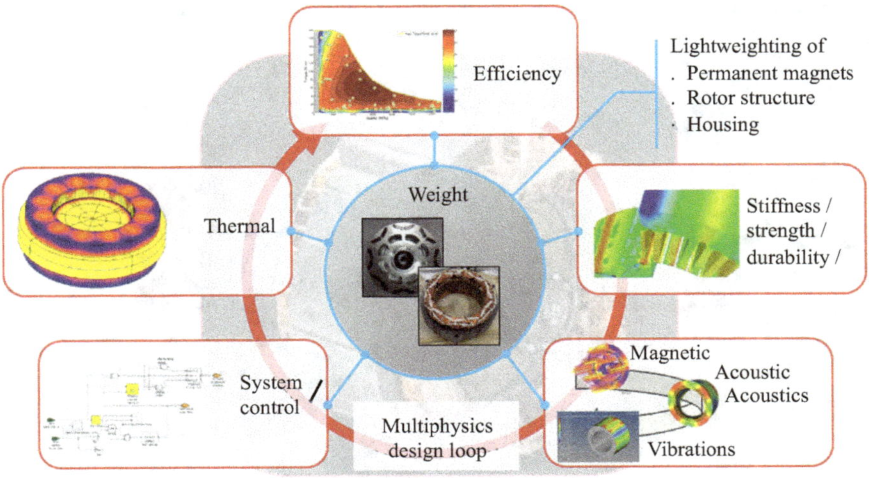

Figure 2.29 Multi-physics design loop

In fact, apart from EMs, there are mechanical, thermal, material and driving circuitry aspects to be considered; to make things even more challenging they are coupled so the problem becomes fully multi-physics. The coupling may not be straightforward to accommodate in the model. One difficulty is the often vastly different time constants (for example of EM and thermal processes). The mechanical stress analysis is likely to require finite-element treatment, but it is unlikely that the mesh suitable for the magnetic solution would be appropriate. Thus, in reality, we need a very complex model to address all the different aspects of the design simultaneously.

Most commercial software these days – including the packages listed in Section 2.5 – offer the flexibility of solving all (or at least some) of the relevant component problems; a good example is the mentioned Altair Flux, as shown in Figure 2.29. It is also worth mentioning that the general purpose COMSOL software has a dedicated multi-physics toolbox as shown in Figure 2.30.

2.18 Benchmarking

New methodologies and algorithms need to be tested and compared; we have already provided such comparisons in Sections 2.12 and 2.14. Numerous mathematical test functions are available devised specifically to challenge and assess the performance of various optimisation routines, refer for example to Sections 2.11 and 2.15, many more are available in literature. Of particular interest and relevance are of course real engineering cases, such as those reported under TEAM problems [40]; we have already looked at TEAM 22 problem.

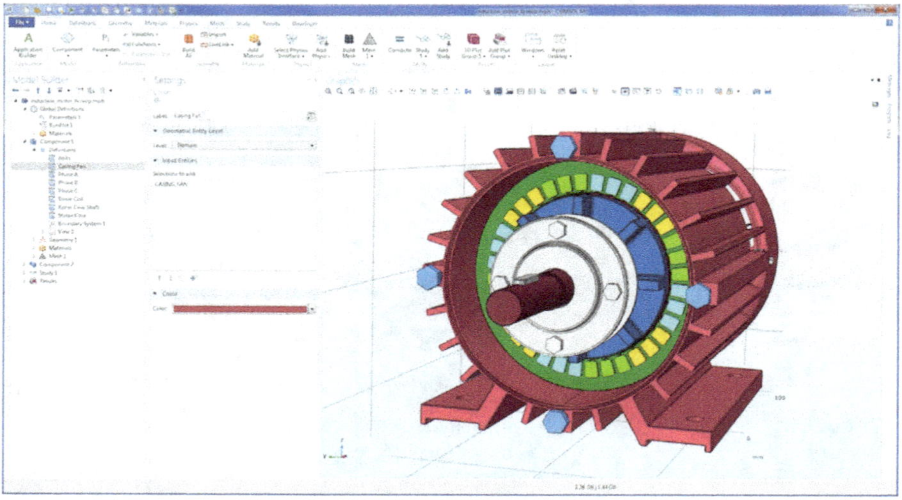

Figure 2.30 Multi-physics in COMSOL

In this context it is probably worthwhile to mention a recent new benchmark problem created specifically for multi-objective Pareto optimisation of devices with EM aspect of performance. Full details may be found in [41,42], a brief summary follows.

Consider an air-cored multi-turn winding (Figure 2.31(a) and (b)). The task is to find the distribution of the turn radii for a given current density that yields the prescribed flux density along the solenoid axis (inside the control region), while reducing the sensitivity of the result to small errors in placing or shaping the turns. Accordingly, the mathematical formulation calls for the minimisation of the field residual (the difference between the postulated and actual value of flux density)

$$f_1(r) = \sup_{q=1,n_p} |B(r(\zeta_l), z_q) - B_0(r_q, z_q)|, \quad l = 1, n_t \tag{2.33}$$

while simultaneously minimising the sensitivity function

$$f_2(r) = \sup_{l=1,n_t} [||B^+ - B(r(\zeta_l), z_q)|| + ||B(r(\zeta_l), z_q) - B^-||], \quad q = 1, n_p \tag{2.34}$$

where B^+ and B^- are the flux density values computed after an expansion (or contraction) of all radii with respect to the unperturbed configuration, that is

$$B^+ = B(r(\zeta_l + \Delta\zeta), z_q), B^- = B(r(\zeta_l - \Delta\zeta), z_q), \quad l = 1, n_t, \quad q = 1, n_p \tag{2.35}$$

where the amount of expansion/contraction is constant and equal to $\Delta\zeta = 0.5$mm. In (2.33) a constant field of value B_p within the controlled region [−5, 5] mm× [−5, 5] mm inside the winding has been considered. A popular NSGA-II algorithm

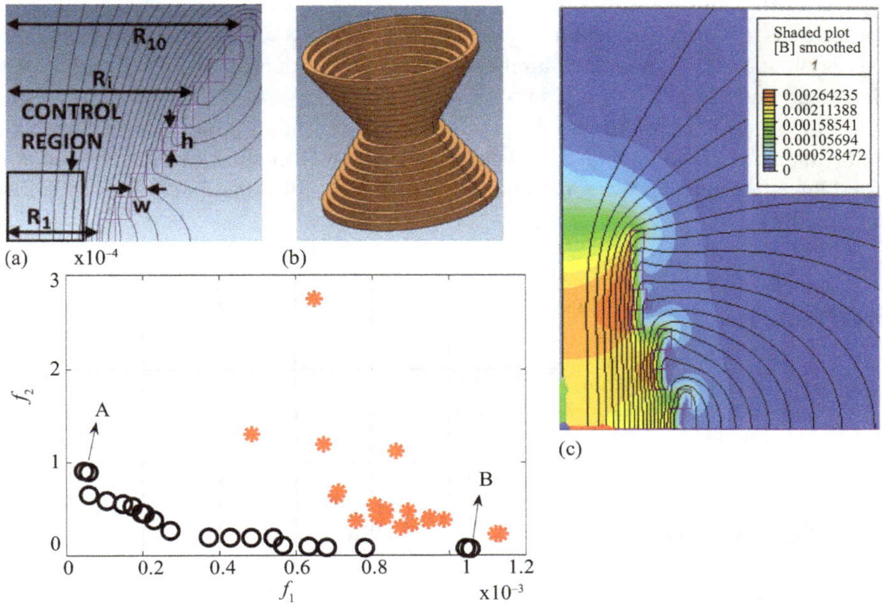

Figure 2.31 New multi-objective Pareto optimisation benchmark problem: (a) geometry, design variables and the control region; (b) a 3D view; (c) typical field solution; (d) results for two objective functions (stars = starting points, circles = final solutions, A and B = end points on the Pareto front)

was used assuming 80 iterations. The starting points were randomly generated in the feasible region. A typical field plot is presented in Figure 2.31(c) and the POF in Figure 2.31(d). Other solutions and comparisons between different algorithms are provided in [41,42]. This benchmark has been designed specifically for multi-objective Pareto optimisation.

2.19 Concluding remarks

The aim of this introductory chapter was to review the current trends in design optimisation, pinpoint the most promising techniques and provide some glimpse into the possible future. There is a vast literature on the subject, new algorithms are proposed all the time and the pace of the developments is breathtaking. This is a very vibrant area of research – involving mathematicians, computer scientists and engineers – and new tools for design optimisation are forthcoming. Inevitably the views expressed here were subjective, but it is hoped that the goal of facilitating the familiarisation with the state of the art has been achieved.

The other aim was to set the scene for the remaining chapters which address far more focused topics and technologies. Optimal designs are mentioned throughout this book and rely on practical applications and implementations of the techniques discussed and reviewed in this introduction.

Finally, if you are interested in the design optimisation issues described here – and want to know more details – please refer to the references I have assembled and cited later. I would also recommend two excellent books listed as [43,44] for further reading.

References

[1] D. H. Wolpert and W. G. Macready 'No Free Lunch Theorems for Optimization'. *IEEE Transactions on Evolutionary Computation*. 1997; 1 (1): 67–82.

[2] G. Bramerdorfer, G. Lei, A. Cavagnino, Y. Zhang, J. Sykulski, and D. A. Lowther 'More Robust and Reliable Optimized Energy Conversion Facilitated through Electric Machines, Power Electronics and Drives, and Their Control: State-of-the-Art and Trends'. *IEEE Transactions on Energy Conversion*. 2020; 35(4): doi: 10.1109/TEC.2020.3013190.

[3] C. W. Trowbridge and J. K. Sykulski 'Some Key Developments in Computational Electromagnetics and Their Attribution'. *IEEE Transactions on Magnetics*. 2006; 42(4): 503–508.

[4] J. K. Sykulski 'Computational Electromagnetics for Design Optimisation: The State of the Art and Conjectures for the Future'. *Bulletin of the Polish Academy of Sciences, Technical Sciences*. 2009; 57(2): 123–131.

[5] https://www.3ds.com/products-services/simulia/products/opera/. [Accessed 28 September 2021].

[6] https://www.mentor.com/products/mechanical/magnet/magnet/.

[7] https://www.ansys.com/products/electronics.

[8] https://www.altair.com/product/flux/new-features/.

[9] https://www.integratedsoft.com/Applications_category/magneticfield.aspx.

[10] http://www.bedl.co.uk/.

[11] https://uk.comsol.com/. [Accessed 28 September 2021].

[12] S. Xiao, G. Q. Liu, K. L. Zhang, *et al.* 'Multi-Objective Pareto Optimization of Electromagnetic Devices Exploiting Kriging with Lipschitzian Optimized Expected Improvement'. *IEEE Transactions on Magnetics*. 2018; 54(3): 7001704.

[13] B. Xia, S. Hong, K. Choi, and C. S. Koh 'Optimal Design of Winding Transposition of Power Transformer Using Adaptive Co-Kriging Surrogate Model'. *IEEE Transactions on Magnetics*. 2017; 53(6): 7203904.

[14] S. Xiao, M. Rotaru, and J. K. Sykulski 'Correlation Matrices in Kriging Assisted Optimisation of Electromagnetic Devices'. *IET Science, Measurement and Technology*. 2015; 9(2): 189–196.

[15] D. R. Jones, M. Schonlau, and W. J. Welch 'Efficient Global Optimization of Expensive Black-Box Functions'. *Journal of Global Optimization*. 1998; 13(4): 455–492.

[16] S. Xiao, M. Rotaru, and J. K. Sykulski 'Adaptive Weighted Expected Improvement With Rewards Approach in Kriging Assisted Electromagnetic Design'. *IEEE Transactions on Magnetics*. 2013; 49(5): 2057–2060.

[17] https://www.compumag.org/wp/wp-content/uploads/2018/06/problem22.pdf. [Accessed 28 September 2021].

[18] L. Yinjiang, M. Rotaru, and J. K. Sykulski 'Kriging Based Robust Optimisation Algorithm for Minimax Problems in Electromagnetics'. *Archives of Electrical Engineering*. 2016; 65(4): 843–854.

[19] G. Taguchi *Introduction to Quality Engineering*, ASIN: B0744MBRQQ, American Supplier Institute, Inc., 1989.

[20] J. Koehler and A. Owen 'Computer experiments' in: Ghosh S. and Rao C. (Eds.) *Handbook of Statistics*, 13, Elsevier Science B.V., 1996, 261–308.

[21] H. G. Beyer and B. Sendhoff 'Robust Optimization – A Comprehensive Survey'. *Computer Methods in Applied Mechanics and Engineering*. 2007; 196(33–34): 3190–3218.

[22] S. Xiao, Y. Li, M. Rotaru, and J. K. Sykulski 'Six Sigma Quality Approach to Robust Optimization'. *IEEE Transactions on Magnetics*. 2015; 51(3): 7201304.

[23] A. Vasuki *Nature-Inspired Optimization Algorithms*, Elsevier: Chapman and Hall/CRC, 2020.

[24] X.-S. Yang 'Nature-Inspired Optimization Algorithms: Challenges and Open Problems'. *Journal of Computational Science*. 2020; 46: 101104.

[25] S. Xiao, M. Rotaru, and J. K. Sykulski 'Correlation Matrices in Kriging Assisted Optimisation of Electromagnetic Devices'. *IET Science, Measurement and Technology*. 2015; 9: 189–196; doi: 10.1049/iet-smt.2014.0194.

[26] L. Yinjiang, S. Xiao, M. Rotaru, and J. K. Sykulski 'A Kriging-Based Optimization Approach for Large Data Sets Exploiting Points Aggregation Techniques'. *IEEE Transactions on Magnetics*. 2017; 53(6): 7001704.

[27] F. Murtagh 'A Survey of Recent Advances in Hierarchical-Clustering Algorithms'. *The Computer Journal*. 1983; 26: 354–359.

[28] F. Pianosi, F. Sarrazin, and T. Wagener 'A Matlab Toolbox for Global Sensitivity Analysis'. *Environmental Modelling & Software*. 2015; 70: 80–85.

[29] P. Weicker and D. A. Lowther 'A Sensitivity-Driven Parametric Electromagnetic Design Environment'. *IEEE Transactions on Magnetics*. 2006; 42(4): 1199–1202.

[30] D. H. Kim, K. Jan J. K. Sykulski, and D. A. Lowther 'Design Optimisation of Electromagnetic Devices Using Continuum Design Sensitivity Analysis Combined With Commercial EM Software'. *IET Science, Measurement and Technology*. 2007; 1(1): 30–36.

[31] Schilders W. 'Introduction to model order reduction' in *Mathematics in Industry*, Springer-Verlag Berlin Heidelberg, 2008, 3–32.

[32] S. Clénet, T. Henneron, and N. Ida 'Reduction of a Finite-Element Parametric Model Using Adaptive POD Methods—Application to Uncertainty Quantification'. *IEEE Transactions on Magnetics*. 2016; 52(3): 7000704.

[33] L. Montier, A. Pierquin, T. Henneron, and S. Clénet 'Structure Preserving Model Reduction of Low-Frequency Electromagnetic Problem Based on POD and DEIM'. *IEEE Transactions on Magnetics*. 2017; 53(6): 7205204.

[34] T. Henneron, A. Pierquin, and S. Clénet 'Surrogate Model Based on the POD Combined With the RBF Interpolation of Nonlinear Magnetostatic FE Model'. *IEEE Transactions on Magnetics*. 2020; 56(1): 7501504.

[35] Y. Sato and H. Igarashi 'Generation of Equivalent Circuit from Finite-Element Model Using Model Order Reduction'. *IEEE Transactions on Magnetics*. 2016; 52(3): 1100304.

[36] H. Sakamoto, K. Okamoto, and H. Igarashi 'Fast Analysis of Rotating Machine Using Simplified Model-Order Reduction Based on POD'. *IEEE Transactions on Magnetics*. 2020; 56(2): 7506104.

[37] K. Lyu, Y. Li, and Z. Zhang 'Attention-Aware Multi-Task Convolutional Neural Networks'. *IEEE Transactions on Image Processing*. 2020; 29: 1867–1878.

[38] A. Khan, V. Ghorbanian, and D. Lowther 'Deep Learning for Magnetic Field Estimation'. *IEEE Transactions on Magnetics*. 2019; 55(6): 7202304.

[39] A. Khan, M. H. Mohammadi, V. Ghorbanian, and D. David Lowther 'Efficiency Map Prediction of Motor Drives Using Deep Learning'. *IEEE Transactions on Magnetics*. 2020; 56(3): 7511504.

[40] https://www.compumag.org/wp/team/. [Accessed 28 September 2021].

[41] P. Di Barba, M. E. Mognaschi, D. S. Lowther, and J. K. Sykulski 'A Benchmark TEAM Problem for Multi-Objective Pareto Optimization of Electromagnetic Devices'. *IEEE Transactions on Magnetics*. 2018; 54(3): 9400604.

[42] P. Di Barba, M. E. Mognaschi, D. S. Lowther, and J. K. Sykulski 'Improved Solutions to a TEAM Problem for Multi-Objective Optimisation in Magnetics'. *IET Science, Measurement & Technology*. 2020; 14(8): 964–968.

[43] P. Di Barba *Multiobjective Shape Design in Electricity and Magnetism*, Springer, 2010.

[44] K. S. Mohamed *Machine Learning for Model Order Reduction*, Springer, 2018.

Chapter 3

Free-form optimal design in electromagnetism exploiting 3D printing

Paolo Di Barba[1] and Roberto Galdi[1]

3.1 Introduction

The success of the additive manufacturing (AM) technology in all its possible forms has generated a considerable interest not only in the field of fast prototyping, but also in the production of devices, previously made with standard technologies. The area of electromagnetic devices has also begun to benefit from this 'revolution', initially intended for mechanical components only. In parallel, the remarkable development of research on materials, which led to the proliferation of a whole set of printing materials added to the 'classic' polymeric bases, contributed to the growth of interest in electromagnetics.

The flexibility of a three-dimensional (3D) printing process makes it possible to fabricate a surface exhibiting at least in principle, any shape which means a substantial increase of degrees of freedom in design. This unique feature can tremendously help the practical realisation of the results of an optimal shape design process, where a surface which is 'optimal' from the viewpoint of a prescribed design criterion could exhibit geometric features hardly implementable by means of standard, old technologies. So far, in fact, this has been the bottle-neck severely limiting the gain in performance of a device designed according to a process of numerical optimisation. In contrast, methods of free-form optimisation (FFO) appear to be particularly suited for exploiting a 3D printing process and manufacturing optimal prototypes.

3.2 Inverse problems and optimal shape design

In engineering science, direct problems are defined as those where, given the input or the cause of a physical process in a device, the purpose is that of finding the output or the effect.

[1]Department of Electrical, Computer and Biomedical Engineering, University of Pavia, Pavia, Italy

Inverse problems, conversely, are those where, given the measured or expected output or effect, one wants to determine the input or the cause; moreover, inverse problems are also those where, given the input and the corresponding output, one tries to understand their interconnection.

The two types of problems, when applied to the same phenomenon or process, represent the two logical ways of conceiving it: from input to output or the other way round. The latter viewpoint is central for design.

In electromagnetics, inverse problems may appear in either of two forms:

- given measured data, which may be affected by noise or error, in a field region, identify, or recover, the relevant field sources or material properties or boundary conditions of the region (identification or parameter-estimation problems);
- given desired fields in a device, or given the device performance based on them, identify, or design, sources or materials or shape of the device, producing the specified performance (synthesis or optimal design problems).

In particular, optimal shape design problems, which are very popular in all branches of engineering, belong to a group of inverse problems where the purpose is to *find the geometry of a device which can provide a prescribed behaviour or an optimal performance.*

Most inverse problems can be numerically solved by minimising a suitable objective function dependent on a design vector and subject to a set of constraints: the relevant formulation has been cast in Section 2.7 for both single- and multi-objective problems. In particular, in a shape design problem design vector x – $x = \{x_i\}, i = 1, n_v$ – represents the geometric variables of the device to be synthesised. This feature in itself makes the dependence of the jth objective $f_j(x), j = 1, n_f$, on x non-trivial. In fact, both the direct problem, through field equations, and the inverse problem, through objective functions, depend on geometry x. As a consequence, since objective f_j is usually a field-based quantity, it depends on x explicitly and also implicitly, by means of the field solution. In general, the following mapping applies

$$\textit{geometry}\{x\} \rightarrow \textit{field } s(x) \rightarrow \textit{objective } f_j(x, s(x)),\ j = 1, n_f$$

Accordingly, the minimisation problem reads

$$\text{find } \inf_x f_j(x, s(x)), x \in \Omega \subseteq R^{n_v},\ j = 1, n_f \tag{3.1}$$

where $s(x)$ is the solution of the direct problem corresponding to the given design vector x:

$$L(s(x)) = 0 \tag{3.2}$$

while L is a linear differential operator like Laplace's operator ruling Poisson's equation of potential, or 'curl' and 'div' operators ruling Maxwell's equations; the case $n_f > 1$ corresponds to a multi-objective problem.

In a problem of shape design, in fact, two aspects are always involved: the optimal synthesis of field s which takes place in the device and the optimal design of device geometry x; formulation (3.1) points out that these two aspects are tightly interconnected.

The situation is even more complicated, because $n_c \geq 1$ inequality constraints might be prescribed on the field; in other words, a set

$$C = \{s(x)|g_k(x, s(x)) \leq c_k,\ c_k \in R,\ k = 1, n_c\} \tag{3.3}$$

where $g_k(x)$ is a constraint function, can be defined. In this case, the minimisation problem reads like (3.1) subject to (3.3), i.e. $s(x) \in C$ and to the field equation (3.2).

For future reference, in a field-based design problem, notation $f_j(x)$ without specifications will stand for formulation (3.1).

The form of the jth objective function f_j suggests a way to classify shape design problems. It can represent the discrepancy between computed and prescribed quantity or the value of a local quantity (e.g. a field component in a part of the device) or, more generally, some characteristics of the device, like weight or volume or cost, and so on. If, for instance, $r(x)$ is the residual vector representing the discrepancy between desired and actual value of the quantity to be optimised, computed in n_s points, then

$$f_j(x) = \sqrt{[r(x)]^T r(x)} \tag{3.4}$$

or

$$f_j(x) = sup_l|r_l(x)|,\ l = 1, n_s \tag{3.5}$$

would be possible forms of the jth objective. In the former case, a least-square (or two-norm) optimisation problem is originated, while in the latter case a min–max (or infinite norm) problem is set up for the jth objective.

The solution to (3.1) is quite troublesome: in fact, function f_j may be neither differentiable nor convex; from the numerical viewpoint, f_j could be a non-smooth function. Moreover, the function evaluation in (3.1) or the constraint evaluation in (3.3) is costly, because any function call requires at least a solution to the field equation (3.2), which in an iterative procedure might be a non-linear one too. This is the main source of insidiousness for field-based inverse problems, which calls for a trade-off among accuracy, runtime and storage.

From the numerical viewpoint, the solution of optimal design problems requires, as a rule, a module for calculating the field, associated with a module performing the minimisation of an objective function. In traditional computer-aided design, these two modules are linked in a way which simulates a trial-and-error procedure. So, starting from an initial design, the field analysis is obtained by means of a numerical method. Then, a check is made whether the device has the desired properties; if not, some variables of the model are updated and the field analysis is repeated, up to a degree of designer satisfaction. This procedure is cumbersome and time-consuming.

The goal of automated optimal design, conversely, is to obtain a tight integration of the module devoted to the field analysis with the module containing the minimisation algorithm, so that the design process is fully automated. This means that the two modules are linked together in a loop. Usually, the analysis of a field can be performed either by methods originating from Maxwell equations (finite-difference, finite-element method (FEM)), or by integral methods amenable to Green's theorems (boundary element method (BEM)). In turn, numerical minimisation can be achieved by means of deterministic or evolutionary methods; the combination of any method for analysis and any method for minimisation gives origin to a variety of iterative procedures for solving an optimal design problem.

Nowadays, most commercially available codes devoted to electromagnetic field analysis are based on the FEM: they proved, in fact, to offer a general-purpose and flexible tool of field simulation. In particular, commercial codes are equipped with an advanced user interface, which enables the designer to develop a model in two or three dimensions by means of graphical operations only. These features make the simulation environment rather friendly and easy to use; so, in practice, FEM has become the most popular one, mainly in an industrial centre for research and development. In this respect, reference is made to Section 2.5.

3.3 State of the art

From the methodological viewpoint, it might be stated that traditional methods of optimal shape design are based on parametric models: this makes it possible to identify improved design solutions by exploiting a variety of optimisation methods, e.g. evolutionary algorithms [1,2]. On the other hand, however, the finite dimensionality of the search space, which is dictated by the number of design variables, severely limits the exploration capability of the search and makes the solutions to be local minima. Moreover, more likely than not, a method of parametric optimisation gives rise to non-smooth shapes which are impractical for technology requirements. In contrast, a parameter-free method able to synthesise optimal shapes, and handling one or more design criteria at a time, would substantially enhance the exploration and preserve the smoothness of the optimal geometry [3]. Eventually, from the application viewpoint, benchmark problems dealing with the design optimisation of electrostatic devices would set guidelines useful for the device designer.

In a parametric model, the modification of the device geometry is achieved by changing the value of the geometrical parameters. During the optimisation process, several combinations of the parameters are checked, and the set offering the lowest value of the objective is treated as an optimum. In order to reduce the number of parameters and find smooth solutions in parametric modelling, the use of non-uniform rational basis-splines (NURBS) curves or surfaces, e.g. the Bézier ones [4], can be adopted.

The main drawback of parametric optimisation is the fact that designers need to have *a priori* precise conception of the possible optimum shape in order to set up the appropriate computer-aided design (CAD) model, and this requires a high level of engineering expertise. As a consequence, unpredicted innovative shapes can hardly be obtained from the parametric approach, because only incremental improvements can be achieved. Besides the traditional methods of optimal shape design, which are based on parametric models, several methods of topology optimisation (TO) were developed [5]. In contrast to the parametric design, TO methods operate by varying the distribution of a given material within a bounded design domain. These methods are thus able to introduce or remove holes and blocks of materials in the region under investigation, changing the volume and exploring new possibilities for the design. The most typical principle of TO considers the normalised densities of materials in each element, varying from 0 (empty) to 1 (filled), as the optimisation variables. During the optimisation procedure, the intermediate densities are penalised, in order to obtain final designs, which are free of intermediate (grey) materials. These approaches are known in the literature as the density methods, or optimised material distribution techniques. However, lack of physical interpretation for intermediate densities, as well as a strong mesh-dependence, has redirected the attention of the research community towards other formulations, e.g. homogenisation-based methods. The heavy reliance on sensitivity-based optimisation, which may often result in solutions that represent local optima, is another limitation of TO-based methods.

In contrast, in a parameter-free approach, commonly known as FFO, the new design is reached by modification of the numerical mesh (in fact, position of the boundary nodes) discretising the surface to synthesise [6,7]. During the optimisation process, the coordinates of nodes are updated and the new shape of the device is achieved. In this case, there is no need to preselect any *a priori* geometrical parameters to be modified. As a consequence, a CAD model does not need to be edited during the analysis. This substantially simplifies the data flow during the optimisation process and, in principle, allows for a wider design freedom (Figure 3.1).

There are various optimisation codes already licensed and available. As far as general purpose codes for design exploration and optimisation are concerned, the Insight, by Dassault Systèmes [8], has to be mentioned as one of the most popular. It offers multi-objective, multi-discipline optimisations. However, it is not always straightforward to link it with the CAD or FEM system which might be available at the designer side, especially in the case of proprietary systems.

Figure 3.1 Independence of an FFO loop on the CAD model

On the other hand, there are some codes, like Abaqus Topology Optimisation Module (ATOM), fully integrated with a given (and single) simulation package. ATOM delivers powerful solutions for performing topology and shape optimisation for single parts and assemblies, but it works mainly with an ABAQUS solver.

Finally, there are also codes focused on non-parametric optimisation, like TOSCA – offering technologies for structural and flow optimisation. TOSCA may work with industrial finite-element analysis standards (Abaqus, Ansys, Nastran) and CFD codes (STAR-CCM+ and Fluent). It provides basically two main optimisation scenarios (topology, shape), with some variants. To the best of authors' knowledge, in the area of dielectric design, the PolOpt code [9] was the first parameter-free optimisation system commercially available.

Shape optimisation has also been a subject of collaborative research projects between industry and academia. A good example is the CASOPT project [10], which includes a comprehensive study of CAD-based parametric optimisation [11] as well as the mesh based [9] and the multi-resolution FFO [12]. The last approach is based on a formulation of the adjoint problem, which delivers the gradient information used for refining the mesh by a subdivision technique. An attractive feature of the multi-resolution approach is its flexibility in defining initial geometry that can have a shape differing significantly from the target and is very roughly discretised. However, its application in dielectric design is limited to a few pilot cases: a tool mature enough for industrial usage is still missing.

A simple FFO algorithm for dielectric design inspired by Neuber conjecture in mechanics [13] is here revisited and applied in dielectric design. The electrode surface, to be synthesised, is modelled in terms of a moving boundary: a NURBS-based model is used first. The coordinates of the control points of the NURBS are design variables handled by an evolution strategy algorithm: on one hand, this enables a broad search in terms of continuously deformable boundaries; on the other hand, however, the finite dimensionality of NURBS prevents the search to be global-optimum oriented. A mesh-based model (FFO) is used next: it is based on an objective-function dependent on a kinematic law which drives the moving boundary towards the optimal shape, improving the design criterion (e.g. the average field strength, to be minimised). In practice, the normal displacement deforming the boundary nodes is proportional to the discrepancy between actual and prescribed objective function value (regularisation step). At each iteration, moving nodes are subsequently dragged in the normal direction in such a way to reduce the discrepancy between highest (E_{max}) and lowest (E_{min}) field strength on the electrode surface (homogenisation step). The two-step procedure is repeated up to convergence.

For the sake of a comparison, the model of the device based on a finite number of parameters is also considered; the relevant design problem is solved by means of an evolutionary algorithm. Specifically, the lowest order evolution strategy (EStra) algorithm [14], which implements a parametric optimisation, is compared with a free-form algorithm, which implements a mesh-based optimisation; BEM is the reference for the electric field analysis.

In contrast to traditional parametric formulation, solutions after FFO are not bounded by the finite dimensionality of the design space. Moreover, smooth shapes that can be easily converted into technologically feasible geometries are always obtained. The benchmark of dielectric design proposed in [11] is referred to and a comparison of optimisation results based on three approaches is carried out: parameter-based model, NURBS-based model (both driven by EStra), mesh-based model (driven by FFO).

3.4 A comparative view of methods

In Figure 3.2 an intuitive representation of how three major classes of methods of optimal shape design work is proposed: in the first column the concept of parametric optimisation is shown with evidence of the single design variable (i.e. the ruling parameter) which can modify the shape of a boundary; apparently, the circular line can only change its radius, but it always remains a circular line.

In turn, in the second column the morphing approach to optimisation is represented: here, a reduced set of control points determines the shape of a boundary. In practice the design variables are the coefficients of a spline function which defines the shape: this way, much more freedom in exploring shapes is made possible; nevertheless, however, the exploration capability is still limited by the finite number of coefficients.

Eventually, in the third column the parameter-less or FFO concept is shown: all boundary nodes are moving nodes which can move freely; therefore, the design space exploration is unlimited in principle.

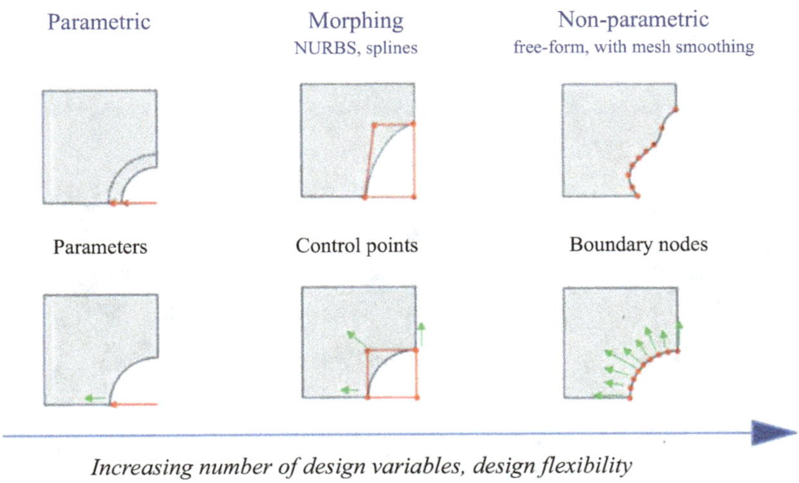

Figure 3.2 A comparative view of modelling methods

Along the horizontal axis of Figure 3.2 both computational complexity and design flexibility increase. In all the aforementioned cases, the movement of nodes is ruled by a minimisation algorithm.

In Figure 3.3, three popular methods to find the minimum of an objective function are shown: according to design of experiment (DoE), a meta-model of the function is generated after regularly sampling the design space by means of a number of experiments, which can be based on either a full-factorial or fractional-factorial scheme. Alternatively, a regression model based on kriging techniques could be utilised (see Section 2.10). Afterwards the *a posteriori* analysis of the function meta-model makes it possible to identify the sub-region of the search space where the minimum is likely to be located.

Subsequently, the iterative search typical of deterministic computing is shown: given an initial solution, only moves which make the objective function decrease are accepted; therefore, the solution converges to a local minimum located nearby the initial one. In order to mitigate the local nature of the search, a multi-start scheme should be applied in spite of a higher computational cost.

Finally, the nature-inspired approach is shown, where moves which make the objective function increase are temporarily accepted in order to substantially strengthen the exploration capability of the algorithm: this way, there is a non-zero probability of converging to the global minimum. Two main classes are devised, namely competitive algorithms, e.g. genetic algorithms, in which the struggle for survival makes the fittest individual survive to the next generation, on the one hand, and cooperative algorithms, e.g. particle swarm optimisation, where agents continuously exchange information each other in order to make the whole swarm converge to the global minimum, on the other hand.

Along the horizontal axis of Figure 3.3 both exploration randomness and computational complexity increase.

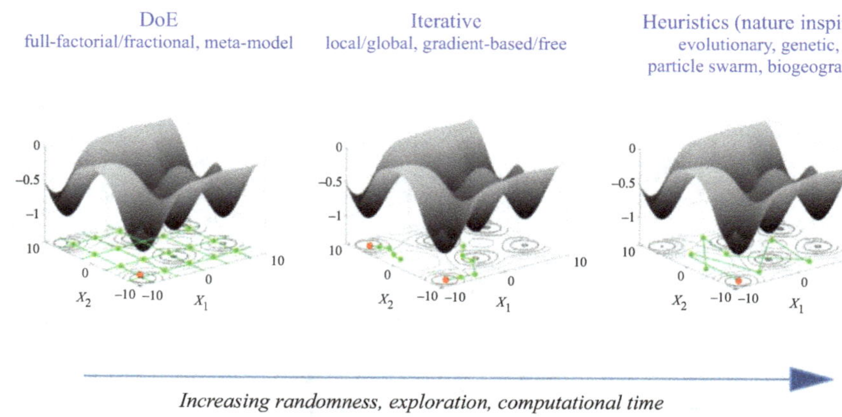

Figure 3.3 A comparative view of optimisation methods

3.5 Free-form optimisation

The idea behind the concept of FFO is amenable to the Neuber conjecture in mechanics [13], according to which the optimum form of a component is achieved when the stresses running along the surface region are uniformly distributed. This means that surface stress or pressure should be homogenised, in order to find out the most effective shape of the component.

Usually a numerical model, amenable, e.g. to the FEM or to the boundary-element method, discretises the component to synthesise. Then a node laying on the component surface can either

- grow→design node is moved outwards (positive amount of displacement)

or

- shrink→design node is moved inwards (negative amount of displacement).

The amount of the displacement is computed as a function of the discrepancy between the actual stress (at the node), and the reference value which might be a prescribed one. Therefore, if the actual stress is higher than the reference – the node will move in one direction, if lower – in the opposite. In the case of a highly stressed area, more materials will be added and the region will grow, while in the opposite situation some materials will be removed and the region will shrink. When repeating this strategy iteratively, the homogenised stress distribution could be achieved. In general, the vector of the displacement is normal to the surface and the amount of displacement is proportional to the difference between actual and reference stress (Figure 3.4). If the node belongs to edges (elements) having normal vectors oriented differently, the displacement direction is calculated as the weighted value.

This simple strategy works pretty well in problems from structural mechanics, where the shape deformation asks for addition or subtraction of structure materials; however, in problems arising from electricity and magnetism the definition of the movement direction is not straightforward.

In the case of a dielectric domain characterised by permittivity ε, surface pressure σ_e at the boundary between two materials depends on the squared value of electric field strength E:

$$\sigma_e = \frac{1}{2}(\varepsilon_2 - \varepsilon_1)\left(E_{t1}^2 + \frac{\varepsilon_1}{\varepsilon_2}E_{n1}^2\right) \tag{3.6}$$

where subscripts t and n stand for tangential and normal, respectively.

In turn, in the case of a magnetic domain characterised by permeability μ, pressure σ_m at the boundary between two materials depends on the squared value of magnetic field strength H:

$$\sigma_m = \frac{1}{2}(\mu_2 - \mu_1)\left(H_{t1}^2 + \frac{\mu_1}{\mu_2}H_{n1}^2\right) \tag{3.7}$$

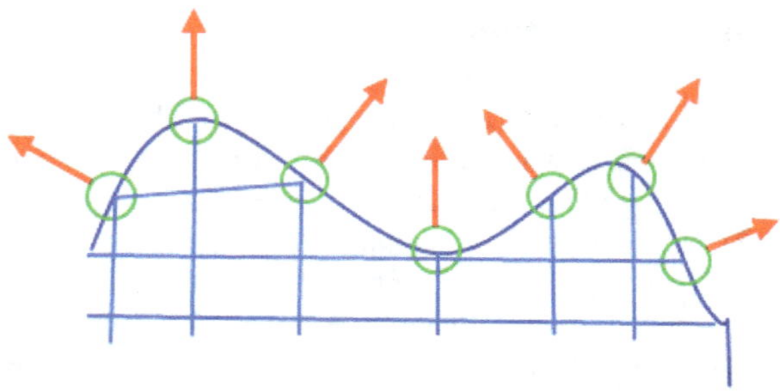

Figure 3.4 *Movable nodes on a boundary surface (green circles) and their normal vector (red arrows). Reprinted from International Journal of Applied Electromagnetics and Mechanics, Vol 60, Paolo Di Barba, Maria Evelina Mognaschi, Tomasz Nowak, Andreas Blaszczyk, Free-form optimisation in industrial dielectric design: A comparative approach, S101–S113, Copyright (2019), with permission from IOS Press. The publication is available at IOS Press through http://dx.doi.org/10.3233/JAE-191109*

From the viewpoint of optimal shape design, dielectric stress or magnetic stress could well be considered as objective functions to minimise with respect to the surface shape, so meeting the criterion of Neuber conjecture. However, the scenario is more complex than in a mechanical domain: for instance, in a dielectric domain, the movement of the design nodes might smooth the external surface equalising the electric field, but at the same time it could also change the distance between components subject to different voltages, in some cases increasing the probability of dielectric breakdown [15].

In the following, an algorithm of FFO for dielectric design is presented.

3.6 A simple algorithm for dielectric design

A simple algorithm of FFO [16] is based on two nested loops, i.e. *homogenisation* and pressure-driven *regularisation*, respectively. The field analysis is assumed to be based on BEM, but also FEM would be applicable.

The basic idea is that the surface to synthesise is considered a moving boundary, discretised by a grid of elements and nodes. Accordingly, in the homogenisation loop, for each surface element the nodes belonging to the element and their position coordinates are recovered as well as the nodal field values. Then, for each node, the normal direction $n = [n_x, n_y, n_z]$ is computed and the normal vector norm is rescaled by a function describing the discrepancy between actual field value at the node and mean field value at the surface. More specifically,

- define the normal velocity v_n attributed to node k as

$$v_n = \frac{\left[\tilde{E}_n - E_n(k)\right]^2 \mathrm{sign}\left[\tilde{E}_n - E_n(k)\right]}{\left[\max\limits_k E_n(k) - \min\limits_k E_n(k)\right]^2} \tag{3.8}$$

where $E_n(k)$is the nodal value of the electric field strength in the normal direction and $\tilde{E}_n$ is the mean electric-field strength:

$$\tilde{E}_n = \frac{1}{m(\Gamma)} \int_{\Gamma} E_n(k) d\Gamma \tag{3.9}$$

with k node index, $m(\Gamma)$ measure (length or area) of moving boundary Γ. In (3.8) max and min refer to the highest and lowest field values on Γ, respectively;
- compute the nodal displacements after scaling n by v_n

$$[dx_h, dy_h, dz_h] = v_n \left[n_x, n_y, n_z\right] \tag{3.10}$$

where subscript h stays for homogenisation.

This way, the coordinates of each node are updated and the node is displaced; at this stage, a smoothing algorithm has to be invoked, in order to suppress possible numerical noise worsening the mesh quality.

The convergence criterion of the homogenisation loop is met when

$$\delta_i < \varepsilon, \ \delta_i = \max_k E_n(k,i) - \min_k E_n(k,i) \tag{3.11}$$

with ε a given threshold, k node index and i iteration index.

The following remark can be put forward: based on (3.8), the boundary movement tends to stop when the nodal field value tends to the average field value, whereas (3.11) keeps the largest field discrepancy as small as possible.

In the subsequent loop, the pressure-driven regularisation is performed and, after a field analysis, the convergence criterion is checked.

In particular, the pressure-driven regularisation is implemented by applying an increment of nodal coordinates calculated as follows:

$$[dx_r, dy_r, dz_r](k, i+1) = \lambda \left| \tilde{E}_n(i) - \tilde{E}_n^*(i) \right| \mathrm{sign}\left(\tilde{E}_n\right) \tag{3.12}$$

with $\lambda > 0$ constant step length, $\tilde{E}_n^*(i)$ the mean electric-field strength averaged over the previous p iterations, namely

$$\tilde{E}_n^*(i) = \frac{1}{p} \sum\nolimits_{l=i-1}^{i-p} \tilde{E}_n(l), 1 < p < i \tag{3.13}$$

$$\tilde{E}_n^*(i) = 0, 1 \leq i < p \tag{3.14}$$

and $\tilde{E}_n$ the discrepancy with respect to the previous iteration, namely

$$\Delta\tilde{E}_n = \tilde{E}_n(i) - \tilde{E}_n(i-1) \tag{3.15}$$

In other words, $\tilde{E}_n(i)$ is the actual mean field at the ith iteration while $\tilde{E}^*_n(i)$ is the mean field within an historical horizon of p iterations. Based on (3.12), in the first group of p iterations of the regularisation procedure, the boundary nodes move in a way to minimise the average field value because according to (3.14) $\tilde{E}^*_n = 0$ is assumed, while in the subsequent iterations the boundary movement tends to make the average field value stable.

Therefore, the kth moving node undergoes a total displacement:

$$\begin{aligned}[dx,\ dy,\ dz](k,i+1) &= [dx_h,\ dy_h,\ z](k,i) + [dx_r,\ dy_r,\ dz_r](k,i+1)\\ [dx,\ dy,\ dz](k,i+1) &= \qquad [dx,\ dy,\ dz](k,i)\\ &\quad +[dx,\ dy,\ dz](k,i+1)\end{aligned} \tag{3.16}$$

Due to the signed distance appearing in (3.8) and (3.12), both outward and inward movements of the node are possible.

In Figure 3.5 the principle flow chart of the FFO algorithm is shown.

Referring to the flow chart, *Homogenisation step* implements (3.8)–(3.10), while *H-convergence* test implements (3.11); in turn, *Regularisation step* implements (3.12)–(3.16), while *R-convergence* test is satisfied when the distance between $\tilde{E}_n(i)$ and $\tilde{E}^*_n(i)$ in (3.12) is small enough.

In general, the proposed algorithm reduces the probability of dielectric breakdown, which represents the most important goal of any optimal shape synthesis in dielectric design.

3.6.1 Dielectric design oriented free-form optimisation

For the sake of an assessment of the proposed method, a real problem of industrial dielectric design was considered [11]. The aim of the optimisation was to control the electric field strength on a shielding ring fitted on an insulating tube of a heat-pipe connected with coolers within an air-insulated circuit-breaker system. The scope is to mitigate the breakdown appearance risk.

The initial shape of the ring was defined by means of three radii (r_1, r_2, r_3) which compose the design vector within feasible region Ω_r. The specific goal was to reduce the electric field strength within a spherical control volume of radius s_r which incorporates the region of interest (Figure 3.6).

The optimisation procedure aimed to redesign the shape of the ring in order to improve its shielding properties. Accordingly, the shape design problem reads as follows: given the applied voltage $U = 150$ kV in DC regime, *find the optimal shape of the ring electrode such that the highest field strength on the electrode surface is minimised and, simultaneously, the homogeneity of the field is maximised.*

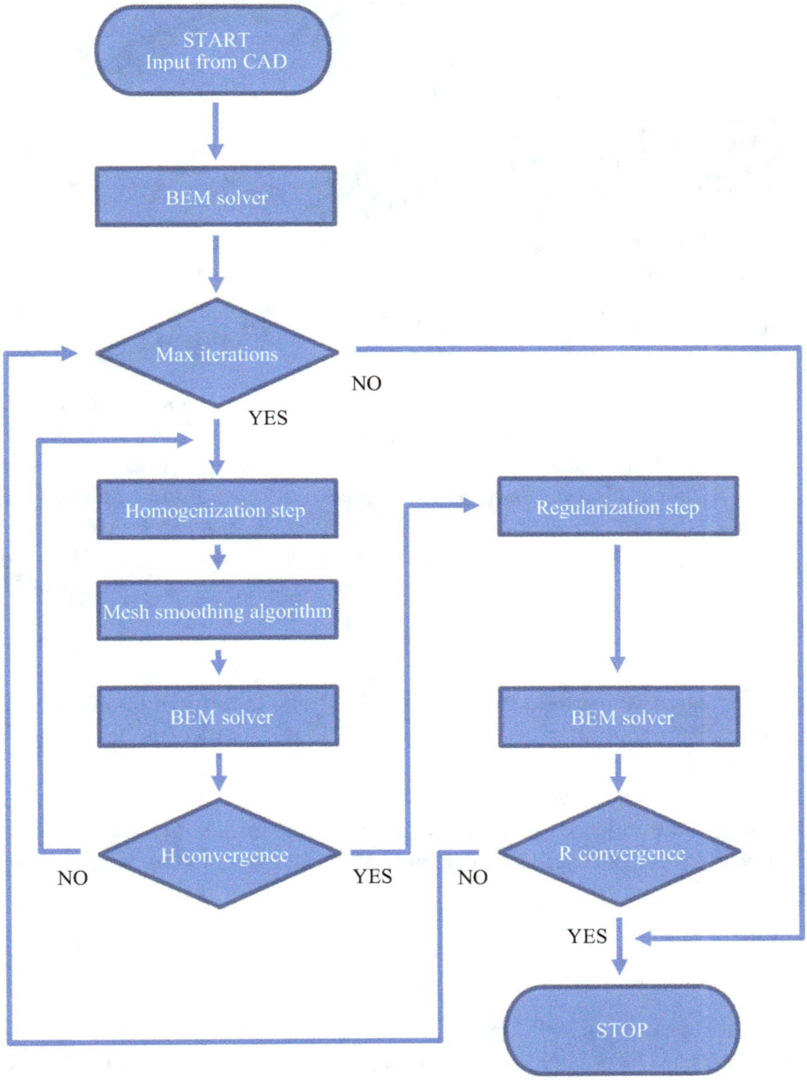

Figure 3.5 Flow chart of a FFO algorithm

To this end, two objective functions are defined: function g governs the homogenisation procedure as follows:

$$g(k) = \max_{k=1,\ n_p} E_n(k) - \min_{k=1,\ n_p} E_n(k),\ P_k(x,y,z) \in \Omega_r \tag{3.17}$$

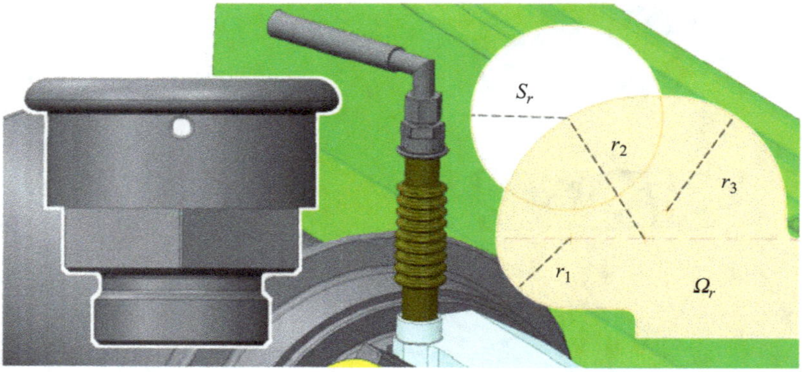

Figure 3.6 Shielding ring: design variables used in the parametric model. Reprinted from International Journal of Applied Electromagnetics and Mechanics, Vol 60, Paolo Di Barba, Maria Evelina Mognaschi, Tomasz Nowak, Andreas Blaszczyk, Free-form optimisation in industrial dielectric design: A comparative approach, S101–S113, Copyright (2019), with permission from IOS Press. The publication is available at IOS Press through http://dx.doi.org/10.3233/JAE-191109

i.e. the discrepancy of the field strength, which is to be minimised in order to have a field distribution as uniform as possible; in (3.17) n_p is the number of movable nodes on the electrode surface.

Moreover, the objective function f, governing the regularisation procedure, is defined: if j is the index of n_p nodes P_j discretising surface Ω_r, then the highest value of electric field strength on the electrode surface, within the control spherical volume of radius s_r, is given by

$$f(j) = \max_{j=1,\ n_p} E_n(j),\ P_j(x, y, z) \in \left\{ \frac{4}{3}\pi(s_r)^3 \cap \Omega_r \right\} \tag{3.18}$$

which is to be minimised with respect to node position $P_j(x,y,z)$.

In the parametric model the design variables are the three radii defining the ring, while in the parameter-free model the design variables are the coordinates of surface nodes (approximately 45,000 nodes).

Three approaches were used to solve the problem in a comparative way, namely:

- an algorithm based on Powell's conjugate direction method [17];
- the proposed FFO method, as described in Section 3.5;
- a heuristic solution based on a trial-and-error procedure.

Moreover, also a simplified axisymmetric geometry, modelled by means of NURBS, was considered. Thanks to the assumed axial-symmetry conditions, it was enough to discretise the ring cross-section by means of a mesh composed of 91 elements, governed by five control nodes; in this case the EStra algorithm was used. All in one, four optimisation cases have been considered.

As far as the field analysis problem is concerned, the 3D numerical model, characterised by approximately 45,000 nodes, was solved by means of boundary element analysis; the latter, in fact, is particularly suited for Laplace field in 3D domains because it allows for handling just a 2D mesh and therefore is cost-effective with respect to finite-element analysis.

3.6.2 Optimisation results

The heuristic optimisation obtained by means of a trial-and-error procedure is considered the reference solution; the resulting geometry is shown in Figure 3.7.

The optimal geometry found by means of the FFO is shown in Figure 3.8.

In Table 3.1, the relevant data of the various optimisations are shown; the corresponding values of the two objective functions are also shown. Specifically, NURBS and FFO identify two complementary solutions that can be classified as non-dominated solutions from the viewpoint of a bi-objective optimal design based on g and f functions [14].

Moreover, it can be noted that the FFO converged with less iterations than the other two optimisations. As far as the computation time is concerned, the method based on NURBS description was the most effective, because this approach exploited a numerical model in two dimensions. In contrast, the least effective method was the parametric optimisation (over two times more long-lasting than

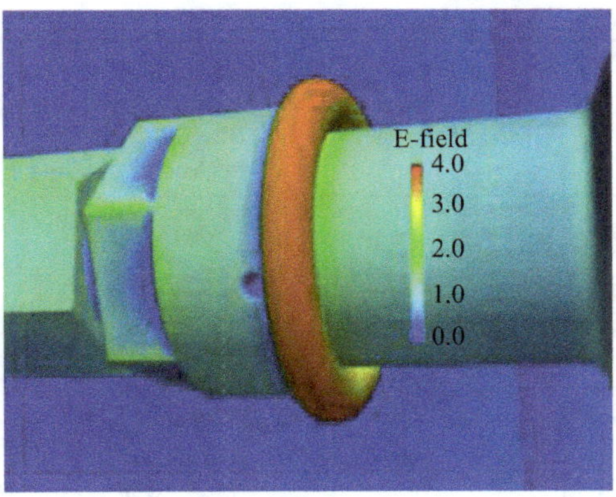

Figure 3.7 Optimal geometries and relevant electric field (kV/mm) maps resulting from trial-and-error solution. Reprinted from International Journal of Applied Electromagnetics and Mechanics, Vol 60, Paolo Di Barba, Maria Evelina Mognaschi, Tomasz Nowak, Andreas Blaszczyk, Free-form optimisation in industrial dielectric design: A comparative approach, S101–S113, Copyright (2019), with permission from IOS Press. The publication is available at IOS Press through http://dx.doi.org/10.3233/JAE-191109

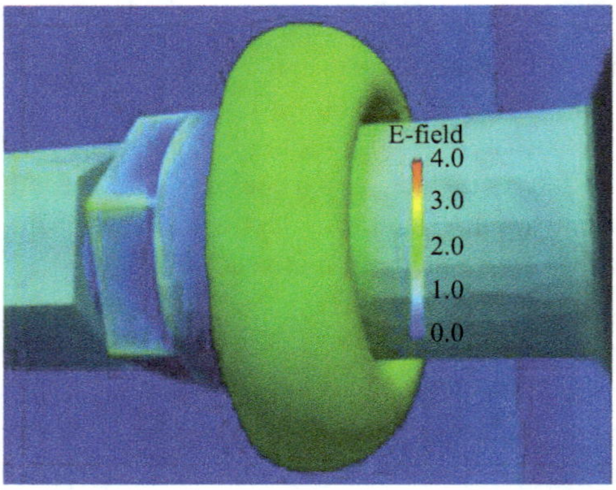

Figure 3.8 *Geometry and relevant electric field (kV/mm) maps resulting from the optimal geometry obtained by means of the FFO optimisation. Reprinted from International Journal of Applied Electromagnetics and Mechanics, Vol 60, Paolo Di Barba, Maria Evelina Mognaschi, Tomasz Nowak, Andreas Blaszczyk, Free-form optimisation in industrial dielectric design: A comparative approach, S101–S113, Copyright (2019), with permission from IOS Press. The publication is available at IOS Press through http://dx.doi.org/10.3233/JAE-191109*

Table 3.1 *Optimisation results*

	Heuristic (trial-and-error solution)	**Parametric optimisation (Powell's method)**	**Free-form optimisation (flow chart in Figure 3.5)**	**2D NURBS (evolution strategy)**
No. of iterations	8	100	80	100
Best iteration	8	87	79	87
No. of unfeasible solutions	0	16	0	0
Computational time (s)	1,296.00	16,200.00	7,560.00	420.00
f (kV/mm)	3.66	3.42	2.84	1.98
g (kV/mm)	2.59	3.08	1.37	1.86

FFO): this is due to the fact that a new mesh model had to be created at each iteration. Moreover, for handling the mesh regeneration, it was necessary to resort to a CAD system with data transfer between CAD system and optimisation algorithms [18]. This was not the case of FFO, where the same mesh topology was used through all the iterations. All calculation jobs were executed on a workstation, equipped with eight cores running at 4 GHz (Intel i7) and 64 GB of RAM.

For the sake of a comparison, a few cross-sectional views of the optimised electrode shapes are shown in Figure 3.9. It can be noted that non-trivial solutions, approximating the optimal one with different degrees of accuracy, are identified.

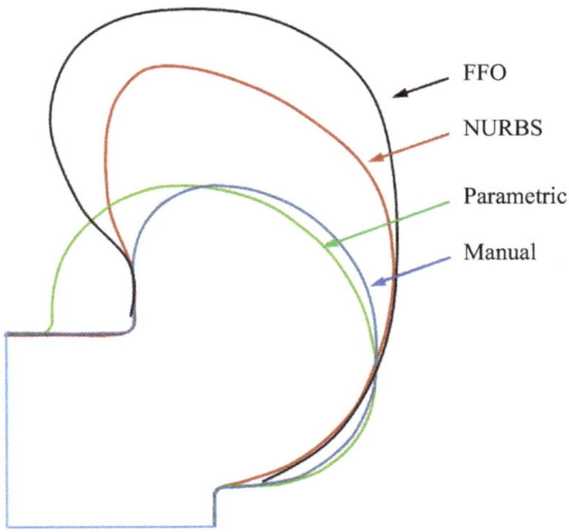

Figure 3.9 Comparison of optimised ring shapes: FFO results, NURBS-based FFO results and Powell method based parametric optimisation results against trial-and-error solution. Reprinted from International Journal of Applied Electromagnetics and Mechanics, Vol 60, Paolo Di Barba, Maria Evelina Mognaschi, Tomasz Nowak, Andreas Blaszczyk, Free-form optimisation in industrial dielectric design: A comparative approach, S101–S113, Copyright (2019), with permission from IOS Press. The publication is available at IOS Press through http://dx.doi.org/10.3233/JAE-191109

In all cases, the use of BEM-based analysis has made the computational costs moderate. The proposed algorithm compares fine with respect to other traditional approaches, e.g. the parametric ones. From the viewpoint of industrial applications, FFO meets also the designer needs: in fact, FFO by its very nature does not require to identify parameterisation schemes, a step that might be cumbersome, especially in the case of complex devices, because inappropriate parametrisation could originate unfeasible results.

3.7 *A posteriori* overview of optimal shape design

For the sake of a comparison, in Figure 3.10 a possible pay-off table comparing methods against models is shown: advantages and drawbacks are annotated showing that FFO is the most preferable approach to optimal shape design irrespective of a peculiar minimisation algorithm which is linked: in fact, the optimal solution is the result of a sequence of smooth modifications of the whole shape, leading to an improvement of performance.

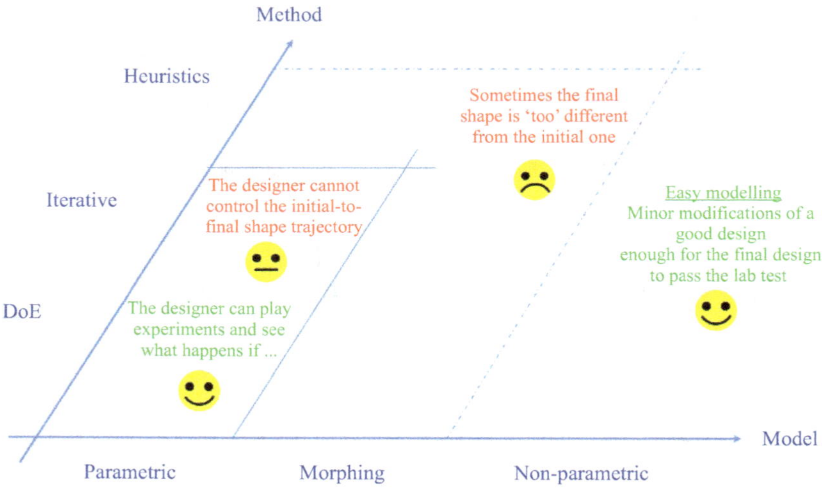

Figure 3.10 A pay-off table of optimisation methods against modelling methods

In contrast, evolutionary computing, which is based on either parametric modelling or morphing models, might lead to optimal designs disruptively different from the prototype: a good feature for innovation, but potentially impractical against available fabrication technologies.

Moreover, deterministic computing is mainly based on parametric modelling and gives rise to a solution which in general represents a local minimum; therefore, the exploration of the search space is severely limited in spite of the designer preferences.

Eventually, the DoEs is felt as a friendly technique because the designer can directly realise the influence of each design variable on the objective function; however, DoE in itself is a technique for investigating the search space through an automated optimal search.

All in one, the fabrication of shapes and geometries identified through numerical methods of FFO looks very promising. Nowadays, it also appears viable thanks to AM technology – sometimes named 3D printing – which makes it possible to create surfaces of virtually any curvature: this unique feature has greatly enlarged the class of achievable prototypes.

On the other hand, however, the stability of a device performance against small perturbations of the device shape is a critical issue. It might well happen that a small drift due to, e.g. fabrication tolerance deteriorates the performance of the device in a substantial way: this is the case when the optimal solution of the design problem is located at a sharp minimum of the objective function (see Chapter 2, Section 2.4).

While classical technology based on subtractive manufacturing has its main limitation in the tolerances of the operating machines, the limitations of AM have yet to be fully investigated. In fact, the performance of a device, designed according to a procedure of optimal shape synthesis and fabricated via 3D printing, is a bench on which to test advantages and disadvantages of this new approach.

3.8 An enabling technology: towards Industry 4.0, level by level

After several unsuccessful attempts, started in 1970 by Dr Hideo Kodama in Japan and reaching as far as the three French engineers Alain Le Méhauté, Olivier de Witte and Jean Claude André who ended their experiments due to lack of funds and perhaps of perspectives, it was in 1980 that the first, industrial, 3D printer projects began.

The first technology to be developed and patented was stereolithography [19].

From US Patent 4575330 [20]: 'Briefly, and in general terms, the present invention provides a new and improved system for generating a three-dimensional object by forming successive, adjacent, cross-sectional laminae of that object at the surface of a fluid medium capable of altering its physical state in response to appropriate synergistic stimulation, the successive laminae being automatically integrated as they are formed to define the desired three-dimensional object. ... the present invention harnesses the principles of computer generated graphics in combination with stereolithography, i.e. the application of lithographic techniques to the production of three dimensional objects, to simultaneously execute computer aided design (CAD) and computer aided manufacturing (CAM) in producing three-dimensional objects directly from computer instructions'.

With these words Charles Hull paved the way for additive printing. After receiving, in 1984, the patent for his idea, which had already been physically realised the previous year, in 1986 Charles Hull co-founded 3D Systems known as the first 3D printing company in the world.

In 1987 the first 3D printer called stereolithography apparatus-1 'SLA-1' was released.

Later in 2014, Hull was inducted into the National Inventors Hall of Fame as the inventor of stereolithography (STL) and received the European Inventor Award.

Parallel to this first successful project research and commercial developments moved very quickly. New companies, founded in various countries, appeared on the market.

In 1989 Steven Scott Crump received the US Patent 5121329 [21], for his idea of fused deposition modelling (FDM), and founded Stratasys company. In this case the production principle was quite different and described in patent as 'Apparatus incorporating a movable dispensing head provided with a supply of material which solidifies at a predetermined temperature, and a base member, which are moved relative to each other along "X", "Y" and "Z" axes in a predetermined pattern to create three-dimensional objects by building up material discharged from the dispensing head onto the base member at a controlled rate'.

Here, too, a reference to well-known computer-based technologies is evident: 'The apparatus is preferably computer driven in a process utilizing computer aided design (CAD) and computer-aided (CAM) software to generate drive signals for controlled movement of the dispensing head and base member as material is being dispensed.'

The first Stratasys printer called '3D Modeller' was produced in 1991.

As a completion, for this first approach to AM, selective laser sintering (SLS) technology and relative apparatus were patented on 5 September 1989 as 'Method and apparatus for producing parts by selective sintering' by Carl Robert Deckard with US Patent 4863538 [22]. Please note that Dickens applied for some other patents in subsequent years, as a result of continuous improvement of his research. His idea was concretised by the foundation of Desk Top Manufacturing, DTM Corp. in 1987.

In Europe, in 1989, the first company founded was Electro Optical Systems, EOS GmbH, in Gräfelfing, Germany. The two founders were Dr Hans J. Langer and Dr Hans Steinbichler. One year later, Steinbichler sold his shares to Langer. In 2001 the company was acquired by 3D Systems.

BMW car company was the first customer to order the EOS rapid prototype system, STEREOS 400. It is curious to know that the first exemplar of the evolved printing system, STEREOS 600, was installed for Mercedes Benz.

In 1994, Ralf Larson with Arcam Ltd, based in London, received a US patent for his ideas of electron beam melting (EBM) [23].

Another company was founded in 1994, ZCorporation. The company based the patented printer and method [24] on an inkjet technology developed at M.I.T. in Boston.

Rami Bonen, Gershon Miller and Hanan Gothait were the researchers who founded, in 1998, Objet Geometries Ltd. The company develops materials and apparatus for PolyJet Technology [25]. From 2011 the company is part of Stratasys universe. One of the founders, Hanan Gothait, in 2005 founded XJet, a company specialised in ceramic and metal printing.

In parallel with the evolution of printing technologies, a series of file formats have been developed for the exchange of data from CAD applications to 3D printers and many, proprietary or OpenSource, software, designed to obtain models increasingly suitable for additive printing, have reached a high degree of maturity.

After the first decades the 3D printing market saw the growth of the companies that first entered the market, by acquisition and huge investment.

The new era of 3D printing started in the early 2000s [26] with the idea of an OpenSource printer project to let as much people as possible to enter AM world. Replicating rapid-prototyper (RepRap) inventor Adrian Bowyer cultivated this idea [27] and invented the 'printer that prints itself' in 2004.

Based on RepRap, in 2009 MakerBot introduced on the market a kit for a DIY printer for 750$. From 2013 the company is a part of Stratasys Ltd.

In Italy in 2012, WASP was founded. The ambitious acronym stands for World's Advanced Saving Project. The first printer developed was the

PowerWASP. It was an economic printer capable of producing artefacts using polymer filaments but was also equipped with a syringe, acting as an extruder, that enabled the printer to use ceramic pastes. The printer set was completed by a milling cutter that could be mounted in place of the extruder so that different materials could be engraved – all for 1.5k €, circa.

WASP continues to develop and produce FDM printers for a variety of materials and in 2015 unveiled the BigDelta WASP, a 12-m high printer used to produce very large-scale artefacts.

In 2013, during the State of the Union address, the President of the United States, Barack Obama, pointed out 3D printing as a technology which carries 'the potential to revolutionize the way we make almost everything'.

A new impulse to the 3D printing market was given by the advent of the crowdfunding platform Kickstarter. Over the course of the first few years, the number of funded 3D printing projects multiplied. An enormous success was *The Micro – The First Truly Consumer Printer*, which, hoping to collect \$50k for the project, raised \$3.4 million in 2014 over the period of a single month. Recently in 2019, *Snapmaker 2.0: Modular 3-in-1 3D Printers* campaign started and after 1 week it raised 1 million \$ and 7,850.866 \$ in 1 year.

Although not all projects became a reality, the substantial funding demonstrated considerable confidence in the possibilities of the 3D printing approach.

During the last 10 years among the most important developments are those involving the application of 3D printing in medicine. In this important field of application, outstanding results have been achieved for the production of prostheses and implants. Currently, research is focusing on bio-materials for tissue reconstruction.

Nowadays, to have an insight of the offers for the industrial market, one can look at aniwaa.com [28]. More than 2,000 different printers are listed.

3.9 An overview of technologies

3.9.1 A technology that could have in the near future a single limit: imagination

During the last decade, the first casting techniques based on filaments and polymer resins have been joined by numerous alternatives. This has been made possible by the development, in addition to printing techniques, of the rapid evolution of printing materials. The research on AM has been driven by the needs of the market in many different sectors: aerospace and defence, automotive, energy, healthcare, art and many others. In each of these sectors, the demand for technology suitable for fast prototyping and, more recently, for real production, has accelerated the work of researchers.

As a result, the race towards evolution has moved in different directions, modifying existing solutions and introducing new ones. It is not uncommon for the same company to produce printers and printing materials guaranteeing users maximum performance.

The differences between the various printing technologies will be, briefly, explored. They differ on the types of materials, the size of the objects that can be produced, the printing times, energy consumption and, last but not least, the cost of purchasing and maintaining printers.

3.9.2 Fused deposition modelling

Stratasys is the owner of the patent for this 3D printing approach.

FDM is, probably, the most diffused technology for AM. This fact depends on the wide range of home printers available at very affordable prices. Adding home and industrial printers, there are more than 1,000 different models available on the market.

The printer works heating, melting and extruding, layer by layer, a thermoplastic-based material, through a nozzle. As the temperature decreases the material solidifies and is ready to be covered by the following layer. Nozzles can be changed in order to have a different diameter dimension and, consequently, different results in time and details. A small diameter allows better details but for converse implies longer printing time. Another parameter affected by nozzle size is the layer height. A larger nozzle diameter, in fact, allows higher layers and shorter printing times to the detriment of quality. It is very important to set up a good balance between all the available parameters to get the desired result. In Figure 3.11 a representation of compromise is shown between time and quality of details.

Taking into account the peculiarities related to complexity and the differences between different models, it is possible to use the graph to identify what is an acceptable compromise in terms of time or filling parameter. A line parallel to the *y*-axis will allow us to 'fix' the time and select the filling percentage from those

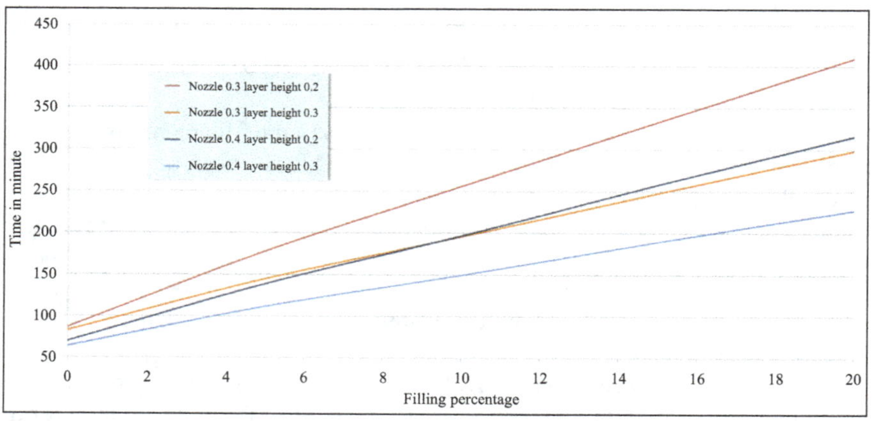

Figure 3.11 Printing time vs filling percentage for the same object using different nozzles and layer heights

available. Drawing a vertical line at a filling percentage, on the other hand, will allow us to choose the quality of the print based on the printing time.

The extruder moves on two axes (2.5D printing) or three axes (3D printing) following the code produced by the slicer. In the case of two-axis movement, the *Z*-axis movement will occur after the layer has been completed, while in real, 3D printing the extruder can deposit material moving on three axes, simultaneously.

A printer could be equipped with more than a single extruder. In this case, it is possible to deposit two different materials, commonly, but not exclusively one of the materials is supposed to be the support material that will be removed at the end of the process. Apart from the removal of the support material, the finished product does not require any special post-processing.

The quality of the component produced via FDM varies greatly depending on the type of printer. For the most advanced device accuracy can be less than $\pm 250\ \mu m$.

3.9.3 Stereolithography and digital light processing

Again, from Charles W. Hull Patent: 'Stereolithography' is a method and apparatus for making solid objects by successively "printing" thin layers of a curable material, e.g. a UV curable material, one on top of the other. A programmed movable spot beam of UV light shining on a surface or layer of UV curable liquid is used to form a solid cross-section of the object at the surface of the liquid.'

This 3D printing approach uses, as base materials, a wide range of wax and resins (epoxy polymers).

The resin is poured inside the production tank and cured using a laser by means of photopolymerisation. The printer works by creating the object in a top-down or bottom-up way. The first method involves the print bed immersed in resin moving downwards, after the creation of each layer. In this case, therefore, the object will always be immersed in resin and its height cannot be greater than that of the tank.

In the opposite procedure the object will be created by moving the print bed from bottom to top. In this way the amount of resin required in the tank, which will be automatically adjusted by the printer, will be less and the size of the tank could also be smaller.

In both cases, it will be necessary to provide support structures for the parts of the objects presenting complex geometry.

The quality of finishes using this technology is very high up to 50 μm for very high-resolution design.

From a mechanical point of view, the operation of a digital light projections printer is similar to the previous one. The difference lies in the source of the light-curing, which in this case is a projector.

While lasers act on a single point at a time the projector covers bigger areas. In this case, the quality of the finish is linked to the resolution of the projector, i.e. the size of the pixels (or voxel, that is a volumetric pixel).

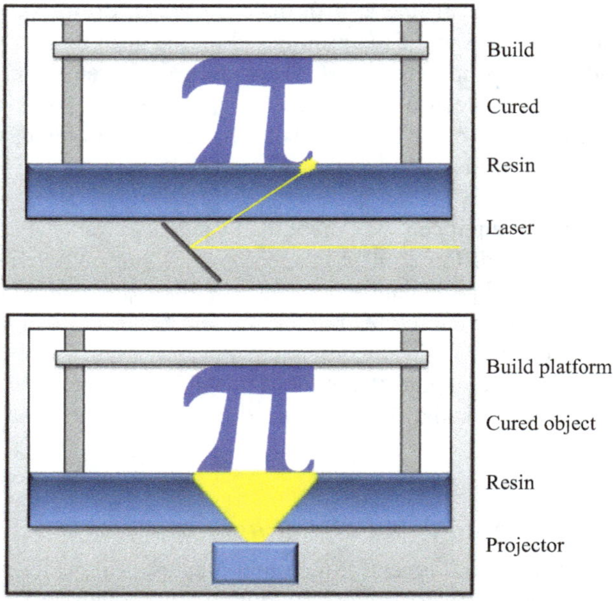

Figure 3.12 SLA (previously shown) and DLP printing process

At the end of the printing process, in both cases, a polymerisation could stabilise the material characteristics and increase the component mechanical strength (Figure 3.12).

3.9.4 Selective laser sintering

The basis of selective laser technologies is the use of powder which is deposited in layers within a production chamber (in vacuum or nitrogen).

During the printing process the powder taken from a feeder is deposited evenly on the printing surface (feeding mode) and, if necessary, smoothed. Then one or more focused laser beams, depending on the type of printer, follow the path required by the printing code, based on the CAD design, solidifying the powder. At the end of the first layer sintering the build platform moves down to the selected height and the process will be repeated until the end of the printing.

At the end of the process the waste material remaining in the printing chamber needs to be removed and, eventually, regenerated to be reused.

The dimension of the printer bed and the printing volume are some of the peculiar characteristics of the printers; indeed, on them depends the number of objects that can be produced simultaneously.

The main difference from SLS, selective laser melting (SLM) and direct metal laser sintering (DMLS), is related to the sintering or melting of the used powder.

In laser sintering the material is heated enough to fuse it at a molecular level. The result is a component whose porosity can be controlled.

In laser moulding the material reaches the melting temperature. Using this technique, the result is a homogeneous component because of the melt process.

3.9.5 Multi-jet fusion

This technology, proprietary of HP, reached the market in 2016.

It is based on polymer printing and infrared light. As for other techniques, the base is a polymeric powder. At the beginning of the process, it is spread to cover the printer bed. For each second, 30 million drops are jetted. In the next step, proper fusing and a detailing agent are jetted to trace the shape, successively light beam, selectively due to the different agents, fuse the material.

One of the advantages of this technology is that the printer works on the whole bed printer area in a single pass. A temperature control, performed at hundreds of points on the printed shape, allows the printer to fine tune the quantity of energy to be provided to correctly fuse the material. HP is working on the chance to use the agents to control parameters such as colours, translucency, mechanical characteristic and electrical conductivity of the printed material.

The use of a detailing agent is something very peculiar. Using this agent on each single voxel, a 3D version of a pixel, which defines the smallest unit of volume that can be printed, Multi-jet fusion (MJF) printers can control in a very precise way its behaviour.

The process is recognised to be very fast in comparison to other approaches.

3.9.6 Material jetting printing

This technique involves the deposition of ink (liquid materials or solid suspensions) layers by layers. The ink is deposited by means of a nozzle and cured exposing it, briefly, to ultraviolet or infrared light. The whole process is safe, if compared with other techniques because it takes place in low-temperature and low-pressure environment.

Possible materials range from polymers to dielectric nanoparticles to conductive nanoparticles. The type of materials strongly differentiates the offer of different companies [29,30].

3.9.7 Electron beam melting

EBM is classified as powder bed fusion technology, as SLS, SLM, DMLS and MJF.

With this technology it is possible to melt metal powder layer by layer. At this scope, it uses an electron beam in a vacuum chamber. EBM is able to process, to melt, the material at multiple points at once. The electrons beam could reach temperatures up to 900 °C.

As for other processes, a layer of metal powder is placed on the build plate, the bed is heated, and the powder is melted by the electron beam.

As the high temperature reaches during the process, the needs of support structure are reduced.

Table 3.2 Technologies vs materials

Technology	FDM	SLA	SLS	DMLS	SLM	MJF	EBM
Material/s	Plastics Composites	Resins	Plastics	Metal alloys	Metals	Plastics	Metals

At the end of the process the pre-sintered powder can be recycled, this translates into important savings, also in relation to the costs of the material used for this process.

3.9.8 Binder jetting

Binder jetting technology, developed at Massachusetts Institute of Technology and in 1993, produces parts using particles of a fine-grained powder. Materials can be polymers or metals.

As for other processes as a first step the powder is spread on a fusion bed then a print head applies the binder agent to the powder. After the layer is solidified, the bed moves down in order to repeat the process for a new layer. So the main difference with other technologies is the use of a binder as a means of solidifying the material. Using two printheads is possible to apply colours to different parts of the printed layer obtaining a full colour print.

3.9.9 3D multi-material printing

Thanks to the evolution of the technology described it is, now, possible to work printing multiple materials at the same time [31]. This result opens a wide range of possibilities, including having objects with different mechanical, thermal and electrical characteristics printed at the same time (Table 3.2).

3.10 Materials vs meta-materials

The ongoing evolution of materials is overcoming the limits of chemical composition by turning towards the world, still largely unexplored, of meta-materials, i.e. materials obtained by assembling elementary materials in such a way that the overall physical properties at the macroscopic level are not observable in any of the component material.

The history of meta-materials, even if the term has been conceived recently, is not recent. In [22], it is possible to find some ideas but, as happened to other areas of research, the advent of AM has led to a broadening of the scope of its possible applications.

Each material in nature has its own chemical composition and consequently certain properties. In the case of meta-materials, to reach peculiar characteristics, there is one more degree of freedom to be used, to control the properties, the structure. In fact, structuring the material according to a specific geometry, it is possible to create characteristics that the material itself, in nature, does not possess. In 3D printing the structure of meta-materials is a repeating pattern realised with polymers or metallic materials. Specifically, the issue of meta-materials based on gradient dielectric structures and split-ring resonator arrays is developed in Chapter 6.

As it happened to other areas of research, the advent of AM has led to a broadening of the scope of its possible applications. The synthesis of printing technology and meta-materials has made it possible, on the one hand, to create, for example, the first prototypes of electro-magnetic devices and, on the other hand, to become aware of the advantages but also the limitations associated with this new production approach [32–34].

3.11 3D printing: a practical implementation

The collaboration among University of Pavia and Lodz University of Technology allowed the authors to explore some aspects of 3D printing, specifically FDM.

The idea came from a line of research at the Institute of Electronics of Lodz University of Technology, in the field of dielectric 3D meta-structures [35]. The goal is to develop an electromagnetic lens, print it and compare the results with the computational ones.

Various geometries have been proposed and simulated. Here we present one of the models printed. Figure 3.13 shows the model in STL format. It is easy to assess the very high number of facets. The model sizes are $50\times50\times50$ mm^3.

One of the problems to be tackled was the, very, long time for the conversion of the STL model in GCODE. GCODE is a text file containing a sequence of commands that have to be sent to the printer to control the movement on the three axes and the behaviour of the extruder. The more detailed the model is, the longer it

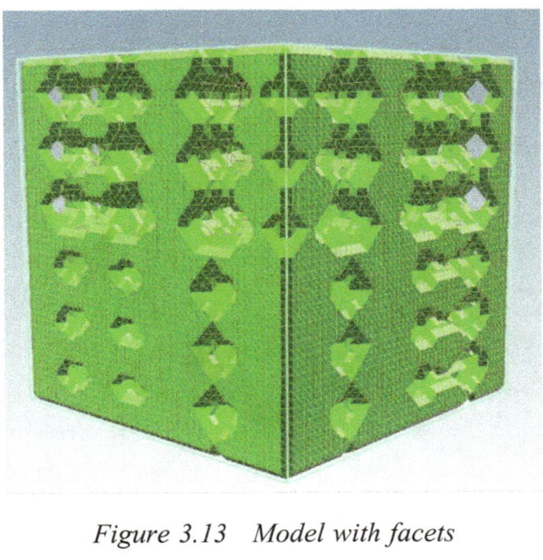

Figure 3.13 Model with facets

takes to write the file. One of the first tests, to solve the problem, was to significatively reduce the density of mesh, and so the facets on the surface of the model. The time was shortened but the quality of the final result appears to have deteriorated. In fact, the model exhibits a very peculiar architecture with many cavities on the surface and inside. As usual the influence of the model structure on the final result was substantial.

In the end, a balance has been found taking into account a matrix of possible variables, the material – different types of acrylonitrile butadiene styrene (ABS), polylactic acid (PLA) and glycol-modified polyethylene terephthalate have been tested, the temperature of extrusion – in a range of ± 30 °C, the speed of the extruder, the nozzle size, the flow rate and model positioning, with regard to the printing process. A re-thinking of the slicing procedure and a fine tuning of the

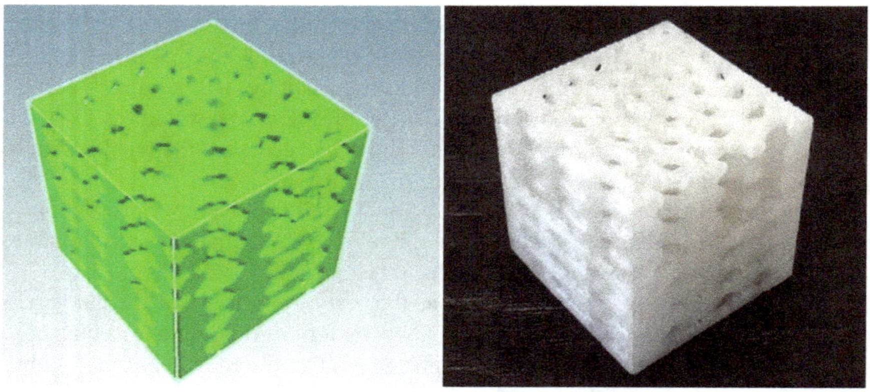

Figure 3.14 Model (left) vs printed (right) component

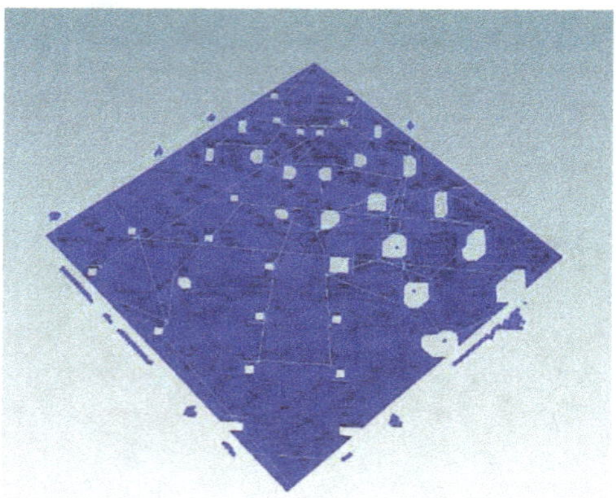

Figure 3.15 Initial, bottom, layer with printing path

parameters help in reducing times enough to get the GCODE, with basically no influence on the quality of the printed part.

The result obtained is, in terms of printing, very positive, also in view of the fact that no support materials were used – and this result is due to the balance mentioned before.

We carried out the printing on an Olivetti 3D-S2 printer, using a single extruder.

The following figures have been added to give a clearer idea of the internal structure of the model.

The good results obtained encourage the investigation of possible future developments (Figures 3.14–3.17).

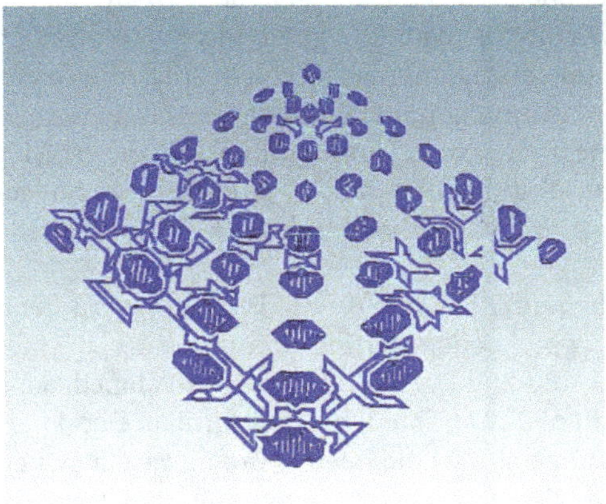

Figure 3.16 *Central layer with printing path*

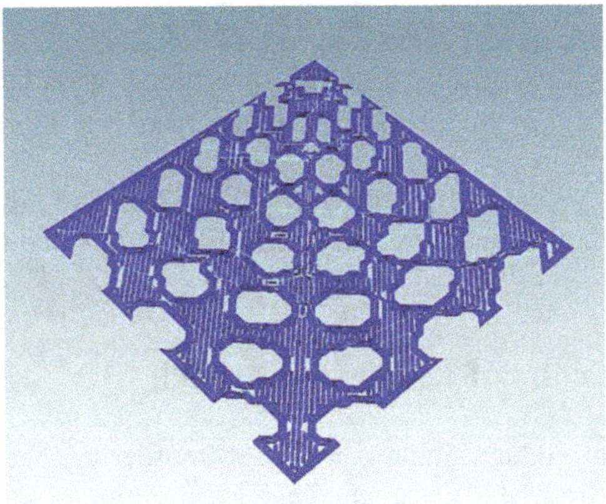

Figure 3.17 *Final, top, layer with printing path*

3.12 Recent FDM experiences in electromagnetism

In the recent past, the number of important additive printing experiences has increased led by experiments of undoubted scientific value that have become propaedeutic to the achievement, more recently, of results of great importance. This has been made possible thanks to new technologies and materials [36].

The goal of the researches in the area of AM, which have as focus electrical devices, in the beginning was to achieve, at least, the same behaviour of devices as obtained through traditional processing. This is the reason why the electrical and mechanical characteristics of materials to be printed has been the object of tests and research [37].

Low electric conductivity for the magnetic material, high electrical conductivity for the winding's material, good insulation attribute and mechanical strength are the parameters against which the quality of a device can be assessed.

Because of low laser radiation absorption and high thermal conductivity, only silver performs better, it was very difficult to use the SLM process for copper [38,39]. The problem is still actual even if, with the usage of powder alloys, it has been partly overcome. In [40] authors review the issues related to pure copper use in 3D printing.

In 2018 at Chemnitz University of Technology, a technique to produce ceramic-insulated copper coils, starting from powder of metallic and ceramic material mixed with binding agents, has been developed, achieving the goal of design and realisation of a 3D printed, high performance coil [41] and its insulation.

The possibility to print a whole device, in this case a switch reluctance motor, using multiple materials represents an important step towards the realisation of other types of devices.

Obviously, printing an entire device with good electrical, magnetic and mechanical properties is not the only goal to be achieved through 3D printing.

The production of single component (rotors, cores, windings, etc.) is also an ambitious objective aimed at achieving lower production costs, shorter production times and better characteristics of the finished product.

In [42], the authors describe the fabrication of coils, using silver ink, to be used in a wireless charging. In the proposed paper a freeform approach of the coil is presented [43].

Moreover, the emergence on the 3D printing market of start-ups bringing knowledge and ideas attracted new investment for the researcher and this will be a benefit to the entire sector.

3.13 Pros and cons of the new approach

Prototyping helps to produce single component in order to check if their quality is the ones expected from the designer. Thus, fast prototyping is a great step in shortening time to market of the projects.

AM approach is constantly evolving and its advantages and weaknesses, in the perspective of FFO, need to be evaluated considering two fundamental factors, materials research and the evolution of printing techniques.

The early additive printing experiences, carried out using advanced versions of CNC machines adapted for FDM, have definitely evolved. As described before, the new printing techniques benefit from the characteristics of new generations of materials and are performed on highly sophisticated printers using different AM approaches.

Various AM techniques for the production of components need to introduce a support material into the object to be printed. This reinforcement is essential in ensuring that the model is produced correctly, especially in the case of several printing layers that form hollows inside the object. It is therefore necessary to take into account the operations of post-processing, removal of the material and finishing of the object, which depend on the type of materials used for printing. There are several types of support materials, available, that differ in the way they are removed from the finished model as well as just different printing techniques that require longer or shorter, and more costly post-processing times. The removal can take place dissolving the support in water and detergent rather than in a chemical solution or by manual removal.

The materials used for AM printing differ depending on the printing method used and so the characteristics of the finished product may or may not come close to those required. These materials are composed of a mix of powder composed by the metal, metallic alloy, ceramic or polymers representing the material to be used for the part and of materials and binding components. The purity of materials makes the difference in maintaining the characteristics of the final component. Other materials, e.g. carbon fibre–filled materials are composed by the infusion of fibre in PLA or ABS and used in FDM printing.

The printing approach based on the printer/material relationship makes it possible to make the full use of the features of freeform design by overcoming the limits imposed by the design method. It is possible, for instance, to print non-planar shapes, thereby optimising devices that must reside in housings and cases with particular shapes.

3.14 3D printing oriented optimal design

At design time of a device, one should take into account the printing technology that will be used to produce it and design the shapes accordingly. If so, SLA, SLS, SLM and DMLS and the other techniques could guarantee the best results even in the case of fairly complex shapes.

Some shapes have embossed parts, holes, bridges, which must be taken into account by correctly introducing the supporting materials.

One of the goals in optimising the design is to have an economically viable process. To this end, one of the fundamental aspects to take into account is the cost of the material used to make the component. In the case of FDM printing, in addition to the cost of the printing material, the need for support structures must also be taken into account. These are necessary in order to avoid the collapse of the de-polymerised material, in correspondence of 'empty' areas and the success of the parts where smaller quantities of filament are deposited, thus allowing the correct realisation of the component. The quantity and quality of this material must also guarantee the simplicity of the necessary procedures in the post-processing phase, during which the removal of the support itself and any finishing and cleaning operations of the component produced will take place.

In the case of prints using powder as the printing material, it is necessary to take into account the material that will remain in the production chamber, which will only be partially recovered and will in any case require a regeneration process before subsequent use.

The geometry of the component to be printed must therefore be designed with these aspects in mind, as well as the printing time, which as with any industrial process must be accounted for, and the amount of energy required for production.

However, optimising the design for additive printing goes beyond the geometry; even the choice of the slicing tool can lead to a significant improvement in the printing code produced and, consequently, in the design goodness parameters expressed before.

The design, often, needs rethinking because of the opportunity offered by AM. Many parts could be designed as a single unit, reaching the so-called part consolidation, and this turns into saving in terms of time and costs.

3.15 Driving the slicing: process-oriented coding

As mentioned earlier, in addition to design ideas, materials and printing techniques, software has also evolved considerably, being a pivotal component of a successful production process. In this aspect, slicing software plays a crucial role.

The production of a component commonly starts with the creation of the design by means of CAD software or by acquisition through 3D scanning. Also, in this case the evolution of acquisition technologies has led to the creation of very sophisticated products capable of acquiring high resolution even large objects. This, second, method is frequently applied to parts to be repaired through a reverse engineering approach.

The final design of the part should take into account, and apply, the possible design optimisations in relation to the type of printing techniques adopted.

The obtained structure has to be translated in a file format that describes the geometry of the 3D object by means of so-called facets. A facet describes a portion of the surface of the object using its normal direction and specifying the three vertices using right-hand-rule rotation.

Please note the analogy with the process of discretisation described in Section 3.4.

There are many different file formats adoptable, here, and we will limit ourselves to outline the main features of the most popular.

STL format was designed in 1987, for 3D Systems. It describes the surface of an object by means of a triangle. This format does not include information about other attributes of the object (e.g. colours, texture, etc.). Information can be saved in ASCII or binary format. ASCII advantage is to be easily understandable while binary format is more compact.

Additive manufacturing file was introduced in 2011, in such a way an evolution of STL. This format, thanks to the fact that it is XML-based, can contain different information, colours, textures, materials and many others. Despite the fact that since 2015 the major players have been supporting this format, its use remains marginal compared to other formats.

OBJ Wavefront format was first introduced by Wavefront Technologies around 1990. It describes objects, including colour and texture information. It can be saved in an ASCII format or in a binary format using MOD extension.

At this step slicing software, the slicer, comes in. As already stated, additive printing produces objects layer by layer, so a tool is needed to create a 2D path to be followed by the printer to create each layer.

Slicing is, for very simple geometry, an automatic process carried out using the default parameter proposed for the printer. In the case of complicated geometry, it is a fundamental step to achieve a good quality of finished product for complicated geometry, moreover, finding a way to adapt the slicing procedure to the model improves production times and costs [32,33].

Each slicer offers different features to control the single aspect of the data to be exported to the file, usually GCODE, that will ultimately control the printer.

Advanced slicing software also offers among its features the possibility of including geometric dimensioning and tolerance information.

Adaptive slicing helps designers and makers to save time keeping a good level of mechanical strength and a good finish quality varying the height of the layers inside a single part.

Many different algorithms are used to achieve a better adaption [34] and in some cases evaluate the surface error and its related costs [35]. Some of them go beyond software proposing a structured approach [36] real methodology to achieve the expected result.

In Figure 3.18, the adaption results applied to the sphere model depicted before. Please note that, to let the detail be clearer, only the last layers are shown.

As the slicer works with the printers, it should support as many of them as possible, moreover besides the common features, some more add-on could make the difference among different software: scripting, a variety of plug-ins, customisable support material insertion opens the possibility to highly customisation code.

It is possible to find on the 'market' good slicer provided under an open source licence as well as commercial products.

As is often the case, there is no perfect product but a range of products that are, to a greater or lesser extent, suited to the needs of the project.

Last but not least the time it takes to produce the code has to be. In the case of complex geometries, it can take several minutes.

Figure 3.18 Facets vs model

3.16 Technology-related sensitivity

In Section 3.2, it has been introduced the weight of the sensitivity from the point of view of the numerical calculation.

The same kind of problem can be declined also from the constructive point of view. Traditional industrial approach had to deal with the problem of sensitivity in terms of working tolerances.

In fact, even projects that are excellent from an engineering point of view have to face the limits of the production machines during their realisation.

With the Industry 4.0 approach and, specifically, the realisations by means of AM of components, we have to take into account the sensitivity related to printing techniques and materials. Since additive printing is based on the deposition of materials in successive layers, each step introduces an opportunity for inaccuracy. Precision of the result is influenced by many elements such as CAD design, printers and print techniques, materials, slicer setting just for a sake of example.

As for the printers, of course, the technical sheet could help in understanding resolution features, but only if the design takes into account all the variables of the printing process, which have already been highlighted, it can lead to the expected result.

Usually the quality of a printed component is evaluated in accuracy, precision and tolerance.

Accuracy can be easily defined as the difference between the real dimension of a digital model and the ones of the printed part. High accuracy means a good result.

The precision parameter evaluates the consistency of the process. Is it possible to replicate the process with the same dimensional results? If so, the precision is good.

Tolerance gives the idea of how much discrepancy from an expected value is acceptable or expected. In Table 3.3, tolerance for different AM technology is expressed. Please note that tolerances and layer height may differ depending on the

Table 3.3 Tolerance and layer height for different technologies

	SLA	FDM	SLS	MJF	SLM	MJP
Layer height (mm)	0.050	0.20	0.100	0.100	0.050	0.020
Percentage (±)	0.500	0.50	0.300	0.300	0.200	0.100
Tolerance mm (±)	0.006	0.02	0.012	0.012	0.008	0.004

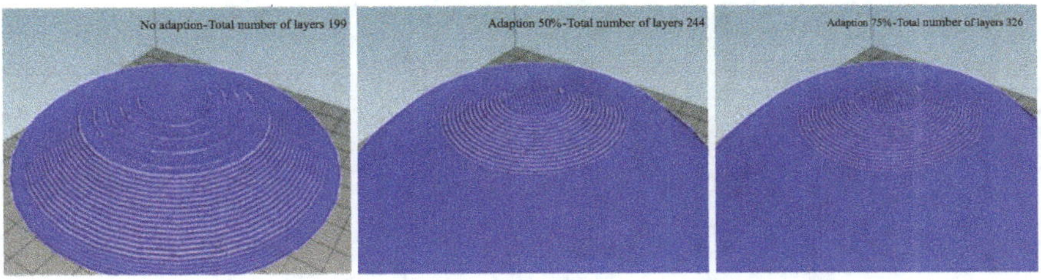

Figure 3.19 Model slicing under adaption increasing percentage

material and printer used. The use of extrusion printers based on FDM has temperature, belt tension and accuracy in printer software calibration and quality of the extruded material among the main causes of inaccuracy.

With respect to powder bed fusion manufacturing, and in particular SLM/SLS processes, four categories have been identified and categorised: laser and scanning parameters, powder material properties, powder bed properties and recoat parameters, build environment parameters, 50 parameters that can affect the quality of printing.

With respect to powder bed fusion printers, and in particular SLM/SLS processes in [44], the authors identify a set of 50 parameters that can affect the quality of printing process and classified them in four main categories: laser and scanning parameters, powder material properties, powder bed properties and recoat parameters, and build environment parameters. It is important to understand the complexity of the printing process in order to optimise it [45] (Figure 3.19).

3.17 Conclusion

The huge conceptual advantage of a combination between optimal shape design methods, for synthesising a device shape, and 3D printing technology, for prototyping the shape, is to create a technique capable of fabricating surfaces characterised by arbitrary curvature and arbitrary distribution of materials. This unique feature has greatly expanded the class of prototypes achievable allowing the implementation of shapes and geometries identified through numerical methods of

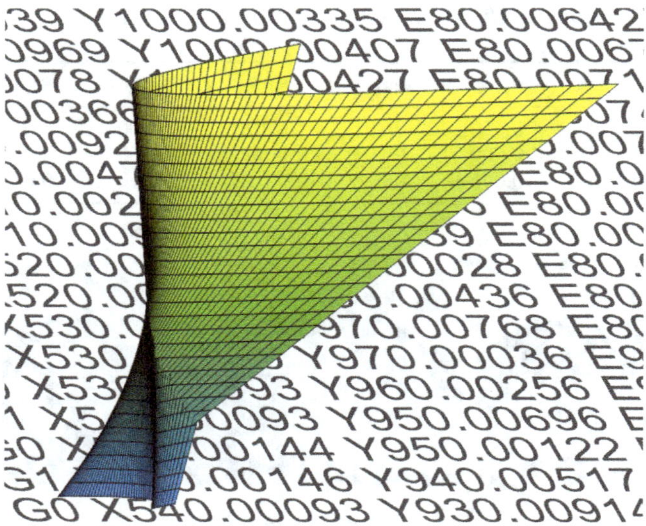

Figure 3.20 A tribute to René Thom

project optimisation. In particular, non-parametric or FFO appears more and more as a concrete objective.

In the end, an artistic picture realised by the authors is proposed (Figure 3.20). The surface represents the swallowtail-like catastrophe according to the popular theory of René Thom [46]: it could well evoke an optimal shape to be printed by means of AM, as the segment of G-code shown in the background suggests. The sophisticated shape and the citation of a code for 3D printing is suggestive enough to capture the spirit of the chapter; in a sense, the title could be 'A tribute to René Thom', who greatly contributed to the development of differential topology, a pioneering topic for understanding the evolution of shapes.

List of acronyms

ABS	acrylonitrile butadiene styrene
DMLS	direct metal laser sintering
FDM	fused deposition modelling
PLA	polylactic acid
RepRap	replicating rapid-prototyper
SLA	stereolithography apparatus
SLM	selective laser melting
SLS	selective laser sintering
STL	stereolithography

References

[1] K. Deb, *Multi-Objective Optimization Using Evolutionary Algorithms*, Chichester, London: Wiley, 2001.

[2] P. Di Barba, F. Dughiero, M. Forzan, and E. Sieni, "Sensitivity-based optimal shape design of induction-heating devices" *IET Science, Measurement and Technology*, vol. 9, n. 5, pp. 579–586, 2015.

[3] P. Di Barba, F. Dughiero, and E. Sieni, "Parameter-free Paretian optimization in electromagnetics: a kinematic formulation" *IET Science, Measurement and Technology*, vol. 7, n. 2, pp. 93–103, 2013.

[4] P. Di Barba, M. Mognaschi, A. Savini, and S. Wiak, "Island biogeography as a paradigm for MEMS optimal design" *International Journal of Applied Electromagnetics and Mechanics IJAEM*, vol. 51, n. S1, pp. 97–105, 2016.

[5] F. Campelo, J. Ramirez, and H. Igarashi, "A survey of topology optimization in electromagnetics: considerations and current trends". Available: https://www.academia.edu/2751679/A_survey_of_topology_optimization_in_electromagnetics_considerations_and_current_trends. [Accessed December 2020].

[6] Z. Andjelić, "A novel approach for free-form optimization in engineering design" *IEEE Transactions on Magnetics*, vol. 49, n. 5, pp. 2010–2014, 2013.

[7] Z. Andjelic, B. Krstajic, S. Milojkovic, and Z. Haznadar, "A procedure for automatic optimal shape investigation of interfaces between different media" *IEEE Transactions on Magnetics*, vol. 24, n. 1, pp. 415–418, 1998.

[8] "Insight & SIMULIA Execution Engine | Dassault Systèmes®". Available: https://www.3ds.com/products-services/simulia/products/isight-simulia-execution-engine/. [Accessed December 2020].

[9] "Home – POLOPT Technologies". Available: http://www.polopt.com. [Accessed December 2020].

[10] CASOPT, EU FP7 Marie Curie IAPP Project, "Controlled Component- and Assembly-Level Optimization of Industrial Devices" ABB Corporate Research, TU Graz, TU München, University of Cambridge, 2009–2013.

[11] B. Dobrzelecki, K. Dudek, T. Nowak, and A. Blaszczyk, "Integration of dielectric design optimization into a product development workflow" in Proc. 1st International Conference on Engineering and Applied Sciences Optimization, OPT-i 2014, Kos Island, Greece, 2014.

[12] K. Bandara, F. Cirak, G. Of, O. Steinbach, and J. Zapletal, "Boundary element based multiresolution shape optimization in electrostatics" *Journal of Computational Physics*, vol. 297, pp. 584–598, 2015.

[13] H. Neuber, "A physically nonlinear notch and crack model" *Journal of The Mechanics and Physics of Solids*, vol. 16, n. 4, pp. 289–294, 1968.

[14] P. Di Barba, *Multiobjective Shape Design in Electricity and Magnetism*, Berlin: Springer, 2010.

[15] Z. Andjelic and S. Sadovic, "Reduction of breakdown appearance by automatic geometry optimization" in Annual Report Conference on Electrical Insulation and Dielectric Phenomena, 2007.

[16] P. Di Barba, M. E. Mognaschi, T. Nowak, and A. Blaszczyk, "Free-form optimisation in industrial dielectric design: a comparative approach" *International Journal of Applied Electromagnetics and Mechanics*, vol. 60, n. S1, pp. S101–S113, 2019.

[17] M. Powell, "An efficient method for finding the minimum of a function of several variables without calculating derivatives" *Computer Journal*, vol. 7, n. 2, pp. 155–162, 1964.

[18] A. Blaszczyk, J. Ostrowski, B. Samul, and D. Szary, "Simulation toolbox: dielectric and thermal design of power devices" *ABB Review*, n. 3, 2013.

[19] C. W. Hull, "The birth of 3D printing" *Research Technology Management*, vol. 58, n. 6, pp. 25–30, 2015.

[20] C. W. Hull, "Apparatus for production of three-dimensional objects by stereolithography". US Patent 4575330, 11 March 1986.

[21] S. S. Crump, "Apparatus and method for creating three-dimensional objects". US Patent 5121329, 9 June 1992.

[22] C. R. Deckard, "Method and apparatus for producing parts by selective sintering". US Patent 4863538, 5 September 1989.

[23] R. Larson, "Method and device for producing three-dimensional bodies". US Patent 5786562, 28 July 1998.

[24] D. B. Russell, T. Anderson, J. F. Bredt *et al.*, "Method and apparatus for prototyping a three-dimensional object". US Patent 6007318, 20 December 1999.

[25] H. Gothait, "Apparatus and method for three dimensional model printing". US Patent 6259962, 10 July 2001.

[26] A. Savini and G. Savini, "A short history of 3D printing, a technological revolution just started" in ICOHTEC/IEEE International History of High-Technologies and their Socio-Cultural Contexts Conference (HISTELCON), Tel-Aviv, Israel, 2015.

[27] "About – RepRap". Available: https://reprap.org/mediawiki/index.php?title=About&oldid=187994. [Accessed June 2020].

[28] "Aniwaa – Get the right 3D printer or 3D scanner". Available: https://www.aniwaa.com. [Accessed January 2021].

[29] E. Napadensky, "Inkjet 3D printing" in *The Chemistry of Inkjet Inks*, World Scientific, Singapore, 2009, pp. 255–267.

[30] "The 3D inkjet printing process explained". Available: https://www.nano-di.com/blog/2019-the-3d-inkjet-printing-process-explained. [Accessed December 2020].

[31] M. Rafiee, D. Farahani, and D. Therriault, "Multi-material 3D and 4D printing: a survey" *Advanced Science (Weinheim, Baden-Württemberg, Germany)*, vol. 7, n. 12, 2020.

[32] C. R. Garcia, J. Correa, D. Espalin *et al.*, "3D printing of anisotropic metamaterials" *Progress In Electromagnetics Research Letters*, vol. 34, pp. 75–82, 2012.

[33] J. Allen and B. Wu, "Design and fabrication of an RF GRIN lens using 3D printing technology" *Proceedings SPIE*, vol. 8624, 2013.

[34] L. Ren, X. Zhou, J. Xue *et al.*, "Terahertz, RF, millimeter, and submillimeter-wave technology and applications VI" *Advanced Materials Technologies*, vol. 4, 2019.

[35] "Design-and-application-of-metamaterials". Available: http://www.eletel.p.lodz.pl/eng/index.php/research/telekomunikacja/design-and-application-of-metamaterials. [Accessed December 2020].

[36] A. Kallaste, T. Vaimann, and A. Rassãlkin, "Additive design possibilities of electrical machines" in 59th International Scientific Conference on Power and Electrical Engineering, Riga, Latvia, 2018.

[37] N. Trnka, J. Rudolph, and R. Werner, "Magnetic properties of ferromagnetic materials produced by 3D multi-material printing" in IEEE 29th International Symposium on Industrial Electronics (ISIE), Delft, the Netherlands, 2020.

[38] P. Lykov, R. Baytimerov, S. Vaulin, E. Safonov, and D. Zherebtsov, "Selective laser melting of copper by 200 W CO_2 laser" in SAE Technical Paper 2016-01-0333, Detriot, 2016.

[39] P. Lykov, E. Safonov, and A. Akhmedianov, "Selective laser melting of copper" in Materials Science Forum, vol. 843, February, 2016.

[40] T. Tran, A. Chinnappan, J. Lee *et al.*, "3D printing of highly pure copper" *Metals*, vol. 9, n. 756, 2019.

[41] F. Lorenz, J. Rudolph, and R. Wemer, "Design of 3D printed high performance windings for switched reluctance machines" in 2018 XIII International Conference on Electrical Machines (ICEM), Alexandroupoli, 2018.

[42] T. Hou, J. Xu, W. S. Elkhuizen *et al.*, "Design of 3D wireless power transfer system" *IEEE Access*, vol. 7, pp. 94793–94805, Accepted for publication, 2019.

[43] S. Lammers G. Adam, H. J. Schmid *et al.*, "Additive Manufacturing of a lightweight rotor for a permanent magnet synchronous machine" in 6th International Electric Drives Production Conference (EDPC), Nuremberg, 2016.

[44] T. Spears and S. Gold, "In-process sensing in selective laser melting (SLM) additive manufacturing" *Integrating Materials and Manufacturing Innovation*, vol. 5, pp. 16–40, 2016.

[45] M. Van Elsen and J.-P. Kruth, *Complexity of Selective Laser Melting: A New Optimisation Approach*, Leuven, Belgium: Katholieke Universiteit Leuven, 2007.

[46] R. Thom, *Structural Stability and Morphogenesis: An Outline of A General Theory of Models*, W.A. Benjamin, Reading, Massachusetts, 1975.

Chapter 4

Innovative motors and shape optimisation

Lidija Petkovska[1] *and Goga Cvetkovski*[1]

Electric motor is one of the few inventions that shaped electric technology the most. All it started when the famous Danish scientist Hans Christian Ørsted (1777–1851) discovered (1820) that a compass needle, when placed near a DC current carrying conductor, is deflecting from the real North. Thus, Ørsted proved that electric currents create magnetic fields and, for the first time, he connected electricity and magnetism, known as electromagnetism. The path for further discoveries was paved.

The very beginning of an electric motor is found exactly two centuries ago, in the year 1821. After Ørsted's discovery, Michael Faraday (1791–1867) undertook a series of experiments in his laboratory at the basement of the Royal Institution in London [1]. In the field of electromagnetism, Faraday is the best known for his discovery of electromagnetic induction and electromagnetic interaction. However, in 1821 on September 3, he first described the principle of *electromagnetic rotation*, which actually stands behind the *electric motor principle*.

Of course, Faraday's primitive motor did not look anything like the modern motors. Even though it had no practical application, Faraday's invention was the first step in the evolution of electric motors and paved the path for the fast forthcoming developments. Soon after many worldwide known scientists and researchers made immense improvements, inventing a variety of electric motor topologies.

The next milestone in the development of modern electric motor technology is the invention of AC motors with rotating magnetic field. It happened that two great minds Galileo Ferraris (1847–1897) in Italy and Nikola Tesla (1856–1943) in the USA independently and simultaneously conceived the idea of building a two-phase commutatorless motor. After several years of individual work, they both published their discovery in 1888 in Turin and New York, respectively. Unlike Ferraris, yet in 1887 Tesla applied with his motor for US patent, which was accepted and granted a year later, in 1888 [2]. Tesla's discovery of the polyphase rotating magnetic field

[1]Faculty of Electrical Engineering and Information Technologies, Ss. Cyril and Methodius University, Skopje, North Macedonia

principle enabled modern-day three-phase systems for electric power production, transmission, distribution and consumption.

After this major milestone, the evolution of electric motors started to grow rapidly in the twentieth century, becoming a cornerstone of many modern devices. Over the last few decades, through numerous innovative techniques, the electric motor has evolved extensively and continues to play a key and growing role in today's society.

4.1 Trends in electric motor technology

With the development of new technologies, the electrical industry and in particular the production of electro-magneto-mechanical devices has been moving forward. Over the time, many significant changes have taken the place in the electric motor technology. Among them, several more important present trends in the production of electric motors, without giving any priority, are depicted in Figure 4.1.

For many years, the energy efficiency of electric motors was the subject of a voluntary commitment by the manufacturers and the agreement with the customers. However, the globally increasing demand for energy savings gave rise to a range of regulations for determining motor efficiency.

First, in 1998, the European Committee of Manufacturers of Electrical Machines and Power Electronics (CEMEP) established the efficiency classification for three-phase induction motors in three levels, using EFF code and respective numbers. The number 1 was given to the most energy-efficient motor (high efficiency), while number 3 was assigned for the least efficient electric motor (low efficiency). The level 2 was the rating for motors with improved efficiency. In other words, the lowest class number represented the highest motor efficiency.

Intending to harmonise the standards describing the motors' energy efficiency worldwide, the International Electrotechnical Commission (IEC) has developed the new classification globally regulated requirements placed upon the energy efficiency of electric motors. It was in 2008, when these IEC standards for energy efficiency of three-phase low-voltage induction motors were initially developed and proposed in a common manner of collaboration. The used code was IE, and again there were defined three efficiency groups from 1 to 3, but in this case number 1 was assigned to the least efficient motors. The respective classes were labelled: standard IE1, high IE2 and premium efficiency IE3. In 2014, for the

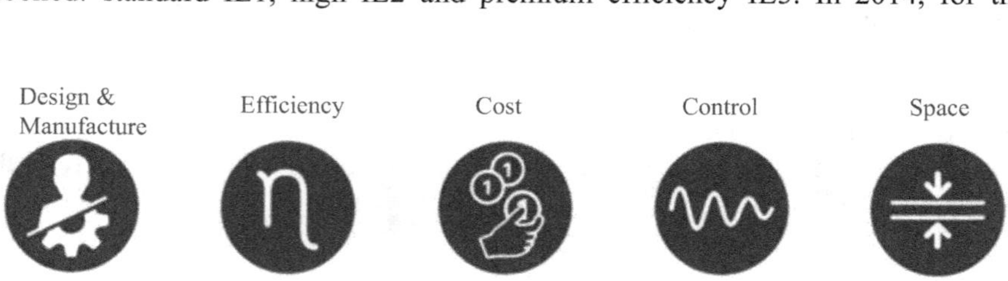

Figure 4.1 Trends in electric motor technology advancements

purpose of expanding the scope, the prior efficiency classification has been updated and an additional efficiency class IE4 has been added. This highest class is reserved for super premium efficiency. The comparison of the earlier and actual efficiency classes for electric motors is presented in Figure 4.2. It is worth to emphasise that under-construction is the next, IE5 motor class, in which would be 20% lesser losses than in IE4 motors [3].

At the present time, the industry for electric motors manufacture has dramatically changed to raise the basic efficiencies on large scale of various electric motors that are fulfilled during the production. The base efficiency of electric motors in general, and of three-phase induction motors in particular – for the number of poles 2–8 and for power range 0.12–1,000 kW – has recorded a dramatic increase in the past decade. Nowadays, even small single-phase induction motors became significantly more efficient. As both the electric motor designers and producers are receptive to the great importance for efficiency increase of electric motors, further innovative advancements have been already undertaken by stakeholders.

Another aspect of motor innovations is the production process. The most recent 3D printing technology developments have added a huge value to the designing, prototyping as well as manufacturing process of electric motors. The broad spectrum of production applications has been unlocked. The laser technology is growing, becoming more workable and affordable, such that additive manufacturing (AM) from 3D printing of only parts of an electric motor has been moved to the printing even of the whole motor structure [4]. The significant reduction of

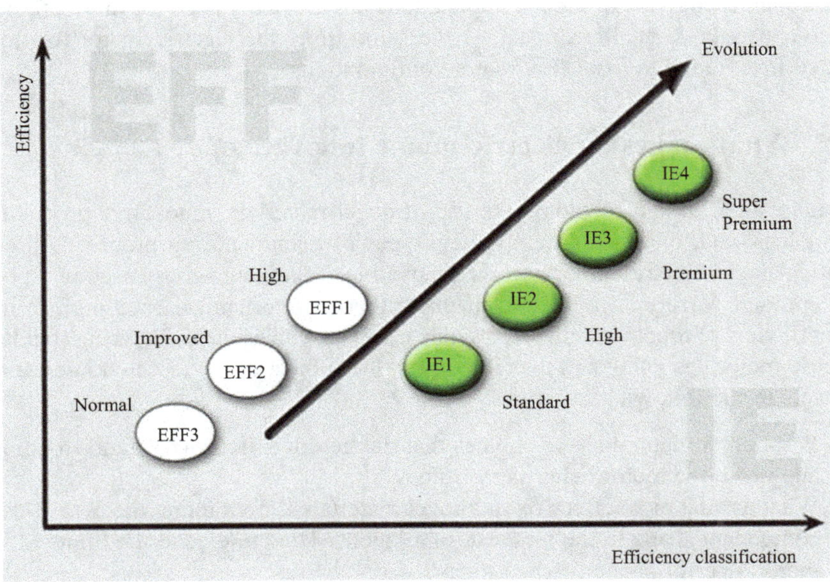

Figure 4.2 Electric motor efficiency classes comparison

time and money consumption, required for the stage of prototyping an electric motor, led to cost-effective massive motor production. Thus, for less money the same quality can be reached, or for the same money more quality motor is obtained.

One concept of the substantial importance for an electric motor performance is speed control. The classical speed control of AC motors, by gear reducers and the number of poles, grace to the fast advancements of variable frequency drives (VFDs) with pulse width modulation, has been sent away. The recent drastic cost reduction and reliability increase of VFDs have driven changes in electric motor designs and the adoption of VFDs to new levels. That fact has brought about to become a standard addition of the inverter as an integral part of the motor structure inside the common housing. As a result of semiconductor technology and microprocessor advances, the VFD has become smarter, more efficient and more accurate. By active modulation of the volt/hertz ratio, a remarkable innovation in the electric motor control took place. This method of motor control, however, causes different electrical stresses on a motor, which necessitated further motor design innovations and led to an inverter-duty rated motor. Current motor innovations, such as in the permanent magnet (PM) AC motor technology, are being refined to bring higher efficiency, cooler running temperatures and incredible speed and torque control accuracy. On the other hand, the rediscovery of synchronous PM and reluctance motors, together with recently added PM-assisted reluctance motors, is becoming evident [5].

Finally, new trends in e-mobility, the developments of EV (electric vehicles) and HEV (hybrid electric vehicles), forced new requirements related to the size and the weight of the drive, resulting in innovations for the embedded electric motors. Consequently, the rise of mobile applications has created a need for more efficient space-constrained smaller drives. At the same time, the electric motor topology drifted from radial to axial flux motor configurations.

4.2 What makes an electric motor innovative?

An innovative design topology, the use of novel materials, innovative production technology and, the most recently, new types of manufacturing process will certainly make an electric motor become innovative. Taking into consideration aspects of the motor performance (electro-magneto-thermal-mechanical coupling), from a point of view of topology, novel materials usage as well as the most recent trends in electric motor designing and manufacturing, the subject of the research interest are answers to the following questions:

- What are structural/shape choices that define novel/improved motor topology and make the motor better performing?
- What are the most effective and novel materials for obtaining the best of each component/shape in the motor design layout? How to extend the limits of the motor's performance reached today?
- How is an innovative motor manufactured? What are novelties in the production of electric motors?

4.2.1 Novel design topologies

After the first fully functional electric motor went to wide application, the invented motor topologies have been relatively unchanged for decades. However, for the petrol crisis of the 1970s in the last century and the more and more restrictive requirements for keeping the environment clean and safe, new movements in the 'energy' world appeared. The green electrical energy production, banned coal and fuel power plants, introducing many alternative energy resources, among which the more important are solar and wind energy. To lower electrical energy consumption, while keeping functionality of appliances, it is required to design novel devices with increased efficiency level. At the same time, the inefficient combustion technology for the cars started to be replaced with electrical or at least combined hybrid technology. All these movements have attracted electric machine designers towards innovative topologies and designs, to meet the new requirements. To visualise an example of innovations in electric motor design and topology, in Figure 4.3 is displayed an expanded view for a PM-assisted reluctance motor.

These meaningful themes challenge the researchers and scientist worldwide to the innovation of electric motors design, by means of their topology. In a single chapter of a book, it is not possible to describe and to present, even a small part of recent innovations, so we recommend finding more information in the listed references.

Commercialisation of innovative motor topologies that have been around for a long time, but have not been broadly found in practice, triggered the growth and penetration of brushless PM motors (PMMs). Some 40 years ago, the major motor sale was for induction and DC brushed motors. Today, induction motors are still important, but brushless PM (BLPM) has grown dramatically, relegating DC

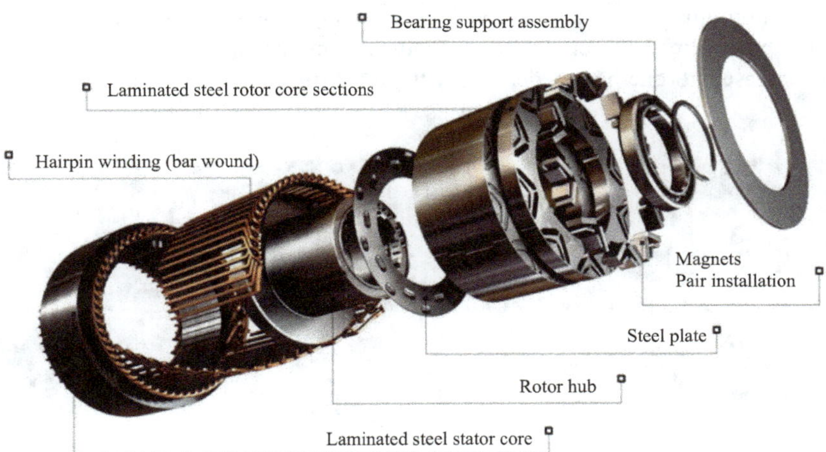

Figure 4.3 Expanded view of innovative PM-assisted reluctance motor

commutator. Thus BLPM motors, along with other topologies such as synchronous reluctance motors – with or without PM assistance – and switched reluctance motors, have increased their application and can be also found in higher volume units. The increase in VFD capability and their cost reduction had also led to a drastic decrease of DC brushed motor applications.

When seeking for a well-synthesised motor for a selected topology, the designer should encompass the most of used materials – iron and copper – and to optimise electromagnetic, thermal and mechanical design, combined with some aspects of noise and vibration. Hence, to provide the best problem solution, the process is always carried out with the participation of experts in motor design, material sciences, electromagnetics, electromechanics, cooling, bearings, etc.

4.2.1.1 Innovative stator winding

In the context of innovative design for AC motors, mostly three-phase, the emphasis is put on the stator windings design innovations and novel production technologies. The growing use of flat (bar) windings – recognised as *hairpin* winding, instead of litz wire windings – with *basket* overhangs, both for wound stator, is leading to better torque density for the same mass and/or motor volume from one side, and to the improved efficiency due to lower copper loss, from the other. Also, the heat dissipation and cooling of the motor, compared to the classical basket winding is much better. Having in mind that the better is motor cooling – the higher is motor efficiency, this fact is certainly of the greatest importance. In Figure 4.4 is presented view for stator winding's overhangs for the both types of stator winding.

The fill factor for litz wire winding is on average 0.35–0.4, while for hairpin flat wire is at least 0.6; with an optimised slot geometry, it even goes beyond this limit. Having at a disposal more copper inside the stator slot with better thermal connections, at unchanged current, allows reducing joule losses in the motor, and hence improving efficiency, as well as reducing over-temperatures. Finally, it means possibility of getting more compact motor with better performance and increased torque/power density, with the same mass and volume of active

(a) (b)

Figure 4.4 Stator-windings overhangs: (a) basket – litz wire; (b) hairpin – flat wire materials – copper and iron. Because of these advantages, the hairpin windings are in growing use and increasingly present in the novel motor designs for EVs and HEVs.

Electric motors with bar windings became one of the most important innovation directions in the diverse fields of EVs drives. Substantial innovations of hairpin windings are undergoing:

- Electrical efficiency of the motor can be further optimised by using parallel pins, packed in a stack, to lower existing AC resistance effects at higher frequencies.
- Additionally, by using uneven pins, more flexible variations of performance and torque classes are possible to be obtained.
- The motor thermal efficiency is optimised by using a highly flexible overhangs cooling for more efficient heat dissipation from the winding head.
- Possibility for hairpin windings to be produced by 3D printing technology is now a new challenging matter in the course.

4.2.2 Novel materials

Regarding the materials used for designing and manufacturing of an electric motor, yet fundamental role is played by the core lamination and PM materials. New developments in electric steel properties to ongoing applications for the production of electric motor cores are presented and compared. The contribution to better motor performance is documented. Here, we are not going deep in material technology, and we suggest interested reader to read Chapter 5 of this book.

Recently, PMMs have been rediscovered and their application in various devices, from small house appliances to large aircrafts, has been visibly extended. Consequently, trends in PM production developments are also worth to be considered.

4.2.2.1 Core lamination materials

Apart from the stator-winding copper loss, another significant loss component that affects the overall motor performance and efficiency is iron loss in the motor cores for the stator and the rotor. Proper selection of the electric steel material for core laminations is an important matter of exploration. Depending on the motor topology, due to electromagnetic circumstances that are specific for the particular motor operation, always one of these two loss components, either stator or rotor iron loss, is possibly neglected.

Speaking about AC motors, induction and PM, usually the rotor core loss is with negligible value, and only the stator core loss is considered. In this regard, a motor designer should encompass new stator topologies, application of novel materials, as well as state-of-the-art manufacturing processes. Stator and rotor cores of an electric motor, depending on the topology, have a variety of shapes and are manufactured from a variety of materials. Lamination examples are presented in Figure 4.5.

Electrical steel is a soft magnetic material, comprising silicon–iron, nickel–iron and cobalt–iron alloys, where iron is the main component. The silicon–iron

Figure 4.5 Lamination examples for stator and rotor cores

(SiFe) alloys with 0.5%–4.0% Si, referred to as silicon steel, lamination steel or transformer steel – with enhanced magnetic properties – are found in a broad use across the applications in variety of electromagnetic devices. The Si added per cent has a marked impact on the lamination properties. Silicon steel can be non-oriented or oriented in dependence of the magnetic properties; also, depending on the thermal treatment is fully processed or semi-processed. Non-oriented fully processed silicon iron is used in the greatest extent. It is available in an array of grades and thicknesses so that this material may be well tailored for various applications. Silicon electric steel is generally specified and selected on the basis of allowable specific iron loss in watts/kg. It has uniform magnetic properties in all directions. The lower silicon alloy grades provide better magnetic permeability and thermal conductivity.

Today, the cobalt–iron (CoFe) electric steel for building stator/rotor core of an innovative motor is more and more often in use. This novel material is characterised by better electromagnetic behaviour, broader use range before saturation and low loss figures, at the same values for magnetic flux densities and frequencies, compared to materials of the silicon–iron (SiFe) class.

With courtesy of the company VACUUMSCHMELZE GmbH & Co. KG (VAC®), Hanau, Germany – one of the world's leading manufacturers of special alloys, especially in the area of magnetism – to illustrate the advantages of innovative high-performance CoFe electric steel in Figure 4.6 are presented comparative characteristics of their products 49%CoFe @ 2%V (vanadium) with commercial names VACOFLUX® and VACODUR®, with reference to 3%SiFe alloy [6].

Thanks to quite high value of magnetic saturation, the cobalt–iron electric steel is generally suitable for applications where high flux density is required, allowing to lower the iron mass, which in return gives decreased iron loss and reduction of the core weight. Other important features of CoFe alloy are high magnetic

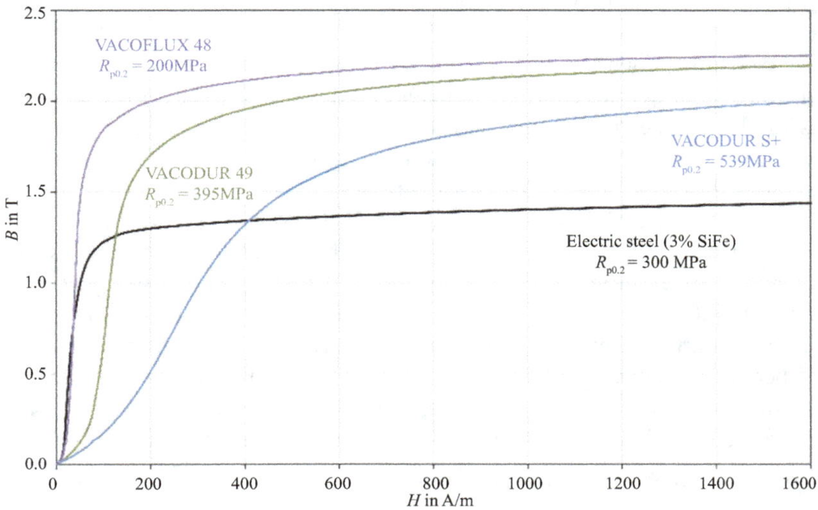

Figure 4.6 Normal DC magnetisation curves of 49%CoFe @ 2%V alloys and a reference material 3%SiFe with corresponding yield strengths

saturation up to 2.4 T, and small specific iron loss 1.5–1.6 W/kg. In addition, due to low coercivity forces the best possible parameters are obtained, and the highest performance electric motors/generators as well as magnetic actuators can be built. Although the mass density is a bit higher, 8.12 kg/dm^3, due to improved magnetic properties, the iron consumption is lowered, and, in turn, the motor dimensions and weight are reduced.

For given power rating, the power density of an electrical machine is determined by the magnetic saturation of the material, and hence the higher magnetic saturation of the lamination is often desired, so that it will result in less material required to transfer the magnetic flux inside the motor magnetic circuit. Further expansion of innovations in electric motor technology with new lamination materials is foreseen.

4.2.2.2 Soft magnetic composite materials

The fast developments of e-traction and renewable power generation, that are characterised with electrical machines of high compactness, higher efficiency and high power/torque density, have driven the research towards not only developing the novel topologies of electrical machines to meet special requirements but also to inventing new materials that will enable their easier manufacturing.

Continual advances in the area of e-mobility and high power density electric motors bring new challenges to the motor constructions. The traditional laminated motor construction is limited to 2D radial magnetic flux to minimise losses in the direction perpendicular to the steel lamination. On the contrary, the topologies of the axial flux motors require very complex 3D magnetising directions where the use of electrical steels is often not practical. In addition, the use of conventional electrical steels is limited to medium frequency values, up to 1,000 Hz. What to do?

The solution is found in the application of soft magnetic composites (SMCs) to manufacture some parts of an electromagnetic device, or even in the whole. By means of the powder metallurgy and advanced AM, SMCs can be formed into complex shapes, allowing three-dimensional magnetic circuits to be designed and easy manufactured. With their help, it is possible to produce more compact innovative and cost-efficient electric motors. The SMCs have many unique characteristics, which can be utilised to improve the performance of traditional electrical machines and to develop innovative designs and topologies with more complex magnetic circuits and increased power density. Sometimes, the geometrical size of the available parts is limited, mainly due to specific compacting processes. The SMCs become interesting at frequency values >500 Hz. However, there is a number of disadvantages of SMCs that should also be taken into account during the electromagnetic design and construction of electro-magneto-mechanical devices.

Advantages of soft magnetic composites

- Higher power density due to existence of 3D magnetic flux in the device.
- Complex shapes are possible to enable a broad range for the product design.
- Lower loss at high frequencies, compared with lamination steel.
- Isotropic physical properties (thermal conductivity and relative magnetic permeability) are independent from the direction.
- Lower eddy current losses due to higher electrical resistance of SMCs.

Disadvantages of soft magnetic composites

- High values of magnetic field strength required to reach magnetic saturation.
- Low magnetic saturation $\approx$1.5 T@10,000 A/m.
- Low magnetic permeability, on average $\mu_r \approx 600$.
- Higher hysteresis losses.
- Inferior mechanical properties.

State-of-the-art applications

The benefits of SMC solutions for both the motor topology design and production processes enabled wide spectrum of state-of-the-art products, which are released by several big worldwide known companies [7]. Soft magnetic powder composites coupled with PM production processes open new possibilities in the design and manufacture of parts for electrical device applications or the whole structures.

With increasing values of operating frequency for electric motors, the use of SMCs can contribute to a substantial decrease of specific core losses of the machine, increasing at the same time the total efficiency [8]. In contrast to laminated cores, the manufacturing process of soft magnetic composites does not influence their final homogenous magnetic properties, which do not change after assembly of the motor.

Somaloy®. is a trademark (TM) of Höganäs AB from Sweden, the world-known producer of soft magnetic composite materials [9]. They have developed a variety of electromagnetic applications. Capabilities of the Somaloy materials have opened

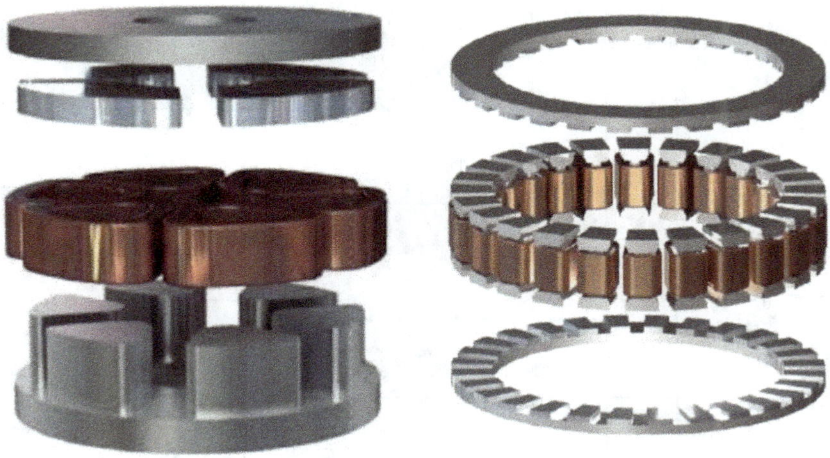

(a) (b)

Figure 4.7 Somaloy innovative axial flux PM motor solutions: (a) single-sided; (b) double-sided

up new opportunities to design innovative motors, produced by novel technology, as elaborated in [10]. They are designed for efficient modular production at low cost, by exploiting state-of-the-art technology for 3D magnetic properties and the net volume shaping. Taking as an example axial-flux electric motors for e-mobility, both single- and double-sided, the proposed solution is pictured in Figure 4.7. Few more interesting papers on practical application of SCMs are listed in [11–14].

Through advanced production technology and 3D shape-making capability, the designed topology is manufactured with reduced size and weight, meeting the high-performance requirements, with less need for machining, i.e. with reduced need for subsequent operations, all that in cost-effective manner.

4.2.2.3 Permanent magnet materials

In the past, the most expensive part for building a PMM was the production of rotor with PMs. It was a serious obstacle to their wider use, besides very good performance characteristics. However, throughout recent years, the magnets offered by various suppliers constantly improve. Nowadays, the PMs available on the market are with higher values of remanence B_r, intrinsic coercivity H_c and energy product, known as maximum energy density $(BH)_{\max}$. The benefit of these advancements is that demagnetisation problems became minor, while the higher motor performance with the same magnet volumes, or the same motor performance with reduced magnet volumes, is enabled.

The use of PMs in favour of the mankind is long-lasting. But their application for building electric motors started somehow less than a century ago, when

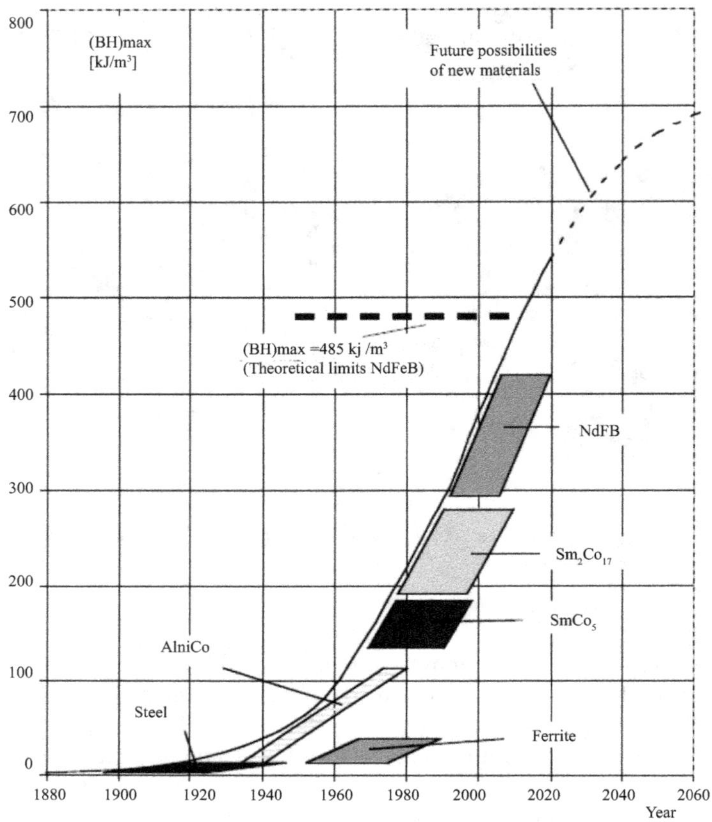

Figure 4.8 Development of permanent magnets by name and energy density

aluminum-nickel-cobalt mixture was sintered under the name AlNiCo, and permanent magnet manufacture became highly effective and commercially viable. Soon after started an intensive research and innovative discoveries that are extended up to todays, and are still in the course. In Figure 4.8 is depicted the historical development of PM materials by decades [15]; corresponding ranges for maximum energy density as per time of their discovery for four distinctive PM materials are depicted. Each generation of PMs is characterised with own features as remanence, coercivity, energy product and price, defining respective advantages and disadvantages. As seen in the figure, the frontiers are open.

Series of various PM materials, with remarkably improved features, have been invented. Typical demagnetisation curves for PMs are presented in Figure 4.9. After AlNiCo-based PMs application started, next were developed cheap ceramic-like PMs, called ferrites. They mutually differ by the value of remanence and

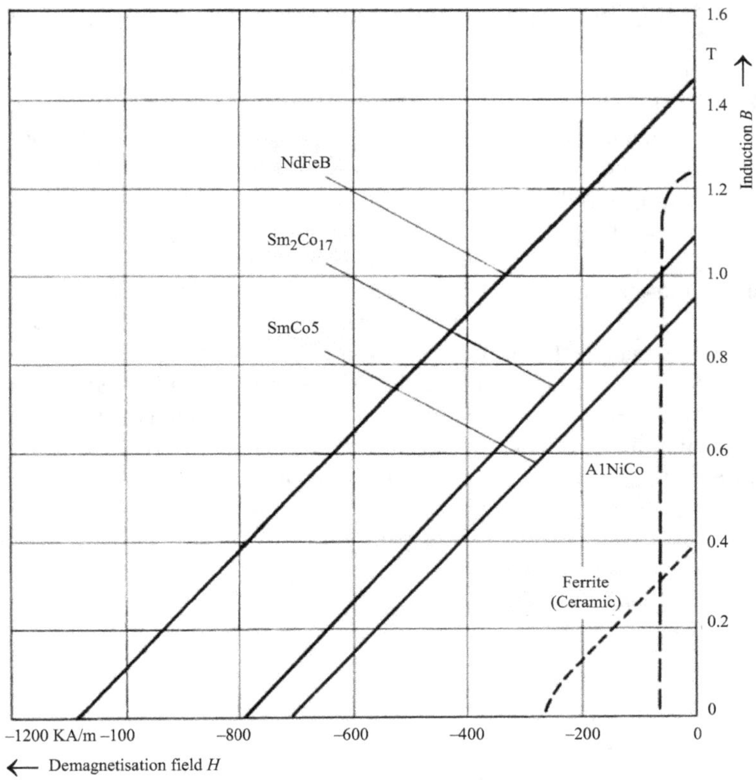

Figure 4.9 Typical demagnetisation characteristics for PM material

coercivity, which stand in inverse correlation. Owing to rare-earth samarium–cobalt (SmCo) PM class, the innovative motor topologies recorded rapid development. Thus, BLPM motor became a combo set with its power supply and control device, dominating in innovative and modern AC electric drives. The only disadvantage was the high cost of PMs production, the fact that challenged future research for PM with features like SmCo and prices as ferrite.

In the last decade of the last century, the sought-after PM material was found in neodymium–iron–boron (NdFeB), the last generation of PMs, which is the best suitable for mounting in electric motor topology. The NdFeB magnets prevailed the others, and are now broadly used, dominating over counterparts. The innovative PM materials, based on NdFeB composition, encompass the best features from the precursors: high magnetic remanence (like AlNiCo), strong coercivity of demagnetisation field (as SmCo) and reasonable price comparable to ferrites. With the help of strong magnetic properties of NdFeB magnets, the same magnetic field is obtained with significantly reduced volume that is often a priority to be achieved. It is of a particular interest when designing electric motors for e-vehicles.

As an illustrating example in Figure 4.10 is presented how the magnet volume reduction is achieved with innovative PM materials [15]. For arbitrary selected PM material from each class with known value for magnetic remanence B_r, there were designed magnet rods to produce the magnetic field of 100 mT flux density at a distance of 5 mm from the surface of the pole. The PM material based on NdFeB, offering the excellent magnetic properties, as well as the highest energy densities currently available, shows dramatic volume reduction.

Electric motors that use PMs are generally more expensive. From an aspect of producers, it is necessary for a design engineer to recognise how important is reducing motor manufacturing cost. Consequently, the motor designer should be able to efficiently select the most suitable PM material, depending on the particular purpose and the specific motor topology, to provide best quality/price.

Outstanding advancements in material composition of PMs, as well as their innovative shaping technology, have resulted in numerous innovative topologies of

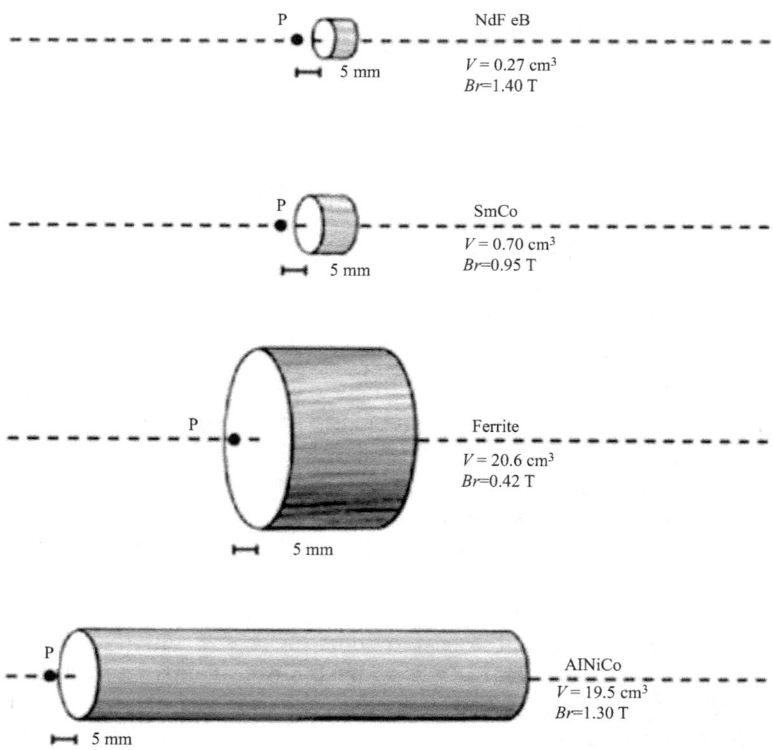

Figure 4.10 Illustrating the volume reduction for various PM material

PMM designs for various applications [16]. PMMs, either radial or axial flux, provide a good compromise of high specific torque and low losses; at the same time, the motor is lighter, safer, quieter and more energy-efficient than any other top drive motor on the market, which justifies its choice for the most automotive applications. Today, PMMs for direct EV/HEV drive are in the forefront of all competitive motors.

4.2.3 Novel production technologies

With growing interest in electrification from clean energy technologies, such as wind power and the use of pure electric mobility in various applications, the demand for the next-generation high-performance electric motors has risen significantly. Electrical machine designer for these applications is facing challenges in terms of meeting very demanding metrics for power density and conversion efficiency [17].

While the innovative electric motors are used in numerous innovative areas, their production has been limited to traditional techniques which, in turn, result in unreliable and not enough efficient products. It is common that electrical machines are usually produced by using traditional manufacturing processes, as laminate punching and packing or powder-based alloy pressing, procedures that often require expensive tooling for extensive machining, or other techniques to remove surplus material. When prototyping a new product, design changes are required, and with conventional manufacturing techniques, an extensive and expensive work should be done. Needless to say, that in every step, the motor must be carefully designed while the used material must be thoroughly selected, to meet the fabrication requirements.

Thereby, it is a substantial motivation of researchers for the exploration of not only advanced novel materials but also innovative production processes for the next generation of compact, high-efficient and light-weight electric motors. Accordingly, it is needed a shift in how these innovative motors are manufactured, and with respect to this, how are designed. It is the AM that can provide this shift. Using 3D computer-aided design (CAD), and having in hand 3D scanners, AM allows the creation of objects with precise geometric shapes, built layer by layer by adding material, opposite to classical manufacturing processes when the excess of material is subtracting. Nowadays, thanks to new technologies and new materials, 3D-printed motors became a reality.

What are challenges? First to explore the new electric motor topologies, then to use advanced optimisation methods for their shape design and synthesis, and finally to exploit state-of-the-art manufacturing of electrical machines. At this stage, it seems the most inspiring is the use of direct 3D printing, widely known as AM of a wide range of products. This subject is elaborated in Chapter 3 and at this point, here are presented only the possibilities of exploiting AM in novel production technologies for innovative design topologies of electric motors. The expected outcome of the feasible applications is to obtain an electric motor that is with compact and lightweight multi-material components, with higher magnetic flux, meaning the higher power density and/or efficiency, easier manufactured and with innovative cooling but, without deteriorating the noise and vibration level [18].

To present days, this promising technology, due to complexity of motor topology, is difficult to be applied on the whole motor structure, although several pioneering research centres throughout the world have reported successful applications. Hence, the AM is used only for particular parts of the motor, e.g. copper windings, or the electric motor is assembled by modular elements fabricated by AM. However, some simple electromagnetic devices, like an inductor, can be performed in whole by 3D printing, and that would result in full multimaterial processing and production.

Another important issue to consider is which materials are feasible for AM? With respect to this, SMCs and SiFe alloys are excellent candidates, while CoFe alloys although with high saturation flux density are still in investigations [19]. In Figure 4.11 is presented expanded view of an axial flux PMM for e-drive, to identify parts of the motor topology and respective materials, for production by AM.

4.2.3.1 Winding production technology

Copper windings in electric motors are of the great importance, because they mostly influence the total loss and define the efficiency. The higher copper content reduces losses and improves the thermal coupling of the winding. By manufacturing copper coils using 3D printing, better performance and higher efficiency are achieved.

The new technology in the field of electric motor topology, as presented before in this chapter, is hairpin windings made of flat copper bars, instead of wound copper wires. They provide 'more copper', increase motor efficiency and improve heat dissipation and motor cooling. The hairpin winding production is easier to be automated than conventional litz-wire wound winding, and hence, they are

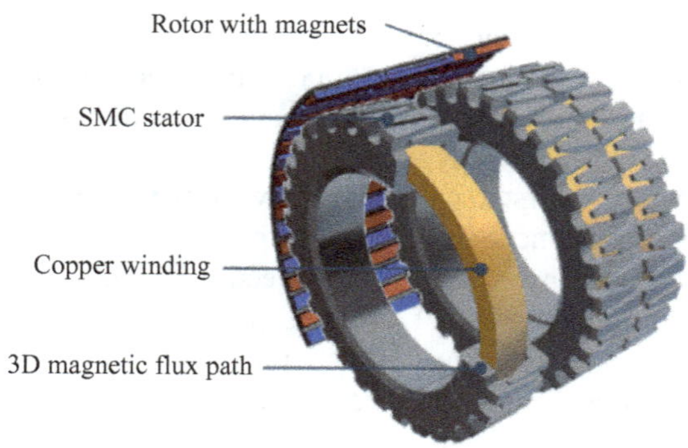

Figure 4.11 Expanded view of AFPM motor

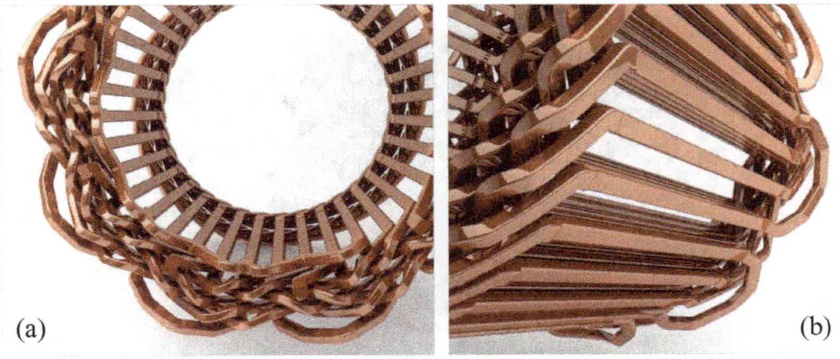

Figure 4.12 Highly efficient 3D-printed hairpin winding: (a) top view; (b) side view

excellent candidates for manufacturing by 3D printing, introducing a new era of electric motor technology. Hairpin winding, produced by Additive Drives group from Halsbrücke (Freiberg, Germany), used when prototyping an e-vehicle, is shown in Figure 4.12. This winding, as stated in [20], is the first real application case and is promising.

4.2.3.2 Electric motors from 3D printer

The earliest practical implementation of 3D printing in electric motor technology was for the purpose of e-mobility, where only some parts of the motor topology have been manufactured, e.g. windings or modular stator core, being used in the stage of prototyping. The 3D-printed functional prototypes can help eliminate expensive early tooling costs, giving engineers ability to make design changes, to iterate quickly and to reduce development time for new products. However, the real challenge for motor designers is producing the whole motor by using AM, the task rather complicated.

The first fully 3D-printed stator for a reluctance electric motor was developed by an enthusiastic group of electrical engineers from Chemnitz University, Germany [21]. This pioneering work was premiered in 2018 at 'Hannover Messe' – the trade fair, and the production phases are shown in Figure 4.13.

Left in the figure is presented the 3D-CAD model of the stator, in the middle is production phase of sintered stator, while right is fully 3D-printed part before heat treatment (Courtesy TU Chemnitz [21]). The stator of the reluctance machine was printed in the Chemnitz University Laboratory using iron for magnetic core, copper for electrical conductors of the winding and ceramics as insulating material, to insulate conductors from each other and from the iron components.

The motor that was created, produced and displayed is a real breakthrough in AM innovative technology for electric motors production, proving the technical principles and demonstrating the feasibility of processes.

Figure 4.13 3D-printed stator of a reluctance machine without winding head

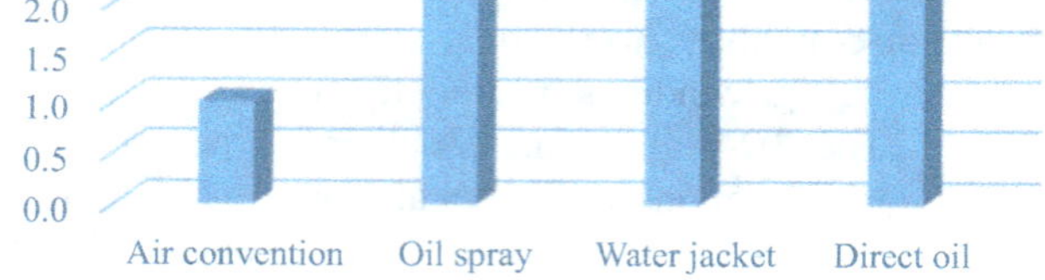

Figure 4.14 Efficiency of various cooling topologies

4.2.3.3 Heat dissipation and cooling

The heat generation in an electric motor is closely connected with the losses: joule loss in the copper of stator windings, eddy current loss and hysteresis loss in the iron laminations of stator and rotor, eddy effect loss in PM material, as well as friction and windage losses in the air gap, due to rotation. All these losses affect the overall motor efficiency. Improving the cooling system and the capability of dissipation, this heat from the motor makes a motor more performing. The better is heat dissipation and the more effective is motor cooling, the higher performance of the motor is achieved.

The main existing cooling topologies for electric motors are free air convection, as the basic one and forced convection outside/inside of the motor, cooling by water jacket and, oil spray or direct oil, as cooling medium. In Figure 4.14 is

Figure 4.15 An example of innovative fully integrated electric motor

presented the comparison of per-unit efficiency for the four cooling topologies, taking as reference the free air convention. This is an important multidisciplinary matter in electric motors technology, and here we are not going deeper to this important issue.

4.2.3.4 Power supply and control trends

The drastic reduction of cost and increase in reliability of VFDs have driven their adoption for electric motors to all new levels. Thus, the further innovations in electric motor technology provide the integration of a motor with power electronics and mechanical transmission system. This perfect mix of components for the best possible motor performance is presented in Figure 4.15. The whole motor enclosure is devised for its connection with the electronic converter for power supply and motor control, and also with the gearbox [22].

4.2.4 Concluding remarks

The innovative electric motors find applications across multiple industry branches, including energy production, e-mobility, manufacturing plants, industrial automation and variety of consumer appliances. To conclude previous headings, few words about expected future of innovative motors' developments and trends in novel production technologies are as follows:

- In general, electric motors and drives will continue to get smaller with higher power densities. Their material and topology will be improved and changed to meet the need for increased efficiency and controllability.
- Induction motors, with significant increase of their efficiency will not go away, still remaining in the widest application.

- Continued growth in variable speed drives, with integrated solutions and other electric motor technologies, will be gaining larger scale of use.
- Brushless PMMs, synchronous reluctance motors without/with PM assistance and even switched reluctance motors, all they based on higher efficiencies and lower cost power electronic devices, will continue to expand and enlarge their use.
- High-speed applications will demand new special motor designs, with embedded VFD, to improve the whole system efficiency and replace traditional mechanical geared drives.
- Due to the increased production of EVs, the new topologies required for automotive industry are expected to drastically drive the demand for innovative electric motors.
- AM with 3D printing will enable designing, prototyping and manufacturing of an electric motor to become easier, faster and more reliable.

4.3 Design optimisation of electric motors

The production of electricity in present days almost entirely relays on electrical generators where on the other hand almost half of the globally produced electricity is used by electric motors. Because of the large energy flow through these machines, the generators and the electric motors have significant impact on the environment. This issue gives the electrical machine designers a good reason to work very hard on the design and production of energy-efficient electrical machines, as well as on innovative design procedures that will involve reliable design techniques, new materials and improved production procedures.

The design of electrical machines has never been an easy task. In general, it is a multi-objective or a many-objective task in which, sometimes, the design has to satisfy rather conflicting objectives, such as high efficiency, low production cost, high torque at less mass of active materials, etc., but by fulfilling various constraints, restrictions and limitations. In practice, such a design process cannot be done by hand, and therefore, at present days, it is usually performed by using powerful computers and applying efficient optimisation algorithms for efficient designing procedures.

In the past, this job was done by various deterministic methods and in many cases they were very successful. However, in optimisation procedures when the objective function is with large number of variables, they did not give satisfactory results or even could not perform at all. The main reasons are as follows: (a) the success of the optimisation depends in large scale on the selection of the starting point, which in some cases cannot be determined exactly; (b) the result can lead to a local optimum instead of a global optimum; (c) some of deterministic methods need determination of the first derivate of the objective function, which in particular problems due to the complexity of the function cannot be done. Important to underline, the deterministic methods perform the optimisation by working on each of the variables separately, which in objective functions with many variables can be time-consuming and at the end the optimisation process might not end at global solution. The problems that have been arisen from those optimisation methods provoked the researchers to seek for better, more reliable and more efficient optimisation methods. Many of them got inspired by nature and its perfect mechanism for adaptation, evolution and reproduction. Other researchers got their inspiration in Darwin's theory of evolution, other in the behaviour of certain species or simply in the mechanisms of the nature. Another group of researchers got inspired by the laws of physics and chemistry and adopted them in successful optimisation methods. That is how the class of optimisation methods is formed generally named 'Nature inspired algorithms'.

4.3.1 Nature-inspired algorithms

Natural processes on planet Earth have been in service and worked perfectly for millions of years in different conditions and environments. Many scientists were inspired by the simplicity and effectiveness of those processes and have decided to borrow them from nature and implement in the artificial-intelligence-computing world. It is estimated that there are about 150 different optimisation algorithms that can be found in literature nowadays and their number is increasing all the time. All of them, in general, can be divided in four groups as presented in Figure 4.16.

Most of those algorithms up to now were used to solve different optimisation problems with great success. In this heading, as an example, the optimal design of a PM synchronous motor (PMSM) by using genetic algorithm (GA), particle swarm optimisation (PSO) and cuckoo search (CS) approach is presented. First, a brief presentation for each of these algorithms, implemented in the optimal designing of the investigated motor, is elaborated.

4.3.1.1 Genetic algorithm

Several researchers while investigating the nature noticed that evolution in nature tends to improve the properties and behaviour of a certain population through the generations. They also have noticed that this behaviour can be mimicked in an optimisation process for different types of linear and non-linear functions. Generally speaking, these evolutionary algorithms refer to a group of optimisation techniques that are inspired by the process of evolution in human and animal life.

GAs are evolutionary algorithms that were first introduced by Holland [23] and his colleagues in the 1960s and the 1970s. Their behaviour is based on Darwin's theory for survival of the fittest species. This method borrows from nature the simplest form of genetic processes of evolution.

The good-quality optimisation results of this algorithm are due to the fact that the method performs the optimisation based exclusively on the value of the objective optimisation function (fitness) and not on the values of the optimisation parameters, as it is the case in the deterministic methods. The optimisation process starts from a randomly generated initial population of solutions (chromosomes). The chromosome of each solution is in fact a vector that can be presented with binary or real numbers [24]. In the algorithm created for optimal designing of a

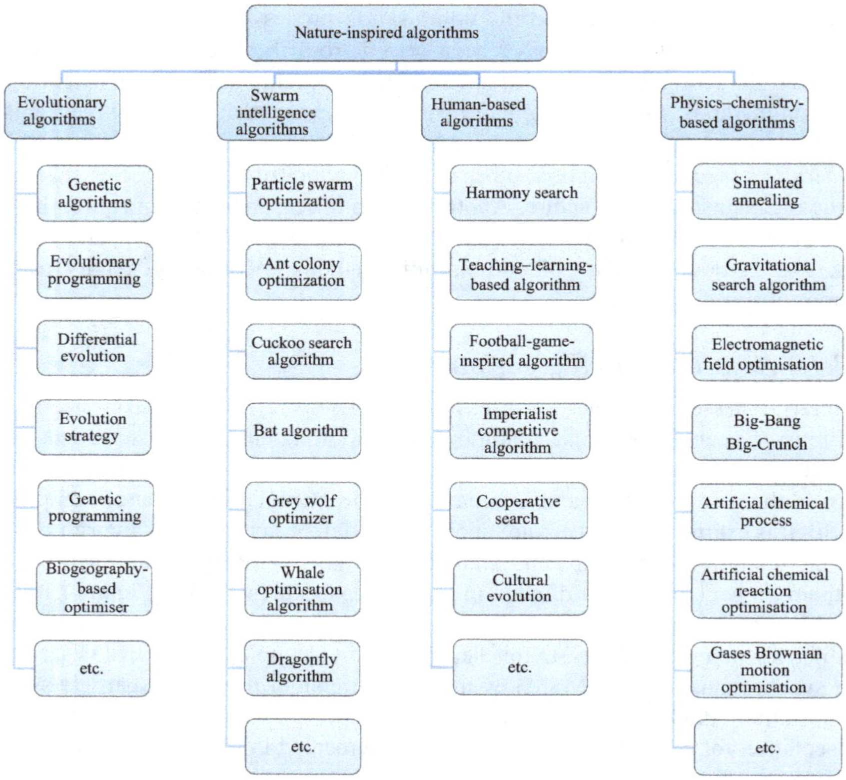

Figure 4.16 Nature-based optimisation algorithms

PMSM, as more suitable for the application, the real number presentation of the variables has been implemented. The principal genetic operators that are in use in the algorithm, like in the nature, involve selection, reproduction, crossover, mutation and elitism. The reproduction on a population is realised by a linear search through a roulette wheel with slots that are weighted in proportion to the string fitness values. For the genetic operator crossover, a uniform arithmetical crossover is applied that guarantees that the values of the new parents will always be in the domain. The mutation is realised through a randomly generated real number for selected variable in its lower and upper bounds. The probability values of these parameters are problem dependent, for each particular application, and should be carefully and thoroughly investigated in order to determine their proper values. The values of all the parameters are always found in the interval between 0 and 1.

As it is the case in nature, the mutation happens very rarely and, therefore, its value is usually taken below 0.1, where on the other hand, the crossover happens more frequently and its value should be 0.6 or above. In this investigation study, the probability of crossover was determined to be 0.85, and the probability of mutation to be 0.07. The last operator that is implemented in the algorithm is elitism that cannot be seen in the nature, and this operator improves the search and ensures that the best solution from one population will not be lost during the process and, for sure, will be transferred to the next generation. In order to additionally improve the GA search, so-called linear fitness scaling is implemented in the algorithm. This procedure adjusts the fitness values of all the chromosomes in such a way that gives the more-or-less equal chance for offspring not only to the top chromosomes but also to the not so good chromosomes. When the genetic operators finish their performance, it is produced a new generation of parents, which have gained new information through the exchange between pairs. The better traits of the chromosomes are carried along to the future generations. This procedure prevents the algorithm from premature convergence of the algorithm and ensures a more reliable search towards global optimum. The stopping rule of the algorithm is usually the number of generations.

The block diagram presentation of the structure for the GA that will be used for the optimal design of a PMSM is shown in Figure 4.17.

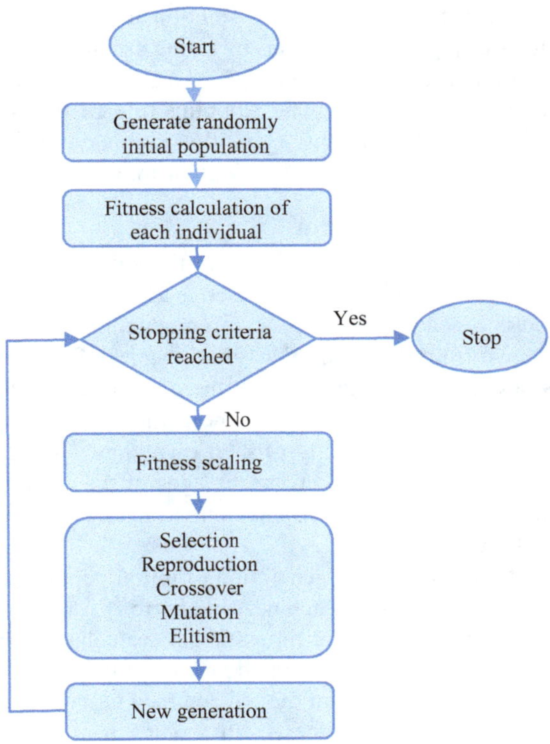

Figure 4.17 Basic structure of a genetic algorithm

4.3.1.2 Particle Swarm Optimisation

PSO belongs to the group of swarm intelligence algorithms, which mimic the swarm behaviour in the nature. This algorithm was introduced by Kennedy and Eberhart in 1995 [25]. Since then, it provoked wide interest in the optimisation community not only for solving general optimisation problems but also for problems of a computer intelligence, as well as shape design optimisation problems.

The PSO algorithm searches the space of an objective function by permanently adjusting the trajectories of individual particles in multi-dimensional working spaces. The individual particles are moved stochastically towards the position with the present velocity of each individual, their own previous best performance and the best previous performance of their neighbours. The effectiveness of PSO algorithm is based on a simple concept, easy implementation, robustness when controlling the optimisation parameters as well as high computational efficiency, compared to the deterministic methods or to some other stochastic optimisation methods. Unlike the other heuristic techniques, PSO has a flexible and well-balanced mechanism to intensify and adjust to both the global and the local exploration abilities.

The PSO algorithm performs the search by using a population of particles, where each particle represents a candidate solution of the problem. In the PSO mechanism, particles change their positions by flying around in a multidimensional search space until a relatively unchanged position has been reached, or until computational limitations are exceeded. The stopping rule in this search is usually defined with the number of iterations. Once this number is reached, the search stops. The flow chart presentation of the PSO algorithm is given in Figure 4.18.

4.3.1.3 Cuckoo search algorithm

Cuckoo search algorithm (CSA) also belongs to the swarm intelligence group of algorithms and was introduced by Yang and Deb in 2009 [26]. The effectiveness of this search method is based on the aggressive reproduction strategy of the cuckoo birds. It is well known from the nature that some species of cuckoo birds lay their eggs in the nests of other birds together with the existing eggs of the nest owner, while the other cuckoo bird species remove the owners' eggs and lay their eggs, thus increasing the hatching probability of their own eggs. On the other hand, the owners of the nests sometimes throw away the eggs of the cuckoo bird or simply abandon the nest. Let us see how this behaviour in nature is implemented in the optimisation algorithm.

The search starts with a random generation of initial population of nests. Then the program calculates the fitness of each nest and performs so-called Levy flight to get new nests' positions for which the fitness is also calculated. The bad nests are removed and replaced by randomly generated new nests in the new positions. Then the program compares fitness values of the old and new nests and defines a new set of nests with improved positions. The nests with worst fitness values are abandoned and they are replaced with new ones using Levy flights. The search continues until

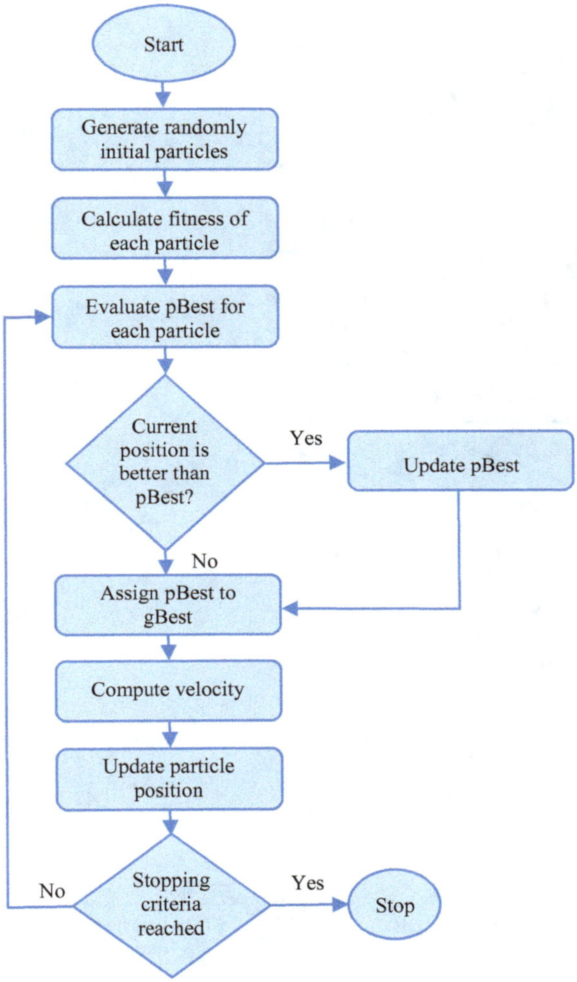

Figure 4.18 Basic block diagram presentation of PSO algorithm

the stopping criteria are reached which is taken to be the number of iterations. Finally, the best selected nest is proposed as a solution of the search. The presentation of the basic steps of the CSA is shown in Figure 4.19.

The authors of this chapter applied the three previously presented methods in the procedure for optimal shape design of a PMSM. First, a brief presentation of the motor is given, as well as only partial presentation of the complex mathematical model of the motor. The equation for the efficiency of the motor, as an objective function in the three optimal design approaches, is derived.

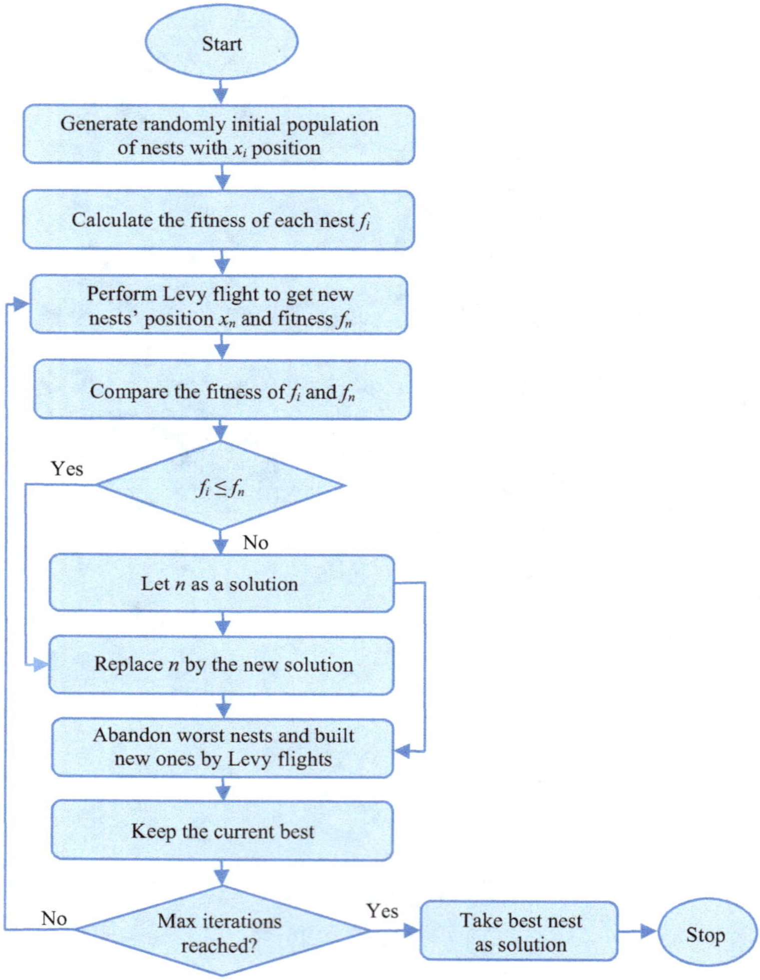

Figure 4.19 Basic block diagram presentation of CS algorithm

4.3.2 Case study

The advancements of PM materials in the last decades have initiated the development of new energy-efficient motors from the PM class of machines. Due to the use of high-quality rare-earth PMs, these motors have higher efficiency, power factor, output power per mass and volume, and better dynamic performance in comparison with squirrel cage induction motors. Like all the other types of machines, they can work either as motors or as generators. In the past, their main application was as motors in various industrial applications. The use as generators increased significantly after their introduction as PM generators in the wind power systems, due to their already demonstrated good performance. As PMMs, in practice, they can be found rating from micro- and milliwatts to kilo- and even megawatts. In particular, PMMs are with radial or axial flux topology, as well as with linear motion and in tubular structures. Regarding the power supply source, they can be supplied with AC or DC voltage and hence, the following motors can be realised as PM brushless synchronous motor, PM brushless DC (PMBLDC) motor and PM brushed DC motor. In the case of radial flux topology, the PMMs can be realised with surface mounted, inserted or buried PMs. Hybrid topologies of PMMs combined with other motor types (reluctance, induction, etc.) can also be found on the market and in practice. These motors are mainly supplied by three-phase inverters that provide good quality start and desirable speed/torque control. There are also some specific PM topologies for direct start electric motors. They have squirrel cage rotor winding, which enables a self-start as an induction motor, and when the rotor speed is reached close to synchronous, due to the presence of the PM field, the motor gets into synchronism and continues working as a synchronous motor. This type of motor is used in applications when there is no need for speed regulation.

4.3.2.1 Surface-mounted permanent magnet – SMPM motor

The optimal design using the three nature-inspired methods is performed on a Koncar-type EKM 90M-6 PMSM, working as an AC motor. The analysed motor is with 36 slots on the stator and a rotor with six surface mounted samarium–cobalt PMs, having the remanent flux density of $B_r = 0.95$ T and coercive field of $H_c = -720$ kA/m. The rated data of the motor are $U_n = 42$ V, $n_n = 1{,}000$ rpm, $M_n = 10$ N m, $m = 3$ phases and the stator winding in Y-connection. The studied motor is supplied and controlled by a built-in three-phase inverter. The outside view of the motor is shown in Figure 4.20.

Figure 4.20 View of the permanent magnet synchronous motor

4.3.2.2 Definition of the objective function

In the optimal design process, the first important task is to select and define the objective function that is going to be optimised. In the most cases, it is recommended to take the motor efficiency as an objective function. But there are also cases where the production cost of the machine, the total volume or the mass, electromagnetic and/or cogging torque as well as the power/torque over mass coefficient are defined as objective functions. In this case study, the efficiency of the PMSM is selected as the objective function.

In general, the efficiency of an electric motor is defined as a ratio of the output power P_2 over the input power P_1, as it is expressed with the following equation:

$$Efficiency = \frac{P_2}{P_1} \tag{4.1}$$

The input and output power are calculated as

$$P_1 = P_2 + P_{losses} \tag{4.2}$$

$$P_2 = T \cdot \omega \tag{4.3}$$

The total power losses of the motor P_{losses} are the sum of the copper losses – P_{Cu}, iron losses – P_{Fe} and additional windage and friction losses – P_{wf}.

$$P_{losses} = P_{\mathrm{Cu}} + P_{\mathrm{Fe}} + P_{\mathrm{wf}} \tag{4.4}$$

The components of additional power losses, at constant speed, are taken as constant and have been determined from experiments, where the copper and iron losses are determined with the following equations:

$$P_{\mathrm{Cu}} = 3R_{\mathrm{ph}} \cdot I_{\mathrm{ph}}^2 \tag{4.5}$$

$$P_{\mathrm{Fe}} = P_{\mathrm{Fe_{SY}}} + P_{\mathrm{Fe_{St}}} \tag{4.6}$$

where P_{FeSY} are iron losses in stator yoke and P_{FeSt} are iron losses in stator teeth.

The phase resistance of the stator winding and the phase current are calculated with the following expressions:

$$R_{\mathrm{ph}} = \rho_{\mathrm{Cu20}} \frac{L_{\mathrm{Cu}}}{A_{\mathrm{Cu}}} \tag{4.7}$$

$$I_{\mathrm{ph}} = \frac{P_{\mathrm{em}}}{3E_{0\mathrm{ph}}\eta_{\mathrm{app}}^*} \tag{4.8}$$

where R_{ph} is stator-winding phase resistance, L_{Cu} is the total length of a single-stator phase winding, A_{Cu} is slot area occupied with the winding, ρ_{Cu20} is copper-specific resistivity at 20 °C, P_{em} is electromagnetic power, $E_{0\mathrm{ph}}$ is induced phase back-EMF and η^*_{app} is apparent efficiency, which is the product of motor efficiency η and power factor $\cos\varphi$.

Each variable in previous equations is calculated with

$$P_{\text{em}} = P_2 + P_0 \tag{4.9}$$

$$P_0 = P_{\text{Fe}} + P_{\text{wf}} \tag{4.10}$$

$$E_{0\text{ph}} = 4.44 \cdot f_n \cdot k_w \cdot N_{\text{ph}} \cdot \Phi_{g0} \tag{4.11}$$

$$L_{\text{Cu}} = N_{\text{ph}} \cdot L_{t_1} \tag{4.12}$$

$$L_{t_1} = 2 \cdot (L + L_{\text{ew}}) \tag{4.13}$$

where: P_0 are no load losses, f_n is rated supply frequency, k_w is winding factor, N_{ph} is the number of turns per phase of stator winding, Φ_{g0} is average air-gap flux at no load, L_{t1} is the length of one turn of stator winding, L is the axial length of the motor, L_{ew} is the length of one side of the stator-winding overhangs.

This is only a partial presentation of the mathematical model for the PMSM that is implemented in the optimal designing procedures. Some of the motor parameters are kept as constant, some are directly dependent on the optimal design variables, while the others are indirectly dependent.

4.3.2.3 Selection of design variables

Besides the definition of the objective function, the selection of the design variables is another main task that the motor designer has to accomplish. Usually the optimisation parameters are problem-dependent and have to be determined very thoroughly. They should correspond with the formulation of the objective function and are used to fully describe the mathematical model of the motor.

In the case of the optimal design of the PMSM, the following design variables are introduced: R_R is rotor core radius, g is air-gap length, l_m is PM height, f_m is PM fraction and L is the axial length of the motor. The visual presentation of the design variables is shown in Figure 4.21.

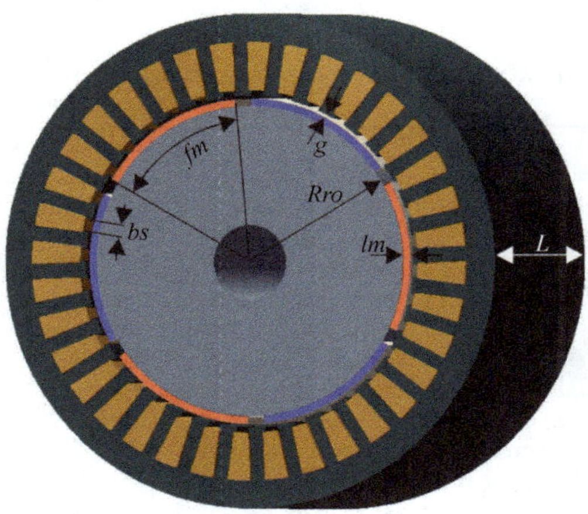

Figure 4.21 Sketch of the design variables

When performing an optimisation procedure, in order to come to a feasible solution, the space for the change of each design variable is defined with respective lower and upper bound. In this research study, the bounds have been selected as variation for ±10% of the reference values, associated with the initial motor model. In this way, lower and upper bounds for each design parameter are determined and presented in Table 4.1. When carrying out previously selected three optimisation algorithms, these values are kept the same.

4.3.2.4 Setting optimisation assumptions

The motor efficiency is selected to be the objective function of optimisation and is subordinated to the five design variables given in Table 4.1. The rotor speed n (ω) and the shaft torque T keep their rated values. Also the number of phases, number of stator slots and the number of PMs are the same as in the reference motor.

In practice, depending on the particular optimisation problem, several additional constraints are required to be introduced. Thus (a) in order to keep the frame size of the motor the same, it is not allowed to change the outer diameter of the stator; (b) to obtain comparable results, materials for PMs, stator/rotor cores and stator windings remain the same.

The corresponding values for the constant parameters are presented in Table 4.2.

Table 4.1 Lower and upper bound of design variables

Design variable	Initial motor model	Lower bound	Upper bound
R_R (m)	0.042	0.0378	0.0462
g (m)	0.0008	0.00072	0.00088
l_m (m)	0.002	0.0018	0.0022
f_m (/)	0.9	0.81	0.99
L (m)	0.09	0.081	0.099

Table 4.2 Reference motor parameters and material properties

Parameter	Description	Value
T (Nm)	Rated torque	10
n (rpm)	Rated synchronous speed	1,000
N_s (/)	Number of stator slots	36
N_{PM} (/)	Number of permanent magnets	6
B_r (T)	Remanent flux density of permanent magnets	0.95
H_c (kA/m)	Coercivity of permanent magnets	−720
μ_{rec} (/)	Permanent magnet recoil relative permeability	1.05
ρ_{Cu} (Ωm)	Specific resistivity of copper at 20 °C	1.733×10^{-8}
γ_{PM} (kg/m^3)	Permanent magnet mass density	8,300
γ_{Cu} (kg/m^3)	Copper mass density	8,960
γ_{FeS} (kg/m^3)	Stator steel mass density	7,650
γ_{FeR} (kg/m^3)	Rotor steel mass density	7,600

Table 4.3 Physical constraints for active materials

Parameter	Description	Value
$B_{SY\text{-}max}$ (T)	Stator back iron maximum flux density	≤ 1.8
$B_{St\text{-}max}$ (T)	Stator teeth maximum flux density	≤ 1.76
J_{max} (A/m^2)	Maximum conductor current density	$\leq 6.5\times10^{-6}$
k_{Cu} (/)	Slot fill factor (conducting packing area)	≤ 0.4

The densities of electric and magnetic loads, which determine copper and iron losses, are important indicators for the heat generation in an electromagnetic device. To reduce these losses and to avoid overheating of the studied motor, respective physical constraints for the active materials (iron and copper) are predefined and presented in Table 4.3. These values are used in the mathematical motor model when calculating the motor dimensions, i.e. the objective function – motor efficiency.

4.3.3 Optimisation results

To analyse the efficacy, advantages and disadvantages of the previously presented algorithms, as well as to come to the relevant facts for the evaluation of the proposed optimisation methods, they all are applied on the studied PMSM topology. The motor efficiency is used as their common objective function, where the design variables with respective bounds remain the same. The previously introduced assumptions and defined physical constraints are also kept the same.

4.3.3.1 Genetic algorithm optimisation

As it was mentioned before, the quality of the GA optimal search basically depends on the values of its parameters such as the number of generations is G, population size is N, crossover probability is p_c and mutation probability is p_m. The values that are assigned are user- and problem-dependent. In order to make a matching selection for the p_c and p_m values, a very complex and detailed analysis of these parameters and their influence on the quality of the GA search, has been performed. The number of generations is determined after a number of attempts with different preset values are performed. After the value of the objective function converges and does not show significant changes across the further generations, it is considered to be a satisfactory criterion to stop the search and to get output results; thus, the computation time and quality of the result remain in good correlation.

In this research study, a large number of GA search procedures have been accomplished. The considered values of the GA parameters for this design problem are population size $N = 20$, crossover probability $p_c = 0.85$, mutation probability $p_m = 0.07$ and the number of generations $G = 15{,}000$. Because of the introduction of the elitism in the algorithm, almost all optimisation attempts give the same output value for the objective function at the end of the search. However, it is not necessarily that the design variables will have the same output values.

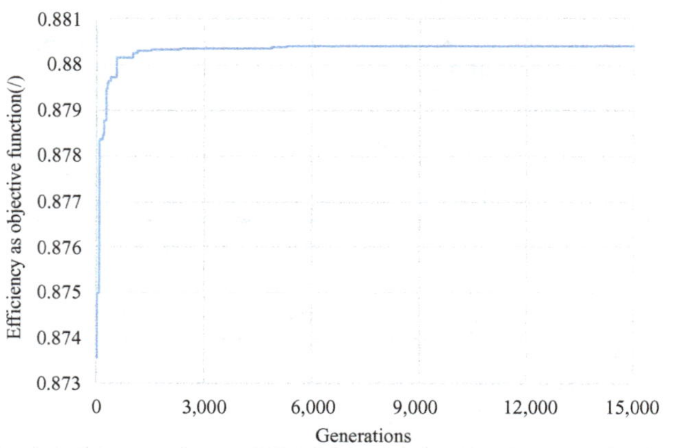

Figure 4.22 Objective function change during generations in GA optimisation

The change of the objective function value during generations for one particular optimisation search is presented in Figure 4.22. From the figure, it can be noticed that the objective function changes rapidly in the first 1,000 generations, then up to 5,000 generation changes of the efficiency are small, and afterwards its change enters a steadier section with slight increase. The last increase of the efficiency is around the 13,000 generations and keeps this value until the end of the search. The changes for the values of design variables during generations are shown in Figures 4.23 and 4.24. From the analysis of the presented diagrams, it can be concluded that during the optimisation procedure only the value of PM fraction exhibits significant changes, while all the other design variables are subject of change only at the beginning of the search and afterwards have almost constant values. The output value of the objective function, i.e. the motor efficiency is found to be

$$\eta = 0.8804 \tag{4.14}$$

In order to evaluate the real quality of GA optimisation (GAO) approach, the next approach for finding the optimal values of design variables that determine the best value of the motor efficiency is the application of PSO.

4.3.3.2 Particle swarm optimisation

The quality of the particle swarm optimal search mainly depends on the values of its parameters such as number of iterations – I, population size – P, cognitive parameter – c_p, social parameter – s_p, constriction parameter – cs_p as well as the velocity – v of the swarm, which is of the big importance for the definition of the particles in this search.

The parameter values, as could be expected, are user- and problem-dependent. Due to the stochastic nature of the algorithm, a large number of PSO approaches are required to be performed. After an extensive investigation and deep analysis of

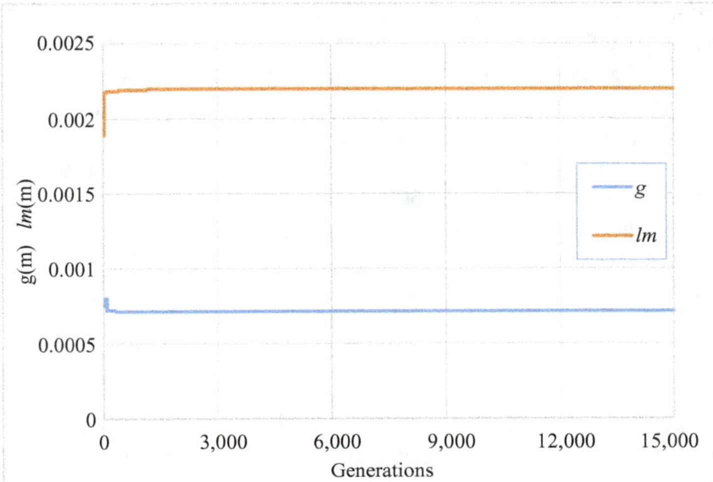

Figure 4.23 *Change of design variables g and lm during generations in GA optimisation*

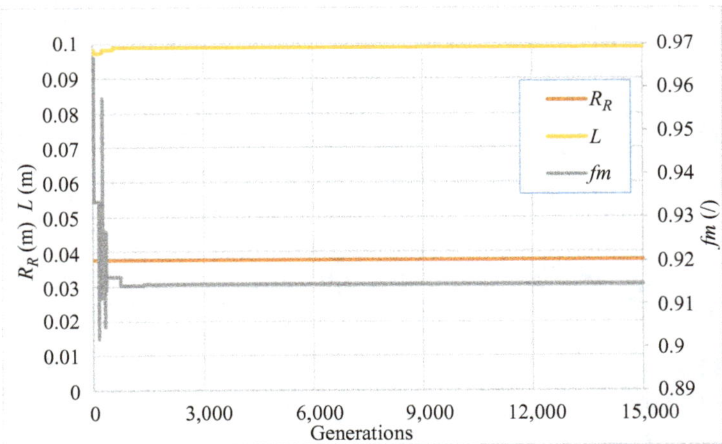

Figure 4.24 *Change of design variables RR, L and fm during generations in GA optimisation*

the output results is carried out, their proper values are selected. The best fitted number of iterations was selected after a numerous PSO attempts, within a wide range of values, have been performed.

For this particular optimal design problem, the values for the optimisation parameters are as follows: population size $P = 20$, cognitive parameter $c_p = 1$, social parameter $s_p = 3$, constriction parameter $cs_p = 0.73$ and the number of generations $I = 500$. The value of the swarm velocity is calculated during the PSO search by using the selected values for the optimisation parameters. The change of the objective function values, i.e. the efficiency values, during the iterations for one particular search, is presented in Figure 4.25. One can easily notice that the best values for the efficiency are reached relatively early, around the iteration $I \approx 120$; although the search continues up to 500 iterations, which are set-up as a stopping criterion, the further values for the objective function remain lower, till the end of the optimisation procedure.

On the basis of the chart from Figure 4.25, for the best efficiency value is found:

$$\eta = 0.8676 \tag{4.15}$$

The output value for the objective function is lower than that obtained from GA search. However, for the assessment of the proposed optimisation approach, one should include values of the other important parameters of an electric motor – like losses, specific electric and magnetic loads, active materials usage, etc. – that are implicitly determined by the output optimal values for the design variables.

At the end of this comparative investigation, the cuckoo search optimisation (CSO) method will be tested on the studied PMM, taken as reference model.

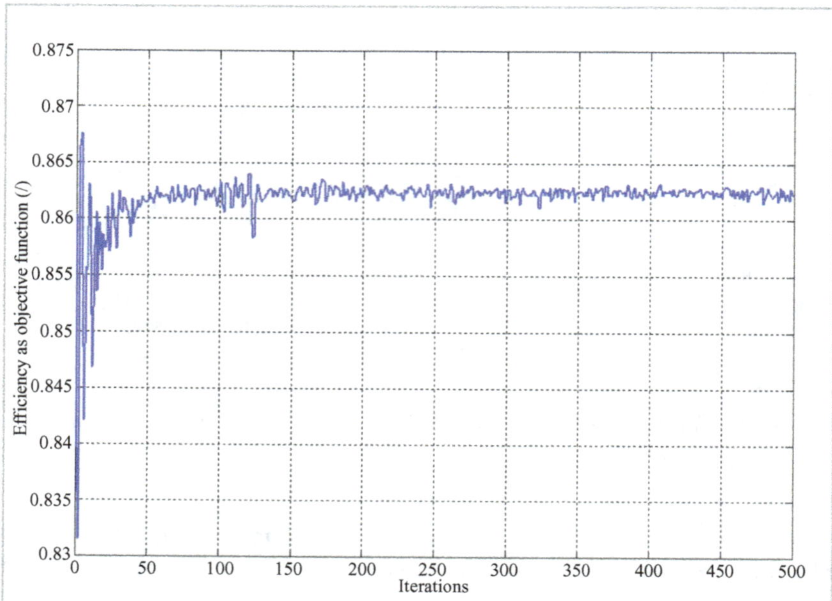

Figure 4.25 Objective function change during iterations in PS optimisation

4.3.3.3 Cuckoo search optimisation

Similar to the previous two optimisation algorithms, the quality of the CS depends on the values of its parameters, where of the greatest importance are the number of iterations – I and the number of nests – N. There are also other parameters of optimisation that determine the quality of the search and influence the output results, but they are not presented and discussed. For more theory on these issues, [26] is recommended. When searching for the optimal motor design by using CS, optimisation parameters are selected to be $I = 100$ and $N = 500$.

The change of the objective function value during iterations, for one particular optimisation search, is presented in Figure 4.26. From the presented diagram, it can be noticed that objective function converges quite rapidly towards the optimal solution and reaches close to the best value after 60 iterations. Afterwards this value improves so slightly that actually it cannot be noticed on the graph. In the last ten iterations, the objective function value practically remains the same, proving that the number of iterations set to 100 is satisfactory.

The output value of the objective function, indeed the optimal efficiency, approaches close to the value obtained from GA search:

$$\eta = 0.8801 \tag{4.16}$$

4.3.4 Comparative analysis

The output values for the selected design variables, as well as the objective function optimal value, being the motor efficiency, compared to the initial values of the reference motor, are presented in Table 4.4. Just a brief look on the last row shows

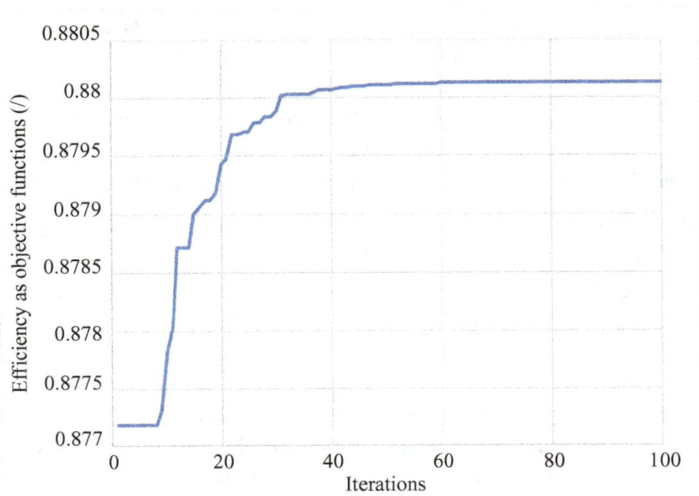

Figure 4.26 Objective function change during iterations in CS optimisation

Table 4.4 Comparison of design variables and objective function

Design variable	Initial model	GAO	PSO	CSO
R_R (m)	0.042	0.0378	0.0412	0.0378
g (m)	0.0008	0.00072	0.00072	0.00072
l_m (m)	0.002	0.0022	0.00195	0.0022
f_m (/)	0.9	0.9147	0.82	0.9143
L (m)	0.09	0.09899	0.09813	0.099
Efficiency (/)	0.8489	0.8804	0.8676	0.8801

that all implemented methods ensure the efficiency improvement, which was set as a target of the optimisation, proving the efficacy of all methods.

In such a way, the GAO provides the highest efficiency increase for 3.71%, the CSO shows close value with 3.68% increase, while with the lowest percentage increase of 2.2 is output result from the PSO search.

From this point, GAO with well-selected optimisation parameters is powerful method for optimal design, or shape optimisation of various types of electrical machines [27]. On the other hand, PSO works well but a designer should pay big attention on the selection of bounds for design variables [28]. Finally, CSO seems to be simple for implementation when optimising electro-magneto-mechanical devices giving results as good as GAO [29].

Related to the design variables figures are similar. The optimal values obtained by GAO and CSO are almost identical, while PSO demonstrates differences. Further analyses, inside the prescribed intervals for the design variables change, show the same trends. The rotor core radius R_R and the air-gap length g are decreased, while the motor axial length L is increased. However, trends for the PM geometry are opposed. While GAO and PSA determine higher PM fraction f_m and PM height l_m, PSO tends to lower their values.

For evaluation, the proposed optimisation methods, as well as an analysis of their advantages and disadvantages, more interesting motor parameters and characteristics are calculated and given comparatively in Table 4.5. Based on the presented values, few remarks and comments are discussed in continuation.

- The higher efficiency is obtained mostly due to lower copper power losses, as they are the major part of the total loss.
- The lower copper loss is achieved by reducing the winding phase resistance, i.e. by increasing a single-wire cross section area and reducing total length of a phase winding, by lower number of turns. The higher current does not have influence.
- The total iron power losses are slightly increased, as the volumes of iron cores of the motor are increased, in favour of lower saturation. However, the overall power losses are significantly lower, which, in turn, gives higher efficiency.

Table 4.5 Comparison of motor parameters and characteristics

Parameter	Description	Initial motor	GAO	PSO	CSO
B_g (T)	Air-gap flux density	0.6097	0.6196	0.5631	0.6198
a_{Cu} (m^2)	Cross section of a single wire $\times 10^{-6}$	2.83	4.459	3.931	4.4
L_{Cu} (m)	Total length of a phase winding	20.97	19.43	19.41	19.44
R_{ph} (Ω)	Stator-winding phase resistance	0.1282	0.0755	0.0755	0.0766
I_{ph} (A)	Phase current	17.70	19.005	19.005	18.915
V_{Cu} (m^3)	Total copper volume $\times 10^{-4}$	1.783	2.599	2.290	2.566
V_{FeSY} (m^3)	Total stator yoke volume $\times 10^{-4}$	2.787	2.885	2.805	2.897
V_{FeSt} (m^3)	Total stator teeth volume $\times 10^{-4}$	1.903	2.516	2.094	2.499
V_{PM} (m^3)	Total volume of PMs $\times 10^{-5}$	4.397	4.869	4.158	4.791
N_{ph} (turns)	Number of turns per phase	54	48	48	48
P_{Cu} (W)	Copper-winding power loss	147.09	99.84	119.54	100.25
P_{Fe} (W)	Iron cores power loss	17.357	20.39	18.53	20.38
P_{tot} (W)	Total motor power loss	186.45	142.23	160.07	142.63
η (/)	Efficiency	0.8489	0.8804	0.8676	0.8801

4.3.5 Concluding remarks

The general conclusion of this research work is that all three optimisation methods are appropriate for application in optimal design procedure for an arbitrary selected case study – a PMSM. The objective function is increased, meaning that the motor efficiency is improved, and thus, the optimisation target is accomplished. There is no doubt that the same approach is applicable for optimal design of other types of electrical machines, in general.

Comparing the value of the objective function and optimal design variables, as was analysed before, GA and the CSA end up at almost the same point with very little differences. The obtained values, compared to the values from particle swarm algorithm, are mutually better corresponding. Hence, both the GAO and CSO methods are quite successful and they both reach the global optimum of the objective function. With these results, we are quite certain that after the optimisation procedure is done, we have in hands the best solution for the selected objective function and for the input parameters. This conclusion is supported by the additional data from all optimisation procedures. Of course, the motor efficiency can be further improved, but more costly, by replacing motor materials with novel ones.

The optimal electric motor design can be further thoroughly investigated by a numerical software based on the finite element method (FEM) analysis. Nowadays, it is a regular practice that an electric machine is numerically tested. Both approaches, optimal designing and numerical analysis, give a powerful tool to a designer when searching for an energy-efficient and effective electrical machine in a relatively short time period. This is very important in today's fast-developing industries and technologies. It is worth to emphasise that this is just one possible approach where the improvement of the motor efficiency is the prime goal. There are also cases where the motor design has to fulfil additional objectives or constrains. In such cases, the multi- or many-objective optimisation approach is highly recommended. We suggest on this issue to read theory in the previous two chapters of this book.

4.4 FEA-based shape synthesis of electric motors

The performance of an electric machine, in general, or an electric motor in particular, depends on many different factors which determine the electromagnetic field properties and the flux distribution in its interior. Among them, the most important are the motor topology with respective geometry, as well as materials used for windings, for magnetic cores and, where exist, for PMs. When the focus is put on designing a particular motor, then the supplying conditions and the load conditions are input parameters, defined and known in advance. Consequently, the main task of an electric motor designer is to get the best motor performance at the least material utilisation, satisfying the given criteria and prescribed physical limitations, i.e. to find the optimal design of a motor [30].

This area of research has been widely exploited, for decades. Recently, the application of classical mathematical approach to the optimisation of a given function, in electric motor design, has been almost abandoned [31]. These mathematical methods, known as deterministic optimisation techniques, have been widely used in the last century. New in this research area, extensively elaborated in Chapter 2, is the development of various optimisation methods inspired by nature, based on the imitation of the behaviour of living beings. Thus, this class of modern optimisation techniques, that have been successfully applied to the electric motor designing, leads to novel motor topologies and better motor performance characteristics.

In general, there are two classes of electric motors' optimisation problems. First one deals with an electric motor designing defined as a full optimisation problem. An engineer in a research and development department of an industrial structure, starting on a 'blank sheet' only with predefined topology, given rated data and certain constraints for a particular motor, carries out *shape synthesis* of the optimal geometry. Usually, the target is at least consumption of materials and minimised production cost to get the most of the motor performance features. In this procedure, a numerical experiment based on finite element analysis (FEA) is widely used. Nowadays, this kind of virtual prototyping became a must issue. Thus, before selecting the motor design, ready for production line, through a series of numerical computations, the performance characteristics can be determined and thoroughly analysed [32].

The second approach can be defined as *shape optimisation* technique, for an existing electric motor to meet and match prescribed requirements. Among the variety of the motor design parameters, the optimisation methodology can be possibly focused on (a) the motor geometry and in particular, dimensions and shapes of stator tooth/slot, stator yoke width related to field saturation, air-gap length, as well as dimensions, shape and magnetisation type of PMs; (b) used materials, where the best quality price is the leading criterion; (c) few determinants related to the motor performance, like no-load back-EMF, load torque value and cogging torque characteristic profile. In this case, through series of electromagnetic field calculations based on the FEM, for the whole range of variation for the selected independent variables, the change of the defined objective function is analysed; the respective design, best fitted to the determined criteria, is found [33].

To conclude, advanced shape optimisation techniques have become a key tool for the design of electric motors. In general, when considering a complex electro-magneto-mechanical assembly, it is often possible to optimise either the whole motor geometry or only the geometry of certain parts, to improve the overall performance of the object. In industrial sector, the shape optimisation of electrical machines is the most economical approach to improve their efficiency and performance [34]. Hereafter, as an example to illustrate FEA-based shape synthesis, a case study for a particular innovative electric motor is presented and described with more details.

4.4.1 Case study background

PMBLDC motors are becoming more popular due to their superior performance, control and energy efficiency. However, when high-energy rare-earth magnets are used, then a significant cogging torque can become a problem [35]. Recently, the topology of PMMs is under reconsideration and new, attractive designs for higher performance motors are developed. A specific class of PMBLDM machines are fractional-slot motors that are characterised by a noninteger ratio of the number of stator slots to the number of rotor magnet poles. Fractional slot motors tend to have a lower cogging torque than motors with integer slot-to-pole ratios. Consequently, fractional-slot winding configurations in PM brushless motors have attracted much attention due to the availability for more compact windings and lower cogging torque.

In general, the drawback of PMMs is the presence of cogging torque often with notable peak values, disturbing the smooth motor running. Hence, both load and cogging torques of PMMs are always a challenging matter for research and analysis [36]. Major factors affecting the cogging torque include PM magnetic field profile, air-gap length, stator openings, ratio of stator teeth and rotor poles and skewing. Relations between these factors are difficult to be defined. Many researchers worldwide tackle the problem of mitigating the cogging torque using a variety of methods, among which are pole shape optimisation, change of the air-gap length between pole tip and rotor, unequal pole shoe thickness, application of dummy slots, or by introducing notches, either in the rotor or in the stator core.

As an approach to design optimisation and shape synthesis of electric motors, a numerical experiment based on FEA is presented. It is applied on the shape redesign for an existing motor, where it is possible to implement non-complex changes only in certain parts of the design aiming to reduce the peak of cogging torque to a

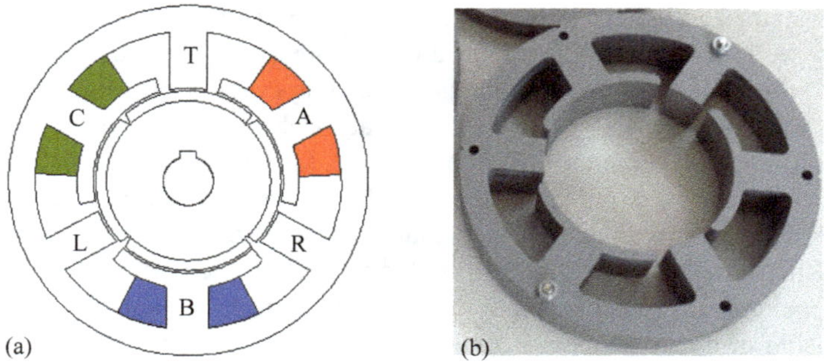

Figure 4.27 Fractional-slot 6/4 PMBLDC motor: (a) outline of the reference motor; (b) stator core topology

minimum possible value. The outline of the studied motor is depicted in Figure 4.27(a), while the stator topology, being the domain of interest, is shown in Figure 4.27(b).

In Figure 4.27(a), the following annotations are introduced: A, B and C are the three large poles, matching with the corresponding phase windings that are inserted in the six available slots; T (top), R (right) and L (left) are three simple small interpoles. This asymmetrical six-slot four-pole topology of PMBLDC motor, with many other new features and advantages, has been adopted as the reference motor model.

The objective of the research herein is to minimise the peak cogging torque, by making simple design adaptations applied to the stator geometry of the reference motor model. Hence, the study is focussed on finding the best shape of stator pole shoe, which provides minimum peak cogging torque while keeping still good performance characteristics of the motor. The method to change stator pole shape based on FEA, as an efficient way to reduce the cogging torque, is reported in [37].

4.4.2 Numerical experiment

The peak cogging torque calculation for the PMBLDC motor with different shape designs is based on FEM, a proved powerful computational tool [38]. The FEM model of the motor is presented in Figure 4.28. The FEA is used to find optimal shape of stator pole shoe and to minimise the peak cogging torque.

The shape optimisation problem is formulated by introducing large pole arc length A and pole shoe thickness T, to be the design variables; the value of the small pole arc length B is subordinated to A and is changed simultaneously with A; the respective initial values are $A_0 = 85°$, $T_0 = 6$ mm and $B_0 = 22°$. The ranges of change for parameters A and T are selected with respect to design limitations. We seek for the optimal solution, i.e. for the pair (A_x, T_y), that gives the minimum value of peak cogging torque by altering the geometry of stator poles according to the two parameters of interest: A and T.

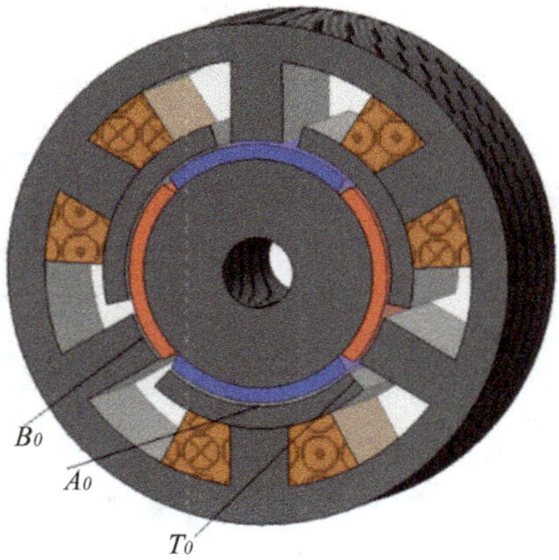

Figure 4.28 FEM model of the PMBLDC reference motor (A_0,T_0)

The following methodology is applied: (a) by sampling the discrete changes of A and T simultaneously, to build as we call *solution matrix* (SM); (b) for each combination, by using FEA, to determine the peak cogging torque and to complete the SM; (c) to present the results on the charts; (d) based on their analysis, to select the best combination of parameters; (e) by applying the fine-tuning, to specify the precise dimensions for the optimal shape of the stator pole shoe.

While tuning, both the ends of large stator pole arc are decreased in m steps, while the small pole arc is bilaterally increased by the same amount. At the same time, the parameter T is varied in n steps. Thus, a discrete two-dimensional solution array $[m \times n]$ is obtained. The sufficiently large number of sample points (A_x,T_y) is needed to ensure the accuracy of the final solution. Even more, in order not to be stacked at a local minimum for peak cogging torque, it is required to analyse the whole array, not only the discretely calculated pairs (A_m,T_n), by performing the fine-tuning of selected variables, inside the area of interest.

4.4.3 Selection of optimal shape design

By employing FEA, an extensive performance analysis of all generated stator designs (20) has been carried out. For each pair of variables (A_m,T_n) from the solution array, the magnetic field calculations are performed when only PMs generate the magnetic field, i.e. at no-load operation and zero current in the windings. Using the output results, the cogging torque profile is developed and the peak values of the cogging torque are determined. The 2D solution array (A_m, T_n) = (5 × 4) is completed and presented in Table 4.6.

Following the analysis of the results shown in Table 4.6, at a first glance, the combination (A_1,T_1) gives the lowest peak of cogging torque. Two separate charts are presented in Figure 4.29: (a) $T_{cog} = f(A_x)$, at T_y = constant; (b) $T_{cog} = f(T_y)$, at A_x = constant. The regions where the minimum peak value of cogging torque takes place are encircled.

Table 4.6 Solution array for peak cogging torque (Nm)

			Pole shoe thickness T_y (mm)			
			$T_0 = 6$	$T_1 = 5$	$T_2 = 4$	$T_3 = 3$
Pole shoe cut for arc A_x (deg°)	A_0	0°	0.1579	0.2835	0.3604	0.1689
	A_1	6 = 2x3°	0.2316	**0.1083**	**0.1233**	0.1282
	A_2	12 = 2x6°	0.2757	0.1904	**0.1468**	0.2103
	A_3	18 = 2x9°	0.3398	0.1941	0.2034	0.1754
	A_0	24 = 2x12°	0.3637	0.3348	0.2833	0.2596

Yet the preselected steps for A and T change were discrete values, it is highly recommended to search inside the intervals. Consequently, there are analysed three 'candidates' for obtaining the minimum peak cogging torque bounded in the table.

4.4.3.1 Fine-tuning of stator poles' design

By interpolation, from the charts in Figure 4.29, it is found that the minimum peak cogging torque appears for $A_m = 6.5°$ and $T_n = 4.8$ mm. Next, three combinations of variables (A_1,T_n), (A_m,T_1) and (A_m,T_n) are studied, so the three new stator pole shape designs are possible; the results of interest are given in respective cells of Table 4.7.

Obviously, the minimum peak cogging torque of 0.0663 N m appears for pole shoe thickness $T_n = 4.8$ mm, when the large pole arc is decreased bilaterally by 6.5°, i.e. $2 \times 3.25°$ each side; the same arc length is added bilaterally to the small pole. Consequently, the selected optimal design of the stator pole shoe is the large pole arc of 78.5°, the small pole arc of 28.5°, while the pole shoe thickness is 4.8 mm.

4.4.4 Analysis of the results

The topology of the motor with new optimised stator core shape is presented in Figure 4.30. To complete the research, the FEA analysis for no-load and rated load operation, for the reference motor (A_0,T_0) compared to the new optimised design (A_{opt},T_{opt}), is carried out. The subjects of interest are cogging torque profiles and electromagnetic torque characteristics.

4.4.4.1 Cogging torque analysis

The assessment of torques, both the cogging and the static electromagnetic, is carried out on the basis of a comparative presentation for the characteristics. In Figure 4.31 are presented cogging torques graphs, for the reference (A_0,T_0) and optimised (A_{opt},T_{opt}) design of BLDCPM motor. An investigation of the cogging torque profiles shows that the period of change remained at 30° mechanical, and the

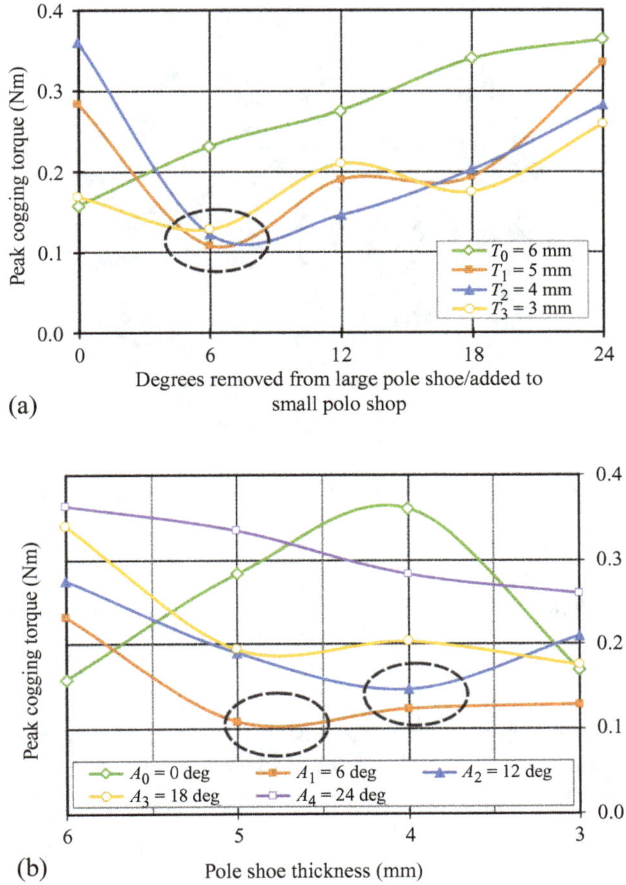

Figure 4.29 Dependence of peak cogging torque on design variables (A_x,T_y): (a) $T_{cog} = f(A_x)$, T_y = constant; (b) $T_{cog} = f(T_y)$, A_x = constant

Table 4.7 Fine-tuning for peak cogging torque (Nm)

		Pole shoe thickness T_y (mm)		
		$T_0 = 6$	$T_1 = 5$	$T_n = 4.8$
Pole shoe cut for arc A_x (°)	$A_0 = 0°$	0.1579		
	$A_1 = 6°$		0.1083	0.0764
	$A_m = 6.5°$		0.0900	**0.0663**

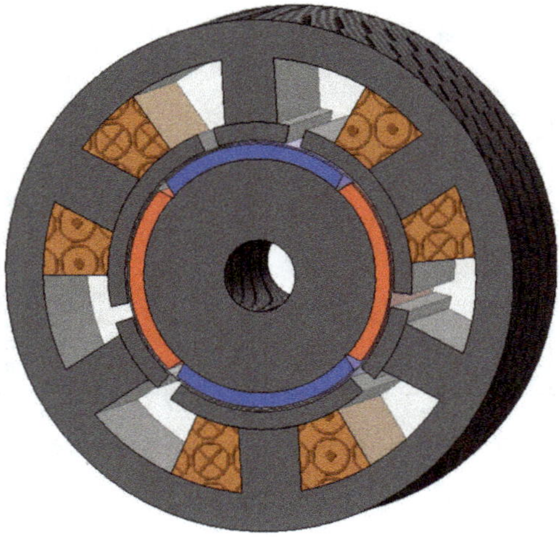

Figure 4.30 Topology of the new PMBLDC motor design (A_{opt}, T_{opt})

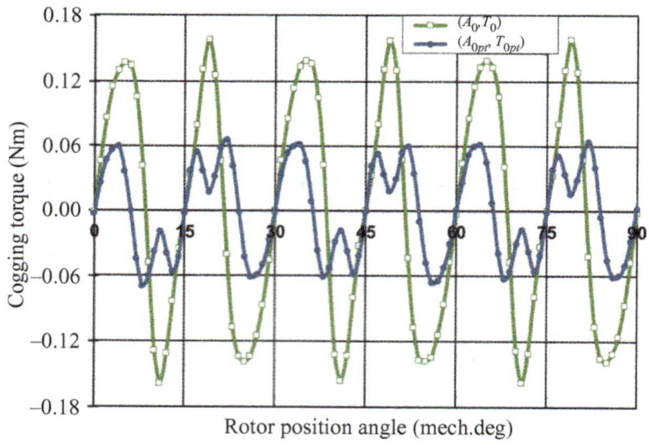

Figure 4.31 Cogging torque comparative characteristics (A_0, T_0) and (A_{opt}, T_{opt})

chart symmetry is same as previous and appears at half period, i.e. at 15° mech., with mirrored values. The peak cogging torque is significantly reduced: from 0.158 N m to a respectable 0.066 N m, which is a mitigation of 2.39 times.

4.4.4.2 Static torque analysis

The analysis of electromagnetic torque characteristics is the next task. For precise electromagnetic torque calculation, by using FEM, accurate global and local field solutions are required. In other words, the high level of mesh discretisation is a

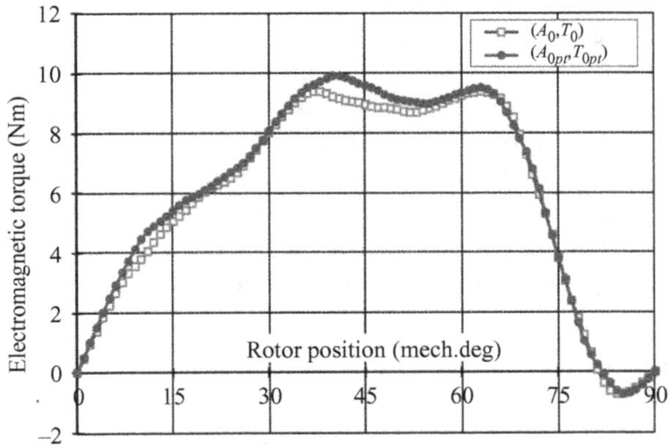

Figure 4.32 Static torque comparative characteristics (A_0, T_0) and (A_{opt}, T_{opt})

prerequisite. The preferred method for calculating the electromagnetic torque is the use of the volume integral for weighted stress tensor in the air-gap domain. The 2D magnetostatic solution problem is performed for half of period, spanned over one pole pitch of 90° mechanical (=180° electrical). The comparative electromagnetic torque characteristics for the reference and optimised motor are presented in Figure 4.32. The higher maximum torque value for (A_{opt}, T_{opt}) is apparent; also, the steeper curve (A_{opt}, T_{opt}) provides more stable operation of the motor under load variations.

Furthermore, the optimised motor design appears to produce much more torque than the reference design. This is very likely to be due to the addition of a component of reluctance torque brought about by an increase in the local saturation within the pole shoes due to a reduction of the pole shoe thickness from 6 to 4.8 mm.

4.4.5 Concluding remarks

The work presented in this heading demonstrates how by using simple design techniques, based on FEM field calculations and respective FEA analyses, it is possible to design and to optimise the shape of selected parts of motor topology. The target to achieve is selected depending on the performance priorities of the studied object. In the case of the fractional-slot PMBLDM motor, the cogging torque profile and its peak value appear to be of the greatest importance. Performing FEA simulations for different design parameters of the stator poles, arc length A_x and shoe thickness T_y, the cogging torque profiles are derived with respective peak values.

In the second step, by fine-tuning of the stator pole shoe geometry, the optimal values for design variables are defined. The presented results show that there is one stator pole arc ratio and pole shoe thickness combination (A_{opt}, T_{opt}) that produces the minimum peak cogging torque. Significantly lower peak cogging torque has been detected, while still good static electromagnetic torque production is retained. In addition, due to the shape change of the stator pole the utilisation of iron is decreased, leading to loss reduction and efficiency improvement.

To conclude, the use of FEM to determine performance characteristics of the studied motor with different shape designs demonstrated to be an efficient strategy. Optimisation of the stator pole shoe of PMBLDC motor, based on the FEA, is proven to be powerful computational tool and good approach for the shape optimisation of certain parts from a given electric motor topology.

4.5 Summary

Innovative electric motors find wide applications across multiple industry branches, including energy production, e-traction, manufacturing plants, industrial automation and variety of consumer appliances. The key question discussed here is *what makes an electric motor innovative?* In the chapter, the present innovations in electric motor technology and the expected future trends are discussed. The utilised materials are explored and presented with the respective features. The possibilities of innovative materials, to meet the need for increased motor efficiency and new production technologies, are described. Several innovative motors and their applications are illustrated and discussed.

The increased production of EVs will drastically drive the demand for innovative electric motor topologies required for e-mobility. The most recent 3D printing technology developments have added a huge value to the designing, prototyping as well as manufacturing process of electric motors. Hence, the novel AM processes with 3D printing production will enable designing, prototyping and manufacturing of an electric motor to become easier, faster, more reliable and more economical.

Shape design and optimisation of electric motors, to meet the prescribed requirements, is a multi-physics problem to which are faced motor designers. To accomplish this task, many different algorithms are possible. In the chapter, few possible methodologies are discussed in details; selected algorithms are implemented and presented with a respective case study.

Selection of the best optimisation algorithm is user- and case-dependent. Three nature-inspired optimisation algorithms, GA, PSO and CSA, are discussed and presented in details. Stator sheet and shape of magnets for an existing PMSM are optimised, and based on the results the evaluation of the methods is presented. For the objective function is selected maximisation of the motor efficiency.

Another approach to the shape synthesis of electric motors, a numerical experiment using FEA, is proposed and discussed in the chapter. As an example, the cogging torque minimisation for a novel brushless DC PMM is presented. By implementing non-complex changes in the design of the stator poles, an original methodology to improve the cogging torque profile is proposed.

New frontiers for the innovative electric motor topologies, for the new material developments and especially for the novel manufacturing processes by exploiting 3D printing, have been already opened to future inventions.

References

[1] 'The National Magnetic Field Laboratory – NSF', Florida State University and University of Florida, FL USA. Available: https://nationalmaglab.org/education/magnet-academy/history-of-electricity-magnetism/museum/faraday-motor-1821 [Accessed 21 January 2021].

[2] 'Tesla Memorial Society of New York', USA. Available: https://www.teslasociety.com/hall_of_fame.htm [Accessed 21 January 2021].

[3] 'Electrical4u'. Available: https://www.electrical4u.net/basic-accessories/what-is-eff1-eff2-eff3-ie1-ie2-ie3-ie4-ie5-efficiency-class/ [Accessed 31 January 2021].

[4] Wu F. and El-Refaie A. M. 'Towards Fully Additively-Manufactured Permanent Magnet Synchronous Machines: Opportunities and Challenges'. *Proceedings of the 2019 IEEE International Electric Machines & Drives Conference – IEMDC'2019*, 2019, San Diego, CA, pp. 2225–2232, https://doi.org/10.1109/IEMDC.2019.8785210.

[5] Pellegrino G.., Jahns Th.M., Bianchi N., Soong, W., and Cupertino, F. '*The Rediscovery of Synchronous Reluctance and Ferrite Permanent Magnet Motors*'. Springer International Publishing, Springer Nature (also Springer-Verlag) AG, Cham, Switzerland, pp. VIII, 136, 2016, DOI: 10.1007/978-3-319-32202-5_1.

[6] 'Soft Magnetic Cobalt-Iron Alloys Vacoflux & Vacodur', Catalogue Edition 2016, VACUUMSCHMELZE® GmbH & Co. KG, Hanau, Germany, pp. 1–20. Available: https://https://www.vacuumschmelze.com/Assets/Cobalt-Iron%20Alloys.pdf. [Accessed 1 February 2021].

[7] 'What Is Soft Magnetic Composite?', St. Marys, PA USA. Available: https://www.horizontechnology.biz/soft-magnetic-composite-resources. [Accessed 01 February 2021].

[8] Zhang B., Andreas S., and Doppelbauer M. 'Development of a Novel Yokeless and Segmented Armature Axial Flux Machine Based On Soft Magnetic Powder Composites'. *Technical paper presented at World PM2016 Congress & Exhibition, 2016*, Hamburg, Germany, *In the World PM2016 Proceedings USB, 2016, European Powder Metallurgy Association (EPMA)*. Available: https://www.worldpm2016.com/.

[9] 'Soft Magnetic Composite SOMALOY 1P, 3P, 5P, Höganäs AB Products Catalogues and material data on-line Somaloy® quick guide', p. 2, July 2015, Publication 1122HOG, © Höganäs AB Sweden.

[10] 'Somaloy® Axial Flux Machine solutions', Brochure by Höganäs AB, Publication 1701HOG, © Höganäs AB Sweden August 2016, pp. 1–4. https://www.hoganas.com/globalassets/download-media/sharepoint/brochures-and-

datasheets—all-documents/somaloy_axial-flux-solutions_1701hog.pdf. [Accessed 2 February 2021].

[11] Jansson P. 'Soft Magnetic Composite Materials Applied to Electric Motors'. *EPE Journal*, 2004, Vol. 14, No. 1, pp. 38–42, DOI: 10.1080/09398368.2004.11463552.

[12] Cvetkovski G. and Petkovska L. 'Performance Improvement of PM Synchronous Motor by Using Soft Magnetic Composite Material'. *Journal IEEE Transactions on Magnetics*, 2008, Vol. 44, No. 11 Part 2, pp. 3812–3815.

[13] Petkovska L. and Cvetkovski G. 'Soft Magnetic Composite Core – A New Perspective for Small AC Motor Design'. *Proceedings of the 11th Spanish–Portuguese Conference on Electrical Engineering – 11CHLIE on CD*, 2009, Zaragoza, Spain, pp. 1–7.

[14] Guo Y. G. and Zhu J. G. 'Applications of Soft Magnetic Composite Material in Electrical Machines: A Review'. *Australian Journal of Electrical and Electronics Engineering*, 2006, Vol. 3, No. 1, pp. 37–46, DOI: 10.1080/1448837X.2006.11464143.

[15] 'Rare Earth Permanent Magnets VACODYM & VACOMAX', Catalogue PD002 Edition 2014, Published by ©VACUUMSCHMELZE® GmbH & Co. KG (VAC), 2014, Hanau, Germany, VACODYM-VACOMAX-PD002_2015_en.pdf, pp. 1–72, https://vacuumschmelze.com. [Accessed 3 February 2021].

[16] Villani M. 'High Performance Electrical Motors for Automotive Applications – Status and Future of Motors with Low Cost Permanent Magnets'. *Proceedings of the 8th International Conference Magnetism and Metallurgy – WMM'18*, 2018, Dresden, Germany, pp. 1–15.

[17] Lamichhane T. N., Sethuraman L., Dalagan A., Wang H., Keller J., and Paranthaman M. P. 'Additive Manufacturing of Soft Magnets for Electrical Machines – A Review'. *Materials Today Physics*, 2020, Vol. 15, paper 100255, pp. 1–23, published on-line https://doi.org/10.1016/j.mtphys.2020.100255.

[18] Halbig M. C. 'Compact Additively Manufactured Innovative Electric Motor – CAMIEM'. *Energy Tech2017 Conference*, Track_6, presentation, 2017, pp. 1–19, https://www.energytech.org/wp-content/uploads/ET2017 Presentations/Track_6/HALBIG_Energy-Tech-2017.pdf. [Accessed 5 February 2021].

[19] Garibaldi M., Gerada C., Ashcroft I., Hague R., and Morvan H. 'The Impact of Additive Manufacturing on the Development of Electrical Machines for MEA Applications: A Feasibility Study'. *Proceedings of the European Conference on More Electric Aircraft – MEA2015*, 2015, Toulouse, France, hal-01178353.

[20] 'Unexpected Efficiency beyond the Wire', Additive Drives GmbH website, Dresden, Germany, Available: https://www.additive-drives.de/en/. [Accessed 6 February 2021].

[21] Rudolph J. and Lorenz F. 'Fully 3D-Printed Electric Motors', Technische Universität Chemnitz, Germany, Press Office and Crossmedia

Communications, Published in University News, 2018. Available: https://www.tu-chemnitz.de/tu/pressestelle/aktuell/8718/en#. [Accessed 6 February 2021].

[22] 'Electric Motor: What is the Real Innovation?'. Available: https://www.electricmotorengineering.com/electric-motor-what-is-the-real-innovation/. [Accessed 7 February 2021].

[23] Holland H. *Adaptation in Natural and Artificial Systems*, Ann Arbor, MI, University of Michigan Press, 1995.

[24] Janikow Z. and Michalewicz Z. 'An Experimental Comparison of Binary and Floating Point Representations in Genetic Algorithms'. *Proceedings of the 4th International Conference on GA*, San Diego, CA, 1991, pp. 31–35.

[25] Kennedy J., Eberhart R. C., and Shi Y. *Swarm Intelligence*. The Morgan Kaufmann Series in Evolutionary Computation, San Francisco, CA, Morgan Kaufmann Publishers, 2001.

[26] Yang X.-S and Deb S. 'Engineering Optimization by Cuckoo Search'. *International Journal of Mathematical Modelling and Numerical Optimization*, 2010, Vol. 1, No. 4, pp. 330–343.

[27] Cvetkovski G., Petkovska L., and Cundev M. 'GA Approach in Design Optimisation of a PM Disk Motor'. *COMPEL – The International Journal for Computation and Mathematics in Electrical and Electronic Engineering*, 2000, Vol. 19, No. 2, pp. 608–614.

[28] Cvetkovski G., Petkovska L., Lefley P., and Ahmed S. 'Optimal Design of Single Phase Permanent Magnet Brushless DC Motor Using Particle Swarm Optimisation'. *COMPEL – The International Journal for Computation and Mathematics in Electrical and Electronic Engineering*, 2014, Vol. 33, No. 6, pp. 1863–1876.

[29] Cvetkovski G., Petkovska L., and Lefley P. 'Cuckoo Search as a Tool for Optimal Design of PM Brushless DC Motor'. *COMPEL – The International Journal for Computation and Mathematics in Electrical and Electronic Engineering*, 2018, Vol. 37, No. 5, pp. 1732–1743.

[30] Di Barba P., Mognaschi M. E., Rezaei N., Lowther D. A., and Rahman T. 'Many-objective Shape Optimisation of IPM Motors for Electric Vehicle Traction'. *International Journal of Applied Electromagnetics and Mechanics*, 2019, Vol. 60, pp. S149–S162.

[31] Jabbari, A., Shakeri M., and Nabavi-Niaki S. A. 'Pole Shape Optimization of Permanent Magnet Synchronous Motors Using Reduced Basis Technique'. *Iranian Journal of Electrical & Electronic Engineering*, 2010, Vol. 6, No. 1, pp. 48–55.

[32] Di Barba P., Mognaschi M. E., Lowther D. A., and Wiak S. 'Field-based Optimal-Design of an Electric Motor: A New Sensitivity Formulation'. *Open Physics*, 2017, Vol. 15, pp. 924–928.

[33] Di Barba P., Mognaschi M. E., Przybylski M., Rezaei N., Slusarek B., and Wiak S. 'Geometry Optimization for a Class of Switched-Reluctance Motors: A Bi-Objective Approach'. *International Journal of Applied Electromagnetics and Mechanics*, 2018, Vol. 56, pp. 107–122.

[34] Petkovska L., Lefley P., and Cvetkovski G. 'Synthesis and Analysis of a High-Performance Low-Cost Permanent Magnet Brushless DC Motor'. *COMPEL – The International Journal for Computation and Mathematics in Electrical and Electronic Engineering*, 2012, Vol. 31, No. 5, pp. 1482–1491.
[35] Pang Y., Zhu Z. Q., and Feng Z. J. 'Cogging Torque in Cost-Effective Surface-Mounted Permanent-Magnet Machines'. *IEEE Transactions on Magnetics*, 2011, Vol. 47, No. 9, pp. 2269–2276.
[36] Ionel D. M., Popescu M., McGilp M. I., Miller T. J. E., and Dellinger S. J. 'Assessment of Torque Components in Brushless Permanent-Magnet Machines through Numerical Analysis of the Electromagnetic Field'. *IEEE Transactions on Industry Application*, 2005, Vol. 41, No. 5, pp. 1149–1158.
[37] Petkovska L., Lefley P., and Cvetkovski G. V. 'Design Techniques for Cogging Torque Reduction in a Fractional-Slot PMBLDC Motor'. *COMPEL – The International Journal for Computation and Mathematics in Electrical and Electronic Engineering*, 2020, Vol. 39, No. 5, pp. 1041–1055.
[38] Mentor Graphics '2D/3D Magnetostatic/Time-harmonic/Transient/Transient with Motion'. User Manual for MagNet, Version 7.9, 2018, p. 131.

Chapter 5

Frontiers and challenges of new ferromagnetic materials

Barbara Ślusarek[1], Marek Przybylski[1] and Sławomir Wiak[2]

The aim of the chapter is to present the current status and development of magnetic materials and methods of their production. These types of magnetic materials are vital parts of many electromagnetic and electromechanical devices. Magnetic materials are mainly applied in electric motors in a wide range of products, from toys to rockets. The application of ferromagnetic materials is constantly expanding.

This chapter is divided into six sub-chapters showing actual production of ferromagnetic materials and ways of developing magnetic materials in the future. Nowadays, we can observe new methods of magnetic materials production and their technology that allows improvement of the parameters of magnetic materials. New magnetic materials could improve and enable the optimization of the parameters of devices with those materials.

5.1 Ferromagnetic materials

Ferromagnetic materials are divided into soft, semi-hard, and hard magnetic materials. Values of coercivity divide magnetic materials into soft, hard, and semi-hard magnetic. Different authors classify ferromagnetic materials by different values of coercivity. Often, soft magnetic materials have a coercivity less than 1 kA/m, semi-hard up to 10 kA/m, and hard magnetic above 10 kA/m [1,2]. Hard magnetic materials are used as permanent magnets, and soft magnetic materials are used as magnetic cores. Semi-hard materials, which have moderately easy demagnetization and magnetization, are applied, for example, as magnetically coupled devices, such as hysteresis brakes and couplings, magnetic flip tabs, anti-theft tags in shops, and magnetic tooling holders [3].

Ferromagnetic materials contain magnetic elements, e.g. Fe, Co, Ni, and Gd. Iron, cobalt, nickel, and their alloys are mainly used as soft magnetic materials. Semi-hard magnetic materials mainly use alloys of Co–Fe–Ni. Permanent magnets

[1]Łukasiewicz Research Network – Tele and Radio Research Institute, Warsaw, Poland
[2]Institute of Mechatronics and Information Systems, Łodź University of Technology, Łódź, Poland

are produced mainly from hard magnetic materials, such as Alnico alloys, ferrite, and alloys of rare-earth elements.

5.1.1 Soft magnetic materials

Soft magnetic materials are specific materials that can be easily magnetised and demagnetised. Soft magnetic materials placed in the presence of a magnetic field are subjected by the magnetic force. Soft magnetic materials that are supposed to be applied in DC magnetic circuits are mainly characterized by a low value of coercivity and high permeability, which amplifies the magnetic flux. Electromagnets need a high value of magnetic flux density in the air gap, and they need materials with a high saturation of the magnetic flux density. For application in AC circuits, a low value of coercivity is needed, causing a small hysteresis loop and loss, and high resistivity, causing a low eddy current loss. Soft magnetic materials for AC applications should have the lowest value of iron loss for mass or volume at a working frequency. Application of magnetic materials in sensors and measurement devices needs a material with the highest permeability [3].

Soft magnetic materials are iron, cobalt, nickel, and their alloys, and ceramic materials, such as ferrites. Due to the intrinsic structure of soft magnetic materials, they are divided into crystalline, amorphous, and nanocrystalline materials [4].

Crystalline materials are classical materials, such as pure elements and their alloys.

Pure iron has very good magnetic properties, such as high saturation magnetization, J_s, or magnetic flux density, B_s, a low value of coercive force, H_c, and very high permeability. The properties of iron depend on its purity. High magnetic parameters are the main reason that iron and their alloys are used in electrical engineering. One disadvantage of iron is its very low resistivity, which causes growth of eddy currents and, in effect, growth of losses when working at changing magnetic flux. It is the reason that, in industry, there are applied alloys with silicon with oriented or non-oriented grains [4].

The different alloys used in industry are alloys of iron with nickel. These types of alloys are known as permalloys. These materials are characterized by a low value of coercive force and a high value of permeability and moderate saturation. They are mainly used in sensors, electromagnets, relays, and low current applications [4].

Another kind of iron alloy applied in industry is an alloy with cobalt. This type of alloy is characterized by a high saturation of magnetization. An example of this type of alloy is Permendur containing Co 50% and Fe 50%. Alloys with a higher value of cobalt are brittle, and different compounds are used as additives [4].

Sendust is another kind of soft magnetic material that is often applied in inductors. It is composed of 85% iron, 10% silicon, and 5% aluminium. Sendust has high magnetic permeability (up to 140,000), low loss, a coercivity up to 4 A/m, good temperature stability, and a saturation flux density up to 1 T [3].

The development of technology allowed for the production of new types of materials called amorphous soft magnetic materials or metallic glasses. They are obtained by rapid cooling of melted material by melt-spinning. The rate of cooling is about 106 °C/s. This kind of production allows Fe–Si–B–C and Co–Ni–Fe–B–Si alloys to be produced. The parameters of amorphous alloys can be improved by curing, especially in the presence of a magnetic field. The temperature of curing cannot exceed the crystallization temperature, which is dependent on the alloy composition [4]. Amorphous materials can be applied in magnetic circuits with higher frequency than electrical steel sheets. The standard thickness of amorphous ribbons is equal to 25–50 μm. The main advantages of amorphous materials are their high permeability, low value of iron loss, and high working frequency. The disadvantage of this type of material is that it has a lower saturation magnetization than crystalline materials [3].

Soft magnetic nanocrystalline materials are obtained by controlled crystallization of amorphous alloys. After crystallization, the material is composed of nanocrystalline grains of silicon with 10- to 20-nm diameters diluted in iron and distributed uniformly in an amorphous matrix. One of the most applied soft magnetic amorphous alloys is the alloy with a chemical composition of Fe 73.5%, Cu 1%, Nb 3%, Si 13.5%, and B 9% [3,4].

Another kind of soft magnetic material is a ferrite material. This type of soft magnetic material is a ceramic material with a spinel structure consisting of 70% iron oxide (Fe_2O_3) and 30% of oxides of different metals, such as MnO, MgO, NiO, and CuO. They are obtained by compression moulding of powder and sintering. This type of material is frequently used in electromagnetic converters due to its low price. It is used especially in high frequency applications due to its high resistivity. The disadvantages of this type of material are its high shrinkage during sintering and its low saturation of magnetization. The magnetic parameters of ferrite elements can be tailored by mixing different kinds of ferrite powders [4].

Nowadays, ferrite nanoparticles have received a lot of attention due to their applications. Ferrite nanoparticles belong to a group of magnetic nanoparticles that have found biomedical to industrial applications. Ferrite nanoparticles are metal oxides with a spinel structure. The common examples for ferrites are MFe_2O_4 (where M = Mn, Fe, Co, Ni, Cu, and Zn), and most of them show superparamagnetic properties at a size below 20 nm in diameter. There are a few methods of ferrite nanoparticle synthesis, such as co-precipitation, hydrothermal, sol–gel, thermal decomposition, solvothermal, sonochemical, microwave-assisted, microemulsion, polyol, electrochemical, mechanical milling, and laser ablation. The quality of the synthesized ferrite nanoparticles is mostly determined by the synthesis technique and the type of precursor and solvent used. Due to the variation in parameters used in the different synthesis techniques, the best and most preferred synthesis has not been identified [5].

Another kind of soft magnetic material is composite. It contains soft magnetic powder, mainly iron, that is bonded by a dielectric, such as resin, glass, or other materials. Dielectric materials play two roles; they insulate each grain from another and they ensure a high strength of the material. This material has high resistivity; however, its saturation magnetization is lower than Fe–Si steels. Losses at low frequency (up to a few 100 Hz) are higher than Fe–Si steels, whereas for higher frequencies, they are lower due to low eddy current loss [6–9].

Tables 5.1–5.4 show exemplary parameters of soft magnetic materials.

Table 5.1 Parameters of chosen crystalline soft magnetic materials [1–3,57]

Kind of material	Composition	$\mu_{r\ max}$ (–)	B_s (T)	H_c (A/m)	ρ (μΩ m)
Iron	99.8% Fe	5,000	2.15	80	0.10
Cobalt	99.8% Co	250	1.79	800	0.06
Nickel	99.8% Fe	600	0.61	60	0.07
Permendur	Fe 50%, Co 50%	5,000	2.45	160	0.04
Fe–Si isotropic alloy	Fe 97%, Si 3%	6,000	2.00	40	0.50
Fe–Si anisotropic alloy	Fe 97%, Si 3%	40,000	2.00	8	0.50
Permalloy	Ni 78%, Fe 22%	100,000	1.10	4	0.16
Sendust	Fe 85%, Si 10%, and Al 5%	120,000	1.00	4	1.00

Table 5.2 Parameters of soft magnetic ferrite materials [2]

Kind of material	$\mu_{i\ max}$ (–)	J_s (T)	H_c (A/m)	ρ (μΩ m)
Mn–Zn	4,000	0.45	20	10^6
Ni–Zn	500	0.30	300	$>10^9$

Table 5.3 Parameters of powder soft magnetic composites [6–8]

Kind of material	$\mu_{r\ max}$ (–)	B_s (T)	H_c (A/m)	ρ (μΩ m)
Atomet EM-1	290	1.4 at 11.9 kA/m	No data	150
Somaloy 1000 3P	950	1.63 at 10 kA/m	217	70
AncorLam	500	1.90	302	80

Table 5.4 Parameters of nanocrystalline and amorphous soft magnetic materials [2]

Kind of material	Composition	$\mu_{r\ max}$ (–)	J_s (T)	H_c (A/m)
Nanocrystalline Finemet	Fe 73.5%, Cu 1%, Nb 3%, Si 13.5%, and B 9%	800,000	1.25	0.5
Amorphous Metglas 2628SC	Fe 40%, Ni 38%, Mo 4%, and B 18%	400,000	0.88	0.5

5.1.2 Semi-hard magnetic materials

The gap between soft and hard magnetic materials is filled by semi-hard ferromagnetic materials. They have a moderate coercivity from 1 to 10 kA/m.

Semi-hard magnetic materials are materials that can keep a magnetic field inside as permanent magnets, but this field is easy to remove by an external magnetic field. Basically, semi-hard magnetic materials are based on cobalt and cobalt alloys [10]. Semi-hard magnetic materials are alloys, such as Fe–Cr–Co–Ni–Mo (also known as Semivac), Fe–Ni–Al–Ti (Sensor-Vac), and Co–Fe–Ni (Vacozet). They have an anisotropic magnetic structure and are produced by Vacuumschmelze company. Vacozet materials are supplied in the form of strips, wires, profiles, and rods. Strips have a thickness of 0.03–0.6 mm and a width up to 240 mm. Wires and rods have a diameter from 0.3 up to 6 mm. As an example, a material made of Co–Fe–Ni called Vacozet 655 has the following parameters: remanence 1.3–1.5 T, coercivity 2.5–4.8 kA/m, and BH_{max} 4 kJ/m^3.

5.1.3 Hard magnetic materials

Hard magnetic materials preserve the magnetization and magnetic field after the removal of the external magnetizing field. After magnetization, they become permanent magnets. Permanent magnets are sources of a static magnetic field. The main hard magnetic materials are divided into four groups depending on the material type or production process.

Depending on the type of hard magnetic material, we can mainly distinguish

- ferrite;
- Alnico;
- alloys of cobalt with rare-earth elements;
- alloys of Nd–Fe–B.

Figure 5.1 shows demagnetization curves of different magnets. As seen in the figure, different permanent magnets have different demagnetization curves. Alnico magnets have the highest remanence, Br, but a low value of coercivity. In turn, sintered Nd–Fe–B permanent magnets have high remanence and coercivity, and sintered Sm–Co permanent magnets have the highest coercivity and moderate remanence. Ferrite magnets have the lowest remanence and moderate coercivity. The gap between ferrite, sintered Nd–Fe–B, and Sm–Co is filled by Nd–Fe–B-bonded magnets.

Hard magnetic ferrites comprise about 80% iron oxide (Fe_2O_3) and 20% barium or strontium oxides. This kind of permanent magnet is characterized by a low value of remanence up to 0.4 T, a moderate coercivity up to 270 kA/m, and BH_{max} up to 32 kJ/m^3. Its cost is low, which is why they are frequently used in electromagnetic and electromechanic devices. Examples of parameters of produced ferrite permanent magnets are shown in Table 5.5.

In turn, Alnico magnets are alloys of Al, Ni, and Co elements with additives, such as Cu, Nb, and Ti. They are mostly characterized by a high value of

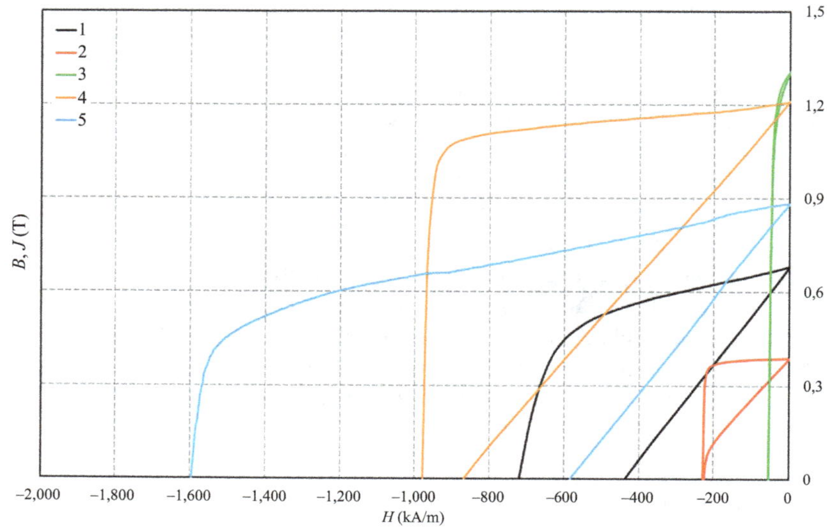

Figure 5.1 Examples of demagnetization curves of permanent magnets: 1 – bonded Nd–Fe–B, 2 – sintered strontium ferrite, 3 – Alnico, 4 – sintered Nd–Fe–B, and 5 – sintered Sm–Co [12]

Table 5.5 Magnetic properties of ferrite permanent magnets [17,58]

Type of ferrite	B_r (T)	H_{cJ} (kA/m)	H_{cB} (kA/m)	BH_{max} (kJ/m^3)
Barium, anisotropic, sintered HF 24/16	0.365	180	175	25.5
Strontium, anisotropic, sintered HF 30/26	0.405	270	250	31.5
Strontium, isotropic, sintered HF 8/26 Sr	0.220	270	140	8.5
Strontium, anisotropic, injection moulded, Plastiform 2051	0.313	249	206	19.2

remanence up to 1.35 T, but their coercivity is low, up to about 150 kA/m, and BH_{max} can be up to 80 kJ/m^3 [11].

Alnico permanent magnets have the most stable magnetic parameters with respect to temperature changes and, therefore, are used mainly in devices that are exposed to changing temperatures. Table 5.6 shows examples of magnetic parameters of produced Alnico permanent magnets.

Alnico permanent magnets are expensive because of the price of cobalt, and that is the reason why their application is essential.

The next group of permanent magnets are permanent magnets from alloys of Co with rare-earth elements, such as samarium. This type of magnet is prepared by different technologies, such as sintering and bonding with plastic, such as

Table 5.6 Magnetic properties of Alnico permanent magnets [11]

Type of Alnico	B_r (T)	H_{cJ} (kA/m)	H_{cB} (kA/m)	BH_{max} (kJ/m^3)
Anisotropic cast Alnico 5–7	1.35	60	58	60
Anisotropic sintered Alnico 5	1.15	50	48	34
Isotropic cast Alnico 2	0.75	46	45	12
Isotropic sintered Alnico 2	0.75	46	45	12
Bonded Alnico BLN8	0.34	107	83	8

Table 5.7 Magnetic properties of Sm–Co permanent magnets [17,28,59]

Type of Sm–Co	B_r (T)	H_{cJ} (kA/m)	H_{cB} (kA/m)	BH_{max} (kJ/m^3)
Anisotropic sintered $SmCo_5$ Recoma 25	1.000	2,400	775	200
Anisotropic sintered Sm_2Co_{17} – Recoma 35E	1.190	1,800	880	265
Anisotropic Nylon12 bonded Sm–Co – injection moulded 2101	0.688	919	477	89.7
Anisotropic compression moulded Sm–Co – BCS-11	0.720	1,200	510	88

Table 5.8 Magnetic properties of Nd–Fe–B permanent magnets [17,27,60–62]

Kind of Nd–Fe–B	B_r (T)	H_{cJ} (kA/m)	H_{cB} (kA/m)	BH_{max} (kJ/m^3)
Anisotropic sintered N55	1.49	876	716	430
Isotropic bonded MQ1-12	0.74–0.80	675–800	470–530	92–99
Anisotropic bonded compression moulded MF-14C	0.98	1,120	600	176
Anisotropic bonded injection moulded RNI-5214	0.79–0.84	1,040–1,120	496–576	116–124
Anisotropic hot pressed and hot forming Neolit NQ 3E	1.27	1,040	915	290
Isotropic hot pressed Neolit NQ 2E	0.80	1,400	560	110

compression moulding or injection moulding. Parameters of this kind of permanent magnet depend not only on the kind of materials but also the applied technology. These kinds of permanent magnets have a high value of remanence up to 1.19 T, a very high value of coercivity up to 2,400 kA/m, and BH_{max} up to 265 kJ/m^3. Examples of parameters of produced Sm–Co permanent magnets are shown in Table 5.7.

The last group of hard magnetic materials are alloys of Nd–Fe–B. They have the main phase intermetallic compound $Nd_2Fe_{14}B$. Nd–Fe–B alloy is a base for the production of permanent magnets with different properties and price. Table 5.8 shows properties of different Nd–Fe–B permanent magnets.

5.2 Methods of magnetic materials production

Magnetic materials are mainly produced by casting, rolling, and powder metallurgy technology. The kind of technology depends on the type of material and application. Soft magnetic materials are produced mainly by casting and rolling. Powder metallurgy, such as sintering and bonding with plastic, are applied in many applications too [12].

The first step in manufacturing electrical steel sheets is casting of an Fe–Si alloy with a defined composition. In 1900, Robert Hadfield found that adding about 3 wt% silicon decreased iron loss in electrical steel by about ten times compared to produced mild steel. The decrease was from 12 to 1.2 W/kg at $B_m = 1.9$ T. Electrical steel is, nowadays, produced with maximum 6.5% silicon. Adding silicon to the alloy increases its resistivity. Higher resistivity decreases iron loss caused by eddy currents. Isotropic electrical steel sheets are produced mainly by casting, hot rolling, and sometimes annealing. Rolling is a kind of metal forming process in which metal is passed through one or more pairs of rolls to reduce the thickness and to make the thickness of the sheet uniform. Fe–Si magnetic steel sheets can be isotropic or anisotropic. Hot rolling without a crystal orientation creates isotropic magnetic steel, whereas hot rolling, cold rolling, and annealing create a crystal orientation called Goss texture with anisotropic magnetic properties. This texture was invented in 1934 by Norman Goss. Silicon steel sheets are produced with different thicknesses, from about 0.2 to 0.65 mm [2]. Recently a group of researchers from Germany developed a new type of non-oriented electrical steel. The new method is based on hot rolling and annealing, cold rolling and then annealing, then cutting and applying the steel in electric machines. This developed method decreases iron loss in electrical steel sheets [13].

Sintered soft magnetic materials are applied mainly in DC circuits with a constant magnetic flux, whereas soft magnetic composites are applied mainly in AC circuits with changing magnetic flux.

Depending on the production technology of hard magnetic materials, we can observe

- casting;
- sintering;
- bonding with plastic (compression moulding, injection moulding, calendaring, and extrusion).

Development of iron alloys with nickel and cobalt called Alnico brought about the development of electric motors with a magnetic field excited by permanent magnets. Alnico permanent magnets are produced mainly by melting and casting of the alloy. Those kinds of permanent magnets can be isotropic or anisotropic in their magnetic structure. In the case of anisotropic permanent magnets, after casting of the alloy and during homogenization, the permanent magnet has to be placed in a magnetic field.

The other method of producing permanent magnets is the method of powder metallurgy, such as sintering and bonding with plastic.

In the case of powder metallurgy, the first step is to obtain magnetic powder. The method of preparing the powder depends on the kind of materials and the further application of the powder. Different types of powders are used in sintering and bonding with plastic.

Ferrite powders are produced by mixing iron oxides or iron compounds with oxides of desired metals, e.g. Mn, Zn, or Ni. The next step is roasting of the prepared mixture and then crushing and milling [14].

In turn, Alnico powders are produced by different methods, such as crushing of Alnico alloy, mixing of Al powder with Al–Fe and Ni–Al alloys, or mixing of Al, Ni, and Co elements [14].

Modern permanent magnets are mainly produced with rare-earth elements, such as Sm or Nd. The first research of magnets with rare-earth elements was carried out on casting permanent magnets, but these trials were not promising. That is why sintering methods were introduced. Sm–Co alloys are mainly obtained by calcio-thermal reaction of oxide powder mixtures (reduction and diffusion processes) or from elements by melting and casting. The next step is crushing of the melted alloy ingot [15]. Nd–Fe–B powders are produced by mechanical synthesis, crushing and milling cast ingots, hydrogen methods, and melt spinning [16].

The second step in manufacturing permanent magnets is densification of the powder, mainly by compression moulding and then sintering. In the case of anisotropic parts, compression moulding is often conducted in the magnetic field. The degree of densification is about 95% in comparison to cast alloy. The sintering process is conducted at a high temperature depending on the kind of alloy. Typically, this temperature is at about 75% of the melting temperature. In the case of Alnico and rare-earth permanent magnets, a protective atmosphere has to be applied.

The next kind of technology for producing permanent magnets is the bonding of powder with glue, such as resins and rubbers. Mainly rigid permanent magnets are prepared by compression or injection moulding, whereas flexible magnets are prepared by calendaring or extrusion.

Compression moulding technology consists of pressing powder with a bonding agent in a die. Compression pressures range from 300 to 1,000 MPa. Compression pressures higher than 1,000 MPa can cause destruction of the green compacts and quick degradation of the pressing tools. The next step in compression moulding of permanent magnets is curing. Curing is conducted at temperatures depending on the kind of bonding material used. For example, for epoxy resin, it is approximately 200°C in atmospheric air for 0.5–2 h depending on the kind of resin. As an example, Figure 5.2(a) shows hydraulic presses and (b) a thermal chamber that are in Łukasiewicz Research Network – Tele and Radio Research Institute.

In the case of injection moulding, the first step in producing ferromagnetic materials is the preparation of the granulate, which comprises thermosetting and thermoplastic polymers with magnetic powder. The maximum value of the amount

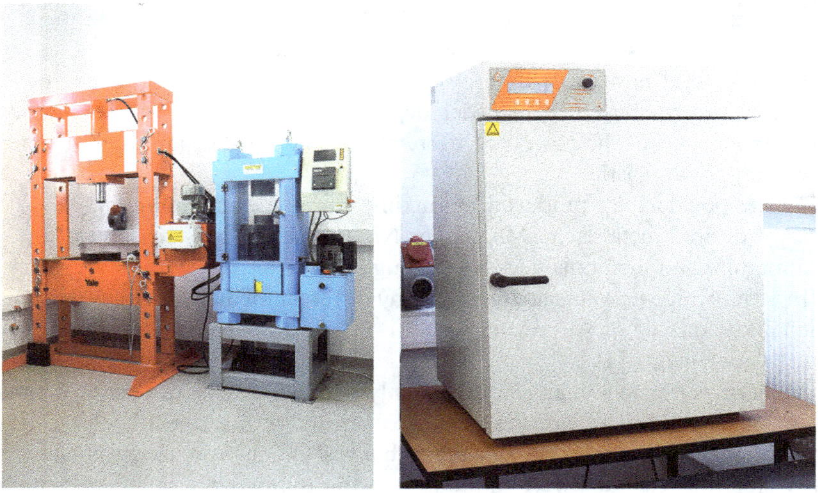

(a) (b)

Figure 5.2 *(a) Hydraulic presses and (b) the thermal chamber in Łukasiewicz Research Network – Tele and Radio Research Institute*

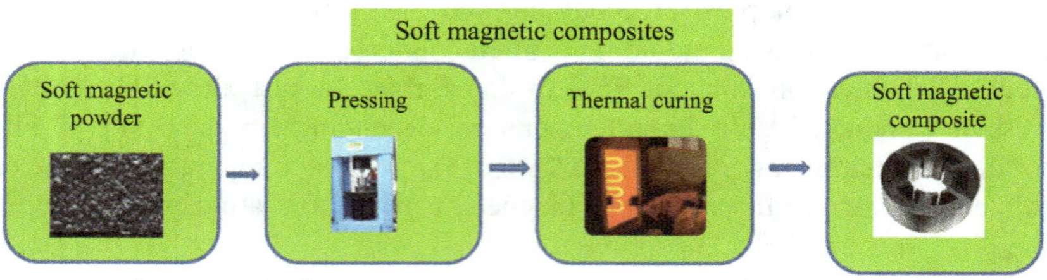

Figure 5.3 *Production of iron soft magnetic composite products*

of magnetic powder is about 60% by volume. The next step in manufacturing is the injection of the heated granulates into the mould. The pressure of the injection is about 60–90 MPa, and the injection temperature depends on the kind of polymer. The main polymers that are used are Nylon 6, Nylon 12, PPS (polyphenylene sulphide), and polyamide. The injection temperature of the permanent magnets with Nylon 6 and PPS is about 180 °C, whereas for Nylon 12, it is 150 °C [17].

In Łukasiewicz Research Network – Tele and Radio Research Institute, powder magnetic materials are made for R&D purposes, or they are commercially manufactured. Figures 5.3 and 5.4 show the paths of producing ferromagnetic elements from powder to finished soft magnetic composite parts and Nd–Fe–B permanent magnets. Figure 5.3 shows an example of an application of compression moulding technology for iron soft magnetic composites, whereas Figure 5.4 shows an example of the production of Nd–Fe–B-bonded permanent magnets.

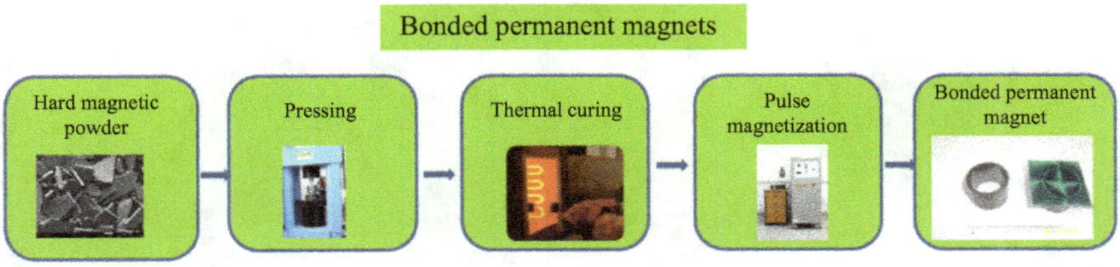

Figure 5.4 Production of Nd–Fe–B-bonded magnets

The obtained composite shown in Figure 5.3 was applied in a magnetic core of a prototype brushless DC motor for a juicer, whereas the compression-moulded-bonded permanent magnet shown in Figure 5.4 was used as a four-pole permanent magnet to the same brushless DC motor.

Powder metallurgy enables the production of ferromagnetic parts from mixtures of soft and hard magnetic powders. This method allows physical parameters of ferromagnetic parts to be tailored, such as their magnetic, mechanical, electrical, and thermal properties.

Permanent magnets can be isotropic and anisotropic in their magnetic structure. Anisotropic permanent magnets must be magnetized in a direction of anisotropy to ensure maximum performance. Isotropic permanent magnets can be magnetized in all magnetic directions and can be magnetized bi- or multipolar with 4, 6, or 8 poles, and so on. Such kinds of magnetization are also the subject of interest in Łukasiewicz Research Network – Tele and Radio Research Institute.

5.3 Application of ferromagnetic materials

Ferromagnetic materials are applied in numerous devices. Soft magnetic materials are used in devices, such as magnetic cores, whereas hard magnetic materials are applied as permanent magnets. Semi-hard magnetic materials are often applied in anti-theft systems.

Figure 5.5 shows main examples of the application of magnetic materials.

Figure 5.5 shows that there are many applications of magnetic materials in our everyday life.

The application of soft magnetic materials depends on the type of devices. Different devices need soft magnetic materials with different parameters. One device may need the highest magnetic flux density, while another may need a maximum initial permeability or maximum working frequency. Sometimes for the application of magnetic materials in AC circuits, the most important parameter is iron loss. Soft magnetic materials in the form of sheets are widely used in electric motors, generators, and transformers. For rotating machines, applied steel sheets are isotropic, whereas in transformers, they are mainly anisotropic.

Soft magnetic ferrites are cheap, which is why they are applied in many devices and applications. The application of soft magnetic ferrites, which are

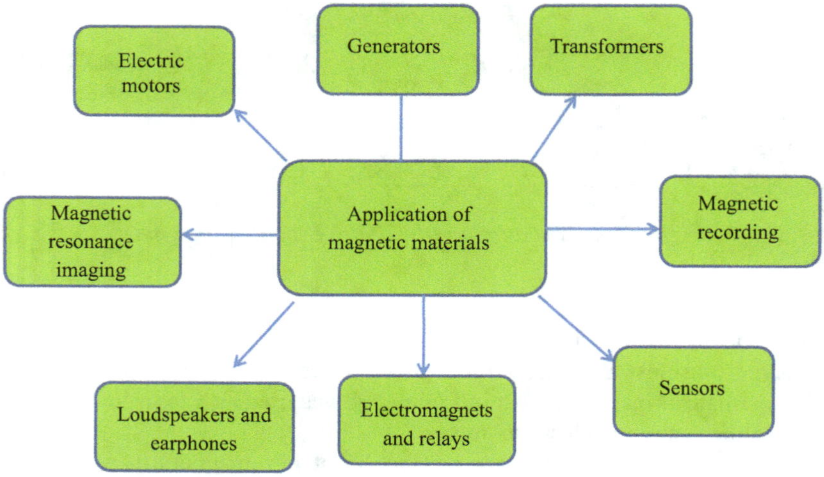

Figure 5.5 Examples of the application of magnetic materials

ceramic ferrimagnetic materials, depends on the permeability and frequency. Soft magnetic ferrites are used for radiofrequency and microwave applications because of their high resistivity. They are used in chokes, inductors, transformers, and antennas [2].

Amorphous materials, also called ferromagnetic glasses, are used at an intermediate frequency from kHz to MHz. They are 50-μm thin ribbons of rapidly quenched amorphous Fe- and Co-based alloys. Magnetic tapes are also applied as magnetic shields [2].

Iron and iron-cobalt bulk magnetic materials are also used as cores of electromagnets, resulting from their high magnetic flux density. In turn, Permalloy materials exhibit high initial permeability and are used in magnetic amplifiers and fluxgate magnetometers [2].

The next group of magnetic materials are hard magnetic materials. Hard magnetic materials are widely used in electric motors and generators as a source of magnetic field excitation. Permanent magnets are used, e.g. in DC motors, brushless DC motors, synchronous AC motors, and hybrid stepping motors. Hard magnetic materials in the form of films on a flexible or rigid substrate are used in tapes for magnetic recording and hard disc drives [2].

Hard magnetic ferrites are very cheap, and they are used in fridge magnets, cheap motors, actuators, and holding devices. They are used frequently in loudspeakers and headphones, cameras, and clocks. Ferrite ring magnets are used in domestic microwaves magnetrons in microwave ovens [2].

Permanent magnets, such as Sm–Co or Nd–Fe–B, are the newest permanent magnets. They are expensive because of the price of the rare-earth elements, but they have very good and diverse magnetic parameters. For example, they are used in MRI (magnetic resonance imaging) and NMR (nuclear magnetic resonance).

Permanent magnets can be used in Halbach cylinders for the creation of a uniform magnetic field in air gaps [2].

Permanent magnets from different hard magnetic materials are also used as a source of non-uniform magnetic fields in, e.g. cathode ray tubes and other electro-optic machines. They are also used in magnetic bearings, couplings, suspensions, magnetic separators, and voice-coil actuators [2].

There have been many attempts in applying powder-soft magnetic composites and bonded Nd–Fe–B permanent magnets in different electromagnetic and electromechanical converters in Łukasiewicz Research Network – Tele and Radio Research Institute.

Figure 5.6 shows examples of obtained Nd–Fe–B-bonded permanent magnet with 36 poles on the outer surface of the cylinder. This permanent magnet was used for trials with a linear and rotary actuator [18].

Another application that was developed at Łukasiewicz Research Network – Tele and Radio Research Institute with Mikroma S.A. company was a magnetic core of a stator made of Atomet EM-1 powder supplied by Rio Tinto company. This stator was applied in a permanent magnet synchronous motor UE-1,5R and was applied in electric tools manufactured by Georges Renault [19]. Figure 5.7(a) shows the stator of this motor, and (b) shows the stator magnetic core.

Powder materials manufactured in Łukasiewicz Research Network – Tele and Radio Research Institute are also shown in Figure 5.8. A brushless DC motor shown in Figure 5.8(a) had a magnetic stator core made of Somaloy 500 powder supplied by Höganäs and a four-pole excitation-bonded permanent magnet made of

Figure 5.6 Permanent magnet with 36 poles on the outer surface of a cylinder

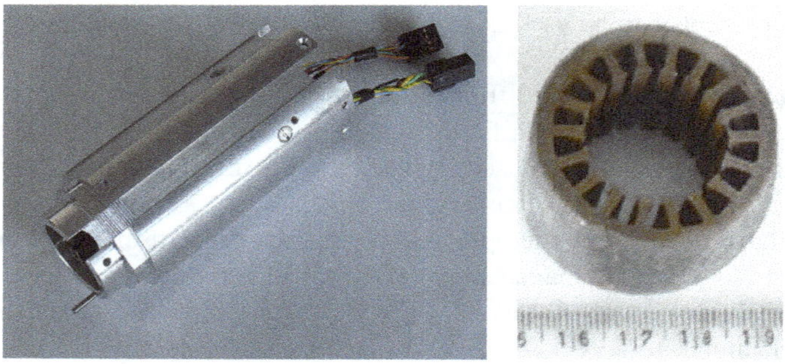

(a) (b)

Figure 5.7 (a) UE-1,5R electric motor for electric tools and (b) soft magnetic composite stator for UE-1,5R motor [19]

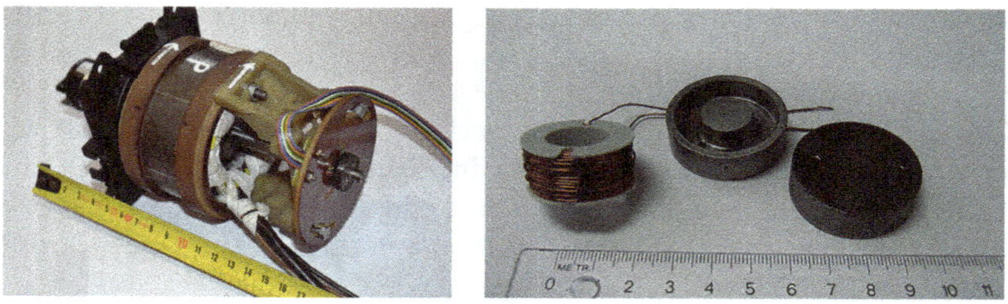

(a) (b)

Figure 5.8 (a) Brushless DC motor with the stator core made of a soft magnetic composite and an Nd–Fe–B-bonded permanent magnet, (b) inductor with the magnetic core made of iron powder

MQP-A Nd–Fe–B powder supplied by Magnequench. This motor was developed for an application in a juicer. In turn, Figure 5.8(b) shows a developed inductor for low and medium frequency applications. The magnetic core was made of Somaloy 500 iron powder supplied by Höganäs company.

Another example of an application of soft magnetic composites is the magnetic core of a switched reluctance motor [20]. Originally, this motor was applied in washing machines called Neptune manufactured by Maytag company. The motor was developed by Emerson Electric Company and has a symbol H55BMBJL-1820. Stator and rotor cores were obtained from Somaloy 700 and Somaloy 700 HR powders, respectively, supplied by Höganäs company. Parts of the composites were cut from larger blanks manufactured in Łukasiewicz Research Network – Tele and Radio Research Institute. These parts were machined and cut using electric discharge machining. Figure 5.9 shows the stator and rotor of the prototype motor. This motor was also optimized by the genetic [21,22] and wind-driven algorithms

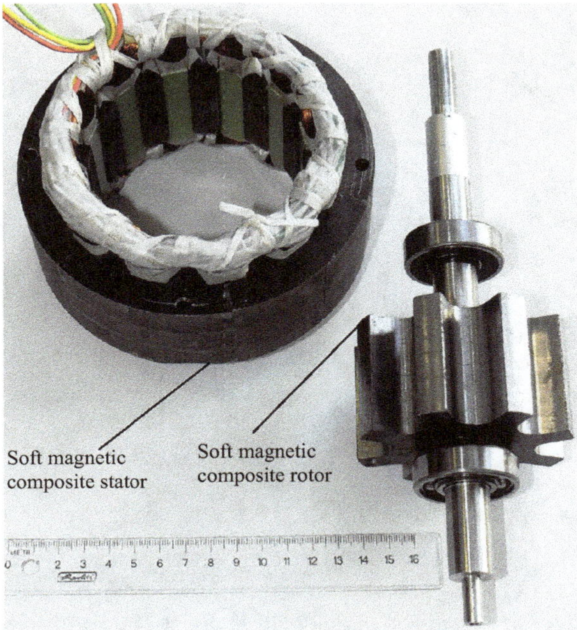

Figure 5.9 Stator and rotor of the prototype switched reluctance motor for a washing machine [20]

techniques [23,24]. Pareto solutions, the lowest losses and the maximum torque, were obtained as a result of geometry optimization.

There were also trials at Łukasiewicz Research Network – Tele and Radio Research Institute on the application of soft magnetic composites in a rotor core of a permanent magnet DC motor with 5-W power [25]. The rotor magnetic core was made of Atomet EM-1 powder supplied by Rio Tinto company. The motor was produced by Mikroma company and was called PRMO-30D. Figure 5.10 shows the rotor of the DC motor.

The next example of an application of a bonded permanent magnet is a magnetic encoder for the ABS (anti-lock braking system) systems for cars [26]. The rubber-bonded strontium ferrite 96 pole permanent magnet is supposed to be applied in an automotive industry. It should replace traditional ABS rings with soft magnetic teeth. Figure 5.11(a) shows the magnetic ring, whereas (b) shows the magnetic poles visualized on a magnetic film.

The next example of an application of soft magnetic composites in an electric motor is a brushless motor with 0.75-kW power and a rotational speed at 18,000 rpm. This motor is produced by Shenzhen Honcar M&C Hi-Tech Co. Ltd company model MBL-92FT-300HA. In this motor, the stator was replaced by a soft magnetic composite. Figure 5.12(a) shows a part of the stator manufactured in Łukasiewicz Research Network – Tele and Radio Research Institute, whereas (b) shows the

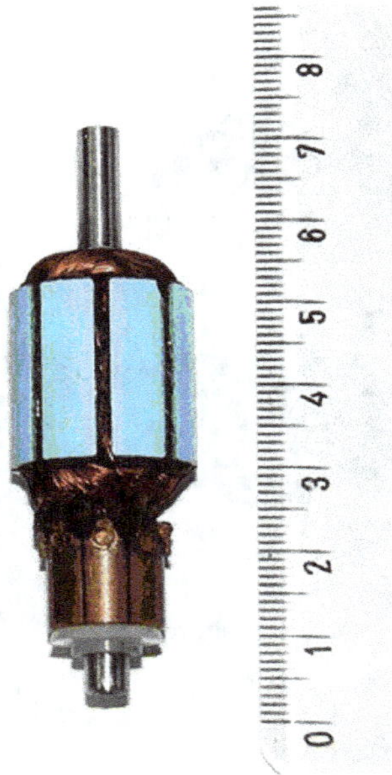

Figure 5.10 *Rotor of permanent magnet DC motor with a soft magnetic composite core [25]*

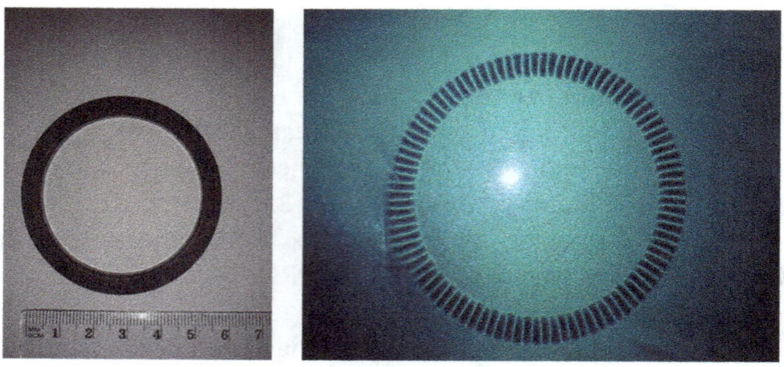

(a) (b)

Figure 5.11 *(a) Permanent magnet ring with 96 poles for the ABS system and (b) visualization of magnetic poles using magnetic film [26]*

structure of the brushless motor. The stator parts were obtained by electric discharge machining from parts that were 63 and 22.5 mm in height. The stator parts were manufactured from Somaloy 500 powder supplied by Höganäs company. The parameters of the newly developed motor were almost identical to the motor produced by the commercial company.

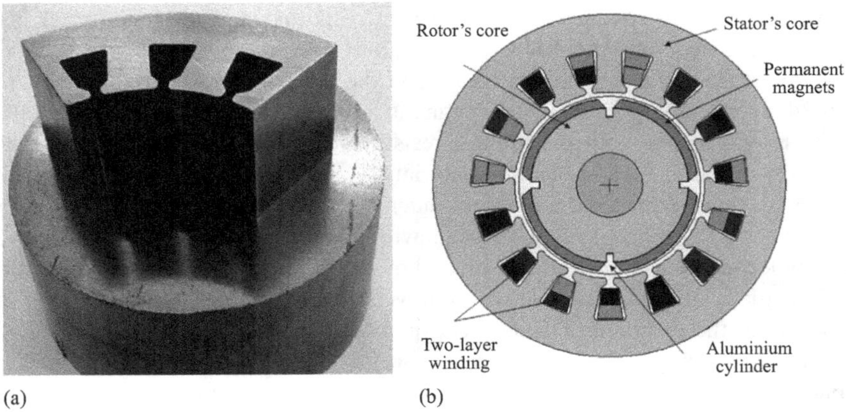

Figure 5.12 (a) A part of the stator and (b) the structure of the BLDC motor [56]

5.4 Frontiers of application of ferromagnetic materials

New types of soft and hard ferromagnetic materials with better properties have not been developed in recent years. The recent works are aimed at improving known ferromagnetic materials. For magnetic materials, physical properties (mainly magnetic) are the most important.

5.4.1 Magnetic properties

For soft magnetic materials, the known barrier is the magnetization saturation of materials. For alloys of 50% iron with 50% cobalt, called Permendur, the maximum magnetic polarization is approximately equal to 2.45 T [2]. Up to now, there have not been any materials that have been developed with higher polarization.

The next important magnetic parameter of soft magnetic materials is its initial and maximum permeability. Initial permeability is most important for applications of the material in low energy devices, e.g. relays. The maximum initial permeability of a material called Supermalloy, with 79% Ni, 16% Fe, and 5% Mo, is about 100,000, whereas the maximum permeability is 10^6 [1].

For soft magnetic materials, coercivity is also a very important parameter. Nowadays, the minimum coercivity of a soft magnetic material is about 0.16 A/m for Supermalloy [1]. Coercivity is dependent on the structure of the material. For an amorphous material such as Metglas 2628SC (40% Fe, 38% Ni, 4% Mo, and 18% B), coercivity is also very low and $H_c = 0.5$ A/m. The same value is obtained for a nanocrystalline material called Finemet. The Finemet material is composed of 73.5% Fe, 1% Cu, 3% Nb, 13.5% Si, and 9% B [2].

For the application of soft magnetic materials in AC conditions, iron loss is a very important issue. Iron loss is composed mainly of hysteresis loss and eddy current loss. In magnetic materials, we can also define additional losses that are the difference between calculated eddy and hysteresis loss and measured total loss. Soft magnetic materials should have the smallest possible hysteresis loss, which is the smallest area of the hysteresis loop, and the smallest eddy currents, which is fulfilled by the composition of a magnetic core with sheets with the smallest thickness or material with the highest resistivity, such as ferrites, iron-dielectric composites, and magneto-dielectric materials.

Iron loss is dependent on the frequency of the magnetic flux. Materials working at 50 Hz are composed of sheets with 0.3- to 0.65-mm thickness, whereas for a higher frequency, such as 400 Hz and more, the thickness is less than 0.3 mm.

The main iron loss parameter for low frequency applications is loss at an amplitude of the magnetic flux density equal to 1 T at 50 Hz and 0.2–0.5 mm of thickness of the sheets. For iron sheets, iron loss is equal to 5–10 W/kg. For isotropic Fe–Si hot rolled steel, it is 1–3 W/kg, and for anisotropic Fe–Si cold rolled, grain-oriented, it is 0.3–0.6 W/kg. The smallest total losses that we can observe are for Permalloy materials. For 50% Ni and 50% Fe, it is about 0.2 W/kg, and for 65% Ni and 35% Fe, it is about 0.06 W/kg [1].

For solid materials that are applied in AC electromagnetic converters, as was mentioned, the resistivity of the material is a very important parameter. Ferrimagnetic ferrites and powder materials have the highest value of resistivity. Soft magnetic yttrium ferrites that are applied in GHz applications are almost insulators. Soft magnetic composites based on iron and organic dielectric have a resistivity up to a few Ω m.

In turn, hard magnetic materials have different values of magnetic parameters compared to soft magnetic materials. For hard magnetic materials, the most important properties are high remanence and coercivity. High remanence and coercivity gives a high density of magnetic energy.

Nowadays, sintered Nd–Fe–B permanent magnets have the highest value of remanence. The highest remanence is equal to 1.49 T for the N55-type magnet [27]. High remanence is desired for high magnetic flux density in the air gap. In turn, high coercivity brings high resistance to demagnetization. The highest coercivity is obtained for Sm–Co permanent magnets (e.g. Recoma 25-type magnet), and it is $H_{cJ} = 2{,}400$ kA/m [28]. The highest magnetic density is obtained for the N55-type Nd–Fe–B magnet, and it is approximately $BH_{max} = 430$ kJ/m^3.

Semi-hard magnetic materials have sufficient properties for customers, and we cannot see much effort for obtaining better and cheaper materials.

5.4.2 Mechanical properties

In many applications, the mechanical properties are also very important. For soft magnetic steel sheets, the most important parameter is tensile strength. The tensile strength for non-oriented sheets is generally higher than for grain-oriented sheets. For non-oriented steel sheet 50HXT780T produced by Nippon Steel Corporation, the tensile strength (UTS) is 847 MPa in the transverse direction of rolling [29], whereas for a grain-oriented sheet produced by Stalprodukt S.A., for ET130-35, it is between 370 and 420 MPa for the transverse direction of rolling [30].

The Vickers hardness (H_v) for non-oriented steel sheet 50HXT780T produced by Nippon Steel Corporation is $H_v = 275$ [29], and for a grain-oriented sheet produced by Stalprodukt S.A., for ETH105-30, it is between 170 and 190 [30]. Companies that produce electrical steel are still looking for electrical steel with higher mechanical properties.

Soft magnetic composites have mechanical properties that are much worse than electrical steel. Producers of soft magnetic powders show, in their catalogues, the mechanical properties of composites manufactured from these powders.

Composites prepared from Somaloy powders produced by Höganäs company have different mechanical properties depending on the kind of powder and the technology of preparing composites from the powder. The maximum mechanical properties exhibit composites manufactured with 3P technology, which cures compacts in a steam atmosphere instead of air. In 3P technology, the pressing tool is heated to 80 °C instead of maintaining it at room temperature. The transverse rupture strength (TRS) is equal to 125 MPa for Somaloy 700 powder manufactured with 3P technology, whereas 40 MPa in the classical technology for the same powder. For 3P technology, for Somaloy 700 powder, UTS = 65 MPa, whereas for the classical technology for the same powder, the UTS is only 15 MPa. In turn, the compressive strength (R_c) for 3P technology is also higher than for the classical technology. For composites manufactured from Somaloy 700 powder, for the same compression pressure of 800 MPa, $R_c = 350$ MPa, whereas for the 3P technology, $R_c = 550$ MPa [7].

A different iron-based composite made from soft magnetic powder is produced by Rio Tinto company, formerly known as QMP. They produce Atomet EM-1 powder and magnetic parts prepared from this powder. For a compression pressure of 800 MPa, TRS = 124 MPa [6]. In turn, GKN Hoeganaes company produces AncorLam powder, and for composites manufactured with 830-MPa compression pressure, TRS = 81 MPa [8].

In the case of Permalloy steel, different mechanical properties are obtained for different compositions. For a Permalloy 80 material called Mumetal with 80% Ni, 5% Mo, and 15% iron, in an unannealed state, TRS = 1,030 MPa, whereas for the annealed state, TRS = 560 MPa [31].

Soft magnetic ferrites are brittle powder metallurgy products, and their main mechanical parameters are tensile and compressive strengths. For typical soft ferrites, UTS = 49 MPa. In turn, $R_c = 441$ MPa [32].

In many cases, producers of soft magnetic materials frequently do not show the mechanical properties of their products.

In the case of hard magnetic materials for permanent magnets, Alnico, hard ferrites, Sm–Co, and Nd–Fe–B have the following properties.

Alnico permanent magnets are mainly produced by casting, but for small magnets, they are produced by pressing and sintering. The main mechanical properties that are presented by the producers of Alnico magnets are Vickers hardness, compression strength, and tensile strength. The maximum Vickers hardness and compressive strengths are similar for the different kinds of Alnico magnets, and $H_v = 700$, whereas $R_c = 400$ MPa. In turn, the tensile strength is different for the different kinds of Alnico magnets. The highest value of tensile strength UTS = 450 MPa [11].

In the case of hard ferrite magnets, the producers show the following properties for sintered magnets: Vickers hardness $H_v = 700$, compressive strength $R_c = 895$ MPa, and tensile strength UTS = 34 MPa [33].

In the case of sintered Sm–Co permanent magnets, the producers show the mechanical properties of magnets, such as Vickers hardness $H_v = 600$, compressive strength $R_c = 1{,}000$ MPa, and tensile strength UTS = 40 MPa [34].

For sintered Nd–Fe–B permanent magnets, the producers also show their mechanical properties, such as Vickers hardness $H_v = 570$, compressive strength $R_c = 780$ MPa, and tensile strength UTS = 78 MPa [35].

5.4.3 Temperature properties

Another limit for the applications of ferromagnetic materials is the Curie temperature (T_c). After exceeding this temperature, magnetic materials lose their magnetic properties and become paramagnetic. The highest Curie temperature is for cobalt ($T_c = 1{,}130$ °C), whereas for iron and nickel, $T_c = 770$ and 358 °C, respectively [1].

Soft magnetic alloys of ferromagnetic elements such as Ni–Fe have the Curie temperature equal to a few 100 °C, depending on alloy composition. As an example, Permalloy 80, called Mumetal, has a T_c of 450 °C [31].

Hard magnetic alloys have different Curie temperatures. Customers frequently need as high a working temperature as possible. The working temperature depends on the Curie temperature and is much lower than the Curie temperature. Alnico alloys have $T_c = 850$ °C, $SmCo_5$ alloys have $T_c = 720$ °C, and $Nd_2Fe_{14}B$ alloys have a $T_c = 312$ °C. Oxide ferrimagnetic materials, such as hard magnetic ferrites, have $T_c = 400$–700 °C [1].

5.4.4 Shape and dimension limitation

Another limitation of ferro- and ferrimagnetic products are their shapes and dimensions. Soft magnetic cores of electromagnetic converters are mainly composed of cast and rolled steel sheets. Parts of magnetic cores are cut or stamped from electrical steel sheets, with their maximum width being about 1.3 m. Tools for cutting and stamping of sheets are expensive and have limited working time. In turn, powder metallurgy products that are magnetically soft and hard allow us to produce more complicated shapes than from sheets and give designers new design possibilities. The dimensions of the produced parts are limited by the pressing force, which influences the pressing pressure. Magnetic parts produced by machining larger elements can meet tight dimensional tolerances of up to a few μm, whereas powder metallurgy products can obtain tolerances up to ± 0.1 mm. Powder metallurgy compacts need pressures up to 1,000 MPa to obtain high density and high physical properties. Such limits influence the maximum dimensions of the compacts. Higher pressures cause Kieck and Riedl cones and the destruction of compacts. For larger products, we can use hot isostatic pressing (HIP) using argon gas as a pressure agent, but the production speed is very low. The maximum weight of HIP products is up to 30 t and 1 m in diameter [36].

5.4.5 Price limitations

One of the most important reasons besides properties that influence the application of ferromagnetic materials is their price. According to MEPS International Ltd. Analysis, soft magnetic materials, such as hot rolled carbon steel, cost about 0.6 $ per kg in October 2020. According to the webpage statista.com, the price of nickel has been quite stable in recent years, and 1 kg of nickel costs about 14 $ in 2019. In turn, the price of cobalt changes rapidly, and according to the webpage tradingeconomics.com, in March 2021, 1 kg of cobalt costs 53 $, whereas in 2018, it was 94 $.

In the case of non-oriented electrical steel sheets type 50W800 produced by Shanghai Zhanzhi Industrial Group Co., Ltd with 0.5 mm thickness and 20–1,250 mm width, the steel costs, in 2020, approximately 0.7 $ per 1 kg.

In the case of hard magnetic materials, the most important factor is the price of the rare-earth elements. The prices of Nd–Fe–B magnets depend on the cost of neodymium oxide. In 2012, the price of Nd oxide rapidly increased from 22 $ per kg in 2010 to 370 $ per kg. Nowadays, the price for 1 kg of neodymium oxide is stable and, from 2016, is about 40 $. The price of Sm–Co magnets depends on the cost of samarium oxide. In 2012, the price of Sm oxide increased rapidly from 3.3 $ per kg in 2010 to 135 $ per kg in 2012. Nowadays, the price for 1 kg of samarium oxide is stable and, from 2015, is 1.8 $ [37]. We can see that the prices are not stable and changes are difficult to predict, which influence the price of permanent magnets in the future.

The price of permanent magnets is very dependent of the kind of materials. The cheapest permanent magnets are ferrite magnets, and the most expensive are the Sm–Co permanent magnets. According to the Internet shop www.magnesy.eu, on 19.03.2021, for example, for permanent magnets with 10-mm diameter and 10 mm in height, the prices are as follows: Alnico LNG37 type costs 1.3 $, the cheapest Nd–Fe–B N38 type costs 0.4 $, and ferrite F30 type costs 0.08 $. In the case of Sm–Co S20-type permanent magnets with approximately the same mass with 6-mm diameter and 20 mm in height, one permanent magnet costs 3.4 $.

Table 5.9 shows the prices for 1 kg of permanent magnet materials calculated from the prices of permanent magnets offered by the Internet shop mentioned

Table 5.9 Prices of permanent magnet material for 1 kg

	Sm–Co S20	**Alnico LNG37**	**Nd–Fe–B N38**	**Ferrite F30**
Price for 1 kg	740 $	230 $	68 $	28 $

earlier. As it can be seen, the price of ferrite magnets is the lowest and is equal to 28 \$ per 1 kg, whereas the price of the Sm–Co material is the highest at 740 \$ per 1 kg. The price of the cheapest Nd–Fe–B permanent magnets is moderate, and mass production enables the price to be decreased. In turn, Alnico magnets are more expensive than Nd–Fe–B and ferrite magnets, but their applications are limited to only where it is essential, especially for stable magnetic parameters with changing temperatures.

5.5 Challenges of new ferromagnetic materials

Nowadays, users of magnetic materials are looking for magnetic materials with better and better physical properties. For some applications, they need the highest magnetic parameters, and for other applications, they need the best mechanical or thermal properties. One of the most important aspects is the price of ferro- and ferrimagnetic materials.

There are two ways for developing ferro- and ferrimagnetic materials. One way is to improve known magnetic materials, and the other way is to develop completely new magnetic materials and production technologies.

In the case of improving existing materials, we can observe changes in the composition of magnetic materials, such as the development of materials without expensive elements, such as rare-earth, or the development of materials with decreased amounts of expensive components. Another way of developing materials is the optimization of the production technology.

New technologies, such as additive manufacturing, are being developed, and they are being increasingly applied to obtain different types of magnetic materials. Additive manufacturing, also called 3D printing, is applied to the production of various magnetic parts. It often uses magnetic particles that are connected to each other during additive manufacturing. It is possible to produce magnetic parts with very complicated shapes.

5.5.1 Soft magnetic materials

In the case of soft magnetic materials, we can see the development of the most popular materials, such as iron–silicon steel sheets. For those materials, producers are looking to decrease iron loss and enhance the magnetic permeability and working frequency. Authors in [13] show that the increase in the electromagnetic properties, especially a decrease in iron loss, is possible by the complex application of interdisciplinary numerical modelling of the production of Fe–Si non-oriented electrical steel. Up to now, the most applied Fe–Si steel sheets have 3% silicon. Adding 6.5% silicon to Fe–Si alloy increases resistivity and decreases magnetostriction to almost zero, causing a decrease in iron loss and enabling the application of such steels in medium-frequency applications. Adding silicon to the alloy causes, unfortunately, an increase in brittleness and prohibits the economical production of iron cores. A higher working frequency enables application of 6.5% silicon in electromagnetic devices and causes minimization of magnetic core sizes and the whole device [38].

In recent years, we can see the development of magnetic materials for high-frequency applications. We can observe new sensors with wires made of new types of alloys, such as Fe–Co–Ni, that are electroplated on Cu–Be wires with 0.1-mm diameter and magnetic amorphous Co–Fe–Si–B wires with 0.15-mm diameter. They are frequently applied in high-frequency applications [39].

Fe–Si steel sheets for low-frequency applications have a thickness as low as 200 μm, whereas for frequencies higher than 50 Hz, the sheets have a 50- to 100-μm thickness. Thinner sheets are manufactured for medium-frequency applications from amorphous and nanocrystalline soft magnetic alloys with very low hysteresis loss, with thickness in the range of 15–30 μm. Amorphous and nanocrystalline materials have, however, a lower saturation magnetization than Fe–Si materials caused by the presence of additive metals, such as Nb and Cu. Further works are focused on the production of nanocrystalline and amorphous sheets with thinner thicknesses and a manufacturing process that increases the heating rates, which increases the saturation magnetization and decreases brittleness. Recently, those materials are more often applied in power transformers [40].

5.5.2 Hard magnetic materials

Nowadays, there are trials to develop permanent magnets without or with less amounts of expensive rare-earth elements [41]. It is connected with a still growing price of rare-earth elements. The greatest amount of rare-earth elements is in China, which is why they can influence the price of the elements. Nd–Fe–B permanent magnets are composed of the ferromagnetic $Nd_2Fe_{14}B$ phase surrounded by a non-magnetic Nd-rich grain-boundary phase. An increase in the magnetic parameters of Nd–Fe–B permanent magnets can be achieved by grain-size refinement, modification of the grain-boundary phase, and the grain-boundary diffusion process. Permanent magnets with smaller grain size exhibit high coercivity. However, small grain size decreases remanence. Smaller size powders of a few micrometer increase the possibility of the powder exploding. Modification of the grain-boundary phase by replacing some of the Nd with other elements, such as Fe, causes an enhancement in the coercivity. In the case of the grain-boundary diffusion process, it is promising to obtain better Nd–Fe–B permanent magnets than those obtained currently. The grain-boundary diffusion process with smaller amounts of heavy rare-earth elements can improve the coercivity of permanent magnets, maintaining remanence [42].

The invention of new permanent magnet materials is connected with the development of new magnetic compounds and the development of methods for creating materials with high coercivity. Research on creating magnets with higher coercivity by the development of manufacturing processes can be observed for Nd–Fe–B and hard ferrite materials [42].

5.6 New technologies of magnetic materials production

Recently, we can observe new technologies that are applied in magnetic materials' production. The broadest development is seen in additive manufacturing methods of magnetic materials. Additive manufacturing is also called 3D printing. Additive manufacturing techniques create 3D objects by successive addition of material. 3D objects are created usually layer upon layer in contrast to the subtraction of material. According to ISO/ASTM 52900 standard, there are seven additive manufacturing techniques: binder jetting, directed energy deposition, material extrusion, material jetting, powder bed fusion, sheet lamination, and vat photopolymerization [43–46]. Some of these have been applied to magnetic materials' fabrication.

Additive manufacturing is mainly applied for permanent magnets. In a smaller range, additive manufacturing is applied for soft magnetic materials' production for research and fast prototyping [47–51].

As an example, the binder jetting technique was applied for obtaining Nd–Fe–B permanent magnets. Binder jetting relies on building a desired object from powder that is glued by a deposited liquid binder. A 3D object is created layer by layer. In the final step, the printed magnet is heated in an oven at 100–150 °C to cure the thermoset binder. The main disadvantage of the obtained magnets is their low density, which causes low magnetic properties and low mechanical strength [52].

The next type of technology that is very promising and that has been used to fabricate permanent magnets is material extrusion. It is one of the cheapest 3D printing methods. In the first step of production, a composite filament with Nd–Fe–B and strontium ferrite magnetic powders is created. Next, the filament is extruded using heated nozzles to melt the filament and create a permanent magnet layer by layer [53].

Direct energy deposition is also used for the production of magnetic elements. The Nd–Fe–B powder is used as a permanent magnet material, and the laser beam source is the source of energy. Here, we have powder as a building material that is melted during deposition. The remanence and coercivity of the obtained Nd–Fe–B magnets decreased [54].

Different kinds of additive manufacturing are still under development. The developed materials suffer from low magnetic and mechanical properties as well as a low speed of production.

5.7 Conclusion

Chapter 5 has shown a general overview of magnetic materials, methods of their production, and their physical properties. Based on research works conducted in Łukasiewicz Research Network – Tele and Radio Research Institute, applications of soft magnetic composites and bonded magnets in electromagnetic and electromechanical devices are shown. In the chapter, the maximum obtained magnetic, mechanical, and thermal properties and other limitations, such as price, are presented. The challenges of new soft and hard magnetic materials as well as new technologies are also presented.

The presented properties of magnetic materials can be useful in designing electromagnetic and electromechanical devices and machines with optimal shapes, taking into account the physical limitations of the materials. New technologies of production, such as 3D printing, enable the development of new devices and machines that are impossible or very expensive to manufacture using conventional techniques.

References

[1] Jiles D., *Introduction to Magnetism and Magnetic Materials*, London, UK: Chapman & Hall, 1991.

[2] Coey J. M. D., *Magnetism and Magnetic Materials*, Cambridge, UK: Cambridge University Press, 2010.

[3] Przybylski M., *Application of Powder Magnetic Materials in Electric Machines*, PhD thesis, in Polish, Poznan University of Technology, Poland, 2012.

[4] Jankowski B., *Influence of Manufacturing Technology on Physical Properties of Powder Soft Magnetic Composites*, PhD thesis, in Polish, Electrotechnical Institute, Warsaw, 2013.

[5] Kefeni K., Msagati T. A. M., and Mamba B. B., Ferrite nanoparticles: Synthesis, characterisation and applications in electronic device, *Materials Science and Engineering: B*, vol. 215, pp. 37–55, 2017.

[6] Rio Tinto Metal Powders catalogue, "Atomet EM-1", 2021.

[7] Höganäs AB catalogue, "Somaloy-Material Data", 2016.

[8] GKN Hoeganaes catalogue, "AncorLam", 2021.

[9] Petkovska L. and Cvetkovski G., "Innovative motors and shape optimisation," in book *Optimal Design exploiting 3D Printing and Metamaterials*, Editors: Di Barba P. and Wiak S. IET Publisher, London, 2022.

[10] Mohapatra J., Xing M., Elkins J., and Liu J. P., Hard and semi-hard magnetic materials based on cobalt and cobalt alloys, *Journal of Alloys and Compounds*, vol. 824, article number 153874, 2020.

[11] Eclipse Magnetics catalogue, "Technical Data Sheet – Alnico Magnets", 2021.

[12] Ślusarek B., *New Powder Magnetic Composites for Electric Machines and Devices*, Tele and Radio Research Institute Monograph, 2012.

[13] Leuning N., Steenties S., Dierdorf J., *et al.*, Continuous process chain modeling of low-loss FeSi sheet for energy-efficient electrical drives, *ICS Newsletter*, vol. 22, no. 3, 2015.

[14] Rutkowski W., Projektowanie właściwości wyrobów spiekanych z proszków i włókien, PWN, 1977.

[15] Schatt W. and Wieters K.-P., Powder Metallurgy – Processing and Materials, EPMA, 1997.

[16] Ślusarek B., Dielektromagnesy Nd-Fe-B, Oficyna Wydawnicza Politechniki Wrocławskiej University, 2001.

[17] Arnold Magnetic Technologies catalogue, "Plastiform Injection Molded Magnets", 2014.

[18] Karbowiak M., Jankowski B., Kapelski D., Przybylski M., and Ślusarek B., Multi-pole magnetization of Nd-Fe-B bonded magnets for rotary linear actuators, *Elektronika*, vol. 10, pp. 147–148, 2011.

[19] Slusarek B., Dlugiewicz L., Gawrys P., and Przybylski M., Brushless high-speed electric motor for electric tools, 14th International Conference Alternating Current Electrical Drives – ACED, pp. 89–92, Yekaterinburg, Russia, 13–16 March 2007.

[20] Przybylski M., Ślusarek B., Di Barba P., Mognaschi M. E., and Wiak S., Temperature and torque measurements of switched reluctance actuator with composite or laminated magnetic cores, *Sensors*, vol. 20, no. 11, article no. 3065, 2020.

[21] Di Barba P., Mognaschi M. E., Przybylski M., Rezaei N., Slusarek B., and Wiak S., Field-based analysis and optimal shape synthesis of switched reluctance motors, *Lecture Notes in Electrical Engineering*, vol. 452, Cham, Switzerland: Springer International Publishing, pp. 71–85, 2018.

[22] Di Barba P., Mognaschi M. E., Przybylski M., Rezaei N., Slusarek B., and Wiak S. Geometry optimization for a class of switched-reluctance motors: A bi-objective approach, *International Journal of Applied Electromagnetics and Mechanics*, vol. 56, pp. S107–S122, 2018.

[23] Di Barba P., Mognaschi M. E., Wiak S., Przybylski M., and Slusarek B., Optimization and measurements of switched reluctance motors exploiting soft magnetic composite, *International Journal of Applied Electromagnetics and Mechanics*, vol. 57, pp. S83–S93, 2018.

[24] Di Barba P., Mognaschi M. E., Wiak S., Przybylski M., and Slusarek B., Wind-driven optimization for the design of switched reluctance motors, Proceedings of 18th International Symposium on Electromagnetic Fields in Mechatronics, Electrical and Electronic Engineering, Lodz, Poland, published on IEEE Xplore, 2017.

[25] Slusarek B., Gawrys P., Gromek J. Z., and Przybylski M., The application of powder magnetic circuits in electric machines, Proceedings of the 18th International Conference on Electrical Machines, published on IEEE Xplore, Vilamoura, Portugal, 6–9 September 2008.

[26] Przybylski M., Ślusarek B., Bednarczyk T., and Chmiel G., Magnetic and mechanical properties of strontium ferrite and Nd–Fe–B rubber bonded permanent magnet, *Acta Physica Polonica A*, vol. 136, no. 5, pp. 685–688, 2019.

[27] Arnold Magnetic Technologies catalogue, "Neodymium Iron Boron Magnet Catalog", 2020.

[28] Arnold Magnetic Technologies catalogue, "Recoma-Sm-Co permanent magnets", 2014.

[29] Nippon steel catalogue, "Electrical Steel Sheets", 2019.

[30] Stalprodukt S.A. catalogue, "Grain Oriented Electrical Steel", 2019.

[31] Xi'an Gangyan Special Alloy Co. Ltd. catalogue, "Permalloy 80/Mumetal", 2021.

[32] Magnetics catalogue, "Ferrite Cores", 2017.

[33] Eclipse Magnetics catalogue, "Technical Data Sheet – Ferrite Magnets/ Ceramic Magnets", 2021.

[34] Eclipse Magnetics catalogue, "Technical Data Sheet – Samarium Cobalt Magnets, SmCo Magnets", 2021.

[35] Eclipse Magnetics catalogue, "Technical Data Sheet – NdFeB Magnets/ Neodymium Iron Boron Magnets", 2021.

[36] Introduction to Hot Isostatic Pressing, "A guide for Designers and Engineers, European Powder Metallurgy Association", 2018.

[37] Hykawy J. and Chudnovsky T., "*Industry Report – Rare Earths*", Stormcrow Capital Ltd., 2020.

[38] Ouyang G, Chen X., Liang Y., Macziwski C., and Cui J., Review of Fe-6.5 wt% Si high silicon steel – A promising soft magnetic material for sub-kHz application, *Journal of Magnetism and Magnetic Materials*, vol. 481, pp. 234–250, 2019.

[39] Kurlyandskaya G. V., Shcherbinin S. V., Volchkov S. O., Bhagat S. M., Calle E., and Pérez R., Soft magnetic materials for sensor applications in the high frequency range, *Journal of Magnetism and Magnetic Materials*, vol. 459, pp. 154–158, 2018.

[40] Azuma D., Ito N., and Ohta M., Recent progress in Fe-based amorphous and nanocrystalline soft magnetic materials, *Journal of Magnetism and Magnetic Materials*, vol. 501, article no. 166373, 2020.

[41] Li D., Li Y., Pan D., Zhang Z., and Choi C.-J., Prospects and status of iron-based rare-earth-free permanent magnetic materials, *Journal of Magnetism and Magnetic Materials*, vol. 469, pp. 535–544, 2019.

[42] Nakamura H., The current and future status of rare earth permanent magnets, *Scripta Materialia*, vol. 154, pp. 273–276, 2018.

[43] ISO/ASTM 52900, "Additive manufacturing — General principles— Terminology", 2015.

[44] Siedlecki M., Zawada A., Przybylski M., and Ślusarek B., Methods of preparing soft and hard magnetic elements by additive manufacturing – review, 15th Selected Issues of Electrical Engineering and Electronics (WZEE), Zakopane, Poland, published on IEEE Xplore, 2019.

[45] Di Barba P. and Galdi R., "Free-form optimal design in electromagnetism exploiting 3D printing," in book *Optimal Design Exploiting 3D Printing and Metamaterials*, Editors: Di Barba P. and Wiak S. IET Publisher, London, 2022.

[46] Lopez-Fernandez X. M., "Industrial design perspectives," in book *Optimal Design Exploiting 3D Printing and Metamaterials*, Editors: Di Barba P. and Wiak S. IET Publisher, London, 2022.

[47] Chaudhary V., Mantri S. A., Ramanujan R. V., and Banerjee R., Additive manufacturing of magnetic materials, *Progress in Materials Science*, vol. 114, article no. 100688, 2020.

[48] Bernier F., Ibrahim M., Mihai M., Thomas Y., and Lamarre J.-M., Additive manufacturing of soft and hard magnetic materials used in electrical machines, *Metal Powder Report*, vol. 75, no. 6, 2020.

[49] Li L., Tirado A., Nlebedim I. C., *et al.*, Big area additive manufacturing of high performance bonded NdFeB magnets, *Scientific Reports*, vol. 6, article no. 36212, 2016.

[50] Domingo-Roca R., Jackson J. C., and Windmill J. F. C., 3D-printing polymer-based permanent magnets, *Materials and Design*, vol. 153, pp. 120–128, 2018.

[51] Goll D., Schuller D., Martinek G., *et al.*, Additive manufacturing of soft magnetic materials and components, *Additive Manufacturing*, vol. 27, pp. 428–439, 2019.

[52] Li L., Post B., Kunc V., Elliott A. M., and Paranthaman M. P., Additive manufacturing of near-net-shape bonded magnets: Prospects and challenges, *Scripta Materialia*, vol. 135, pp. 100–104, 2017.

[53] Palermo E. M., Casaleiz D., Jimenez N. A., *et al.*, Magnetic-polymer composites for bonding and 3D printing of permanent magnets, *IEEE Transactions on Magnetics*, vol. 55, no. 2, 2019.

[54] Sridharan N., Cakmak E., List F. A., *et al.*, Rationalization of solidification mechanism of Nd–Fe–B magnets during laser directed-energy deposition, *Journal of Materials Science*, vol. 53, pp. 8619–8626, 2018.

[55] Yang F., Zhang X., Guo Z., Ye S., Sui Y., and Volinsky A. A., 3D printing of NdFeB bonded magnets with $SrFe_{12}O_{19}$ addition, *Journal of Alloys and Compounds*, vol. 779, pp. 900–907, 2019.

[56] Kapelski D., Jankowski B., Przybylski M., and Ślusarek B., The influence of cryogenic temperatures on the characteristics of a brushless motor, *Technical Transactions—Electrical Engineering*, vol. 1E, pp. 281–289, 2015.

[57] Ripka P., *Magnetic Sensors and Magnetometers*, Norwood, MA: Artech House, 2001.

[58] MS-Schramberg GmbH & Co. catalogue, "Permanent Magnets", 2014.

[59] "Compression molded Sm-Co permanent magnets", MPCO Magnetics Co., Ltd, Catalogue, 2020.

[60] Neo Magnequench catalogue, "Magnequench – Preferred Partner for the Bonded Neo Industry", 2021.

[61] Aichi Steel Corporation catalogue, "MagFine World's Strongest Bonded Magnet", 2018.

[62] Tridelta Magnetsysteme GmbH, "Neolit – The Innovative Material for High Performance Permanent Magnets", Catalogue, 2021.

Chapter 6
Synthesising metamaterials with 3D printing and conductive layers

Sławomir Hausman[1] and Łukasz Jopek[1]

Metamaterials are artificial composite media that exhibit unique electromagnetic (EM) properties. The chapter covers selected aspects of the modelling, design, optimisation, and fabrication of metamaterials based on gradient dielectric structures and split-ring resonator (SRR) arrays. In the first part of the chapter, a synthesis method for gradient metamaterials is presented, and its application to a flat focusing lens is studied. Next, we discuss metasurfaces comprising conductive resonant elements. We present full-wave and equivalent-circuit modelling approaches for resonator arrays and two example designs embracing a flat lens for magnetic resonance imaging (MRI) and an artificial magnetic conductor (AMC). We also report on the progress in frequency-selective metasurfaces in the terahertz band. The last part of the chapter discusses 2D/3D printing of effective dielectric media and proposes a method to improve its accuracy and speed.

6.1 3D gradient dielectric metamaterials

6.1.1 3D gradient dielectric metamaterial anatomy

In order to understand how dielectric gradient metamaterials work, it is helpful to consider how EM waves interact with their conventional constituent homogeneous and isotropic materials. The EM response of any medium is governed by its three parameters, i.e. the permittivity ε, the permeability μ, and the conductivity σ. Most dielectrics used in practice can be assumed to have $\mu = 1$ and often also $\sigma = 0$. It is possible to fabricate composite materials with μ significantly greater than 1, such as polylactic acid with ferroelectric oxide inclusions (PLA/Fe_3O_4), but they will not be considered in this chapter. In general, media are dispersive, and their parameters are frequency dependent, so we should be aware that the permittivity is a function of frequency $\varepsilon(\omega)$. When a harmonic EM wave travels in a dielectric medium, its frequency does not change, while its wavelength is reduced by a factor of $1/\sqrt{\varepsilon_r}$, where ε_r is the relative electric permittivity of the medium. Conventional

[1]Institute of Electronics, Lodz University of Technology, Łódź, Poland

dielectrics have $\varepsilon_r > 1$ and are isotropic. It is possible to design and fabricate gradient dielectric metamaterials (nonhomogeneous or/and anisotropic) exploiting the concept of effective media with a desired effective permittivity achieved by alternating constituent elements of subwavelength size with two permittivity values, e.g. air with $\varepsilon_r = 1$and some other material with $\varepsilon_r \gg 1$, which is illustrated in Figure 6.1 for a 1D-layered case. Even if the constituent materials are isotropic, the resulting metamaterial can be anisotropic with different permittivity values in different directions. This can be achieved by adequately engineering the material's internal structure consisting of meta-particles or unit cells. The unit cell is a fundamental building block of meta-structures and is usually arranged quasi-periodically [1,2].

Each unit cell has its effective relative permittivity, which may be calculated by the following equation:

$$\varepsilon_{reff} = (1 - F)\varepsilon_1 + F\varepsilon_2 \tag{6.1}$$

where ε_1 is the relative electric permittivity of the first material (e.g. air), ε_2 is the relative electric permittivity of the second material, F is the volumetric fill fraction coefficient for the medium with permittivity ε_2.

Examples of 3D unit-cell shapes with various fill fraction F values are shown in Figure 6.2.

When a (meta)material is anisotropic, the EM field lines tend to be refracted towards the direction of the highest permittivity values depending on the strength of

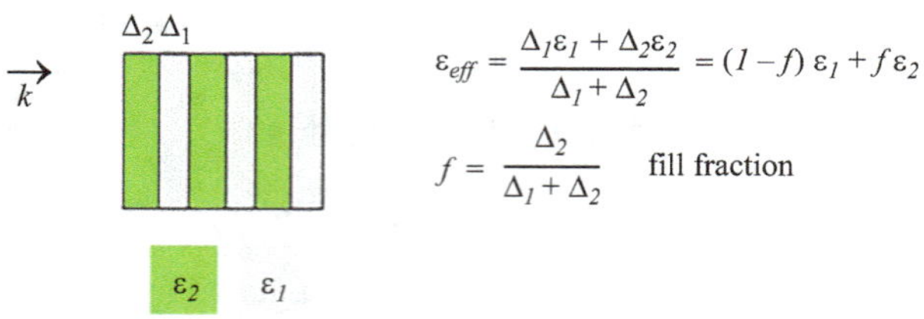

Figure 6.1 Composite 1D effective medium consisting of two dielectrics – effective permittivity ε_{eff} calculation

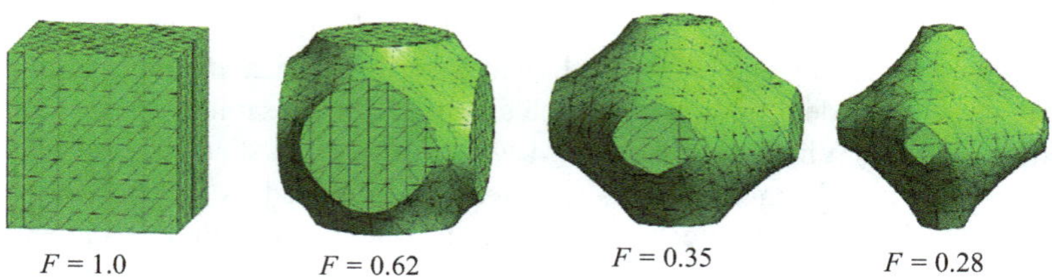

Figure 6.2 Example 3D unit cell with various values of the fill fraction F

the anisotropy [3]. Unit cells with variable fill fraction, which also exhibit anisotropy, may be helpful for complex EM wave shaping but make the design intricate.

6.1.2 3D metamaterial synthesis techniques

In this section, we present an efficient method to design 3D gradient dielectric metamaterials. It is based on the Fourier series expansion and so-called grating functions described in [2]. The primary element of the structure is a unit cell which might be similar to the one in Figure 6.2. The unit cells are placed repeatedly in the structure, but they can have different directional orientations, sizes, and fill fractions, as shown for a 2D case in Figure 6.3.

The scheme which we will present was initially proposed in [2,4]. The method is suitable both for 2D and 3D design. As the input, the algorithm requires a predefined shape on the unit cell and three spatial 'maps' (i.e. spatial 2D/3D functions) of the fill fraction $\vec{F}(\vec{r})$, phase (orientation) $\vec{\Theta}(\vec{r})$ and period (unit cell size) $(\vec{r})$. The unit-cell shape is then represented in the frequency domain by calculating its Fourier series [2]. The number of harmonics $a(p,q)$ in the following equation is truncated to some small value $M = (2N+1) \times (2N+1)$:

$$a(p,q) = DFT_{2D}(f(x,y)) \tag{6.2}$$

Greater values of M result in more accurate unit-cell shape reconstruction at the expense of increased computational complexity and memory requirements (the whole lattice may consist of millions of unit cells). The next step is generating a lattice containing unit cells modified by the fill fraction, orientation, and period maps. Defining these maps to achieve some target functionality of the metamaterial is an important and nontrivial task in itself. The target lattice should not exhibit

Figure 6.3 Example 1D quasi-periodic structure with a triangular unit cell (white triangle). The cells are of different sizes and orientations depending on their position in the lattice

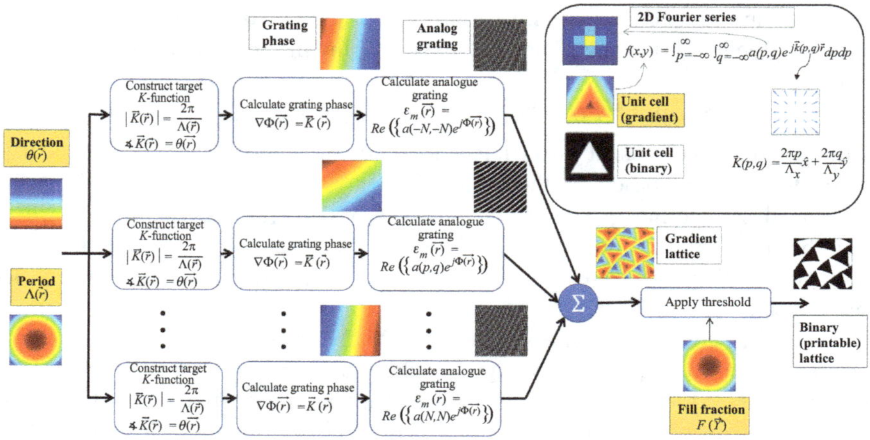

Figure 6.4 Synthesis of 2D gradient dielectric metamaterials using period, fill fraction, and direction functions ('maps')

significant discontinuities to avoid unintended scattering of the EM waves. Generally, all the cell size, orientation, and fill fraction changes should be gradual and not abrupt.

The synthesis algorithm is explained in Figure 6.4. In the process, spatially variant grating phase function $\Phi(\vec{r})$ is used, where $\vec{r}$ is the position vector for the lattice. The $\Phi(\vec{r})$function controls the period and direction of each spectral component of the unit cell within the lattice. $\Phi(\vec{r})$ is derived by solving the gradient equation (6.3). The wave number function $K(\vec{r})$ is given by (6.4)

$$\nabla\Phi(\vec{r}) = \vec{K}(\vec{r}) \tag{6.3}$$

$$K(\vec{r}) = \frac{2\pi}{\Lambda(\vec{r})}e^{j\Theta(\vec{r})} \tag{6.4}$$

where $\Lambda(\vec{r})$ is the period map and $\Theta(\vec{r})$ is the direction map.

Two examples of analogue (non-binary) grating functions are shown in Figure 6.5. A preliminary spatially variant lattice is constructed from the sum of all of the spatially variant gratings.

At this stage of the algorithm, the unit cell is not a binary object but a continuous gradient image with maximum values $\cong$1 near the centre of the triangular cells. To generate the spatially variant binary lattice, spatially variant fill fraction function $F(\vec{r})$is applied.

Two examples of spatially variant lattices are shown in Figure 6.6. In Figure 6.6(a), only the direction and period of the unit cell is altered. Direction varies from 22.5° to 62.5° from the bottom to the top of the lattice, and that is how the unit cells are rotated depending on their position in the structure. Similarly, the unit-cell period varies from 1.0 to 3.0, from corners towards the lattice centre. The

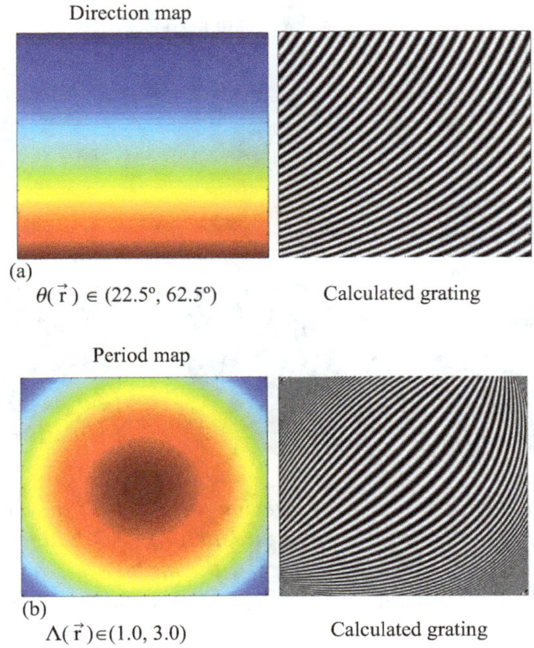

Figure 6.5 Example of spatially varying orientation (a) and period (b) and the resultant gratings

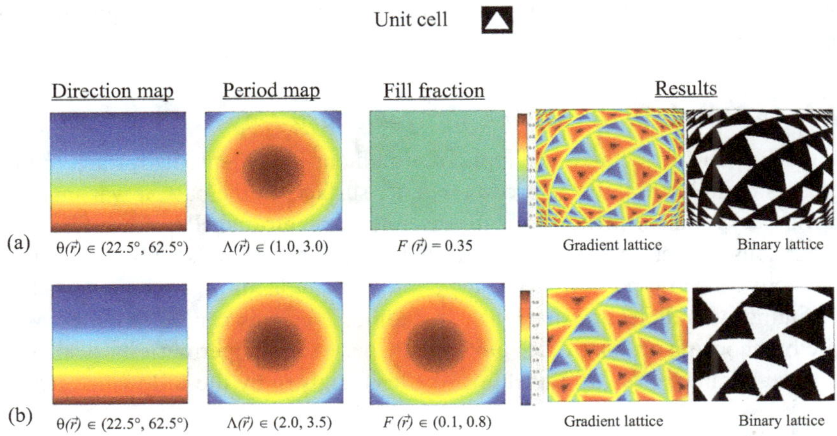

Figure 6.6 2D spatially varying lattices with a triangular unit cell for some example period, fill fraction, and direction functions with fill fraction map constant (a) and variable (b)

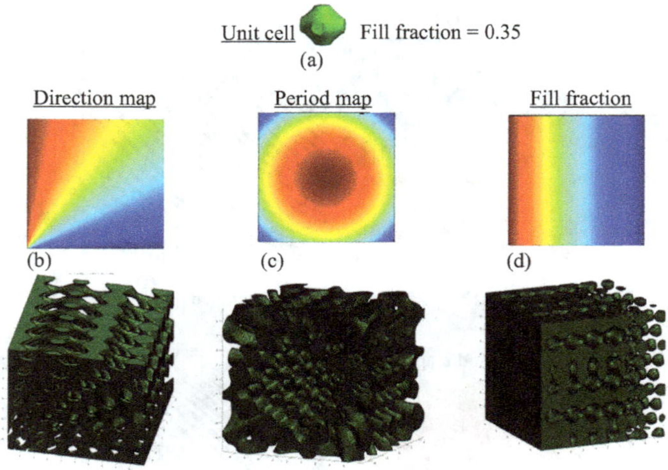

Figure 6.7 Examples of 3D spatially varying lattices with a symmetric unit cell (a), 3D lattice with varying direction (b), 3D lattice with varying period (c), and raw 3D lattice with varying fill fraction (parts of the lattice are 'floating' and therefore not printable) (d)

fill fraction, in this case, is constant for the whole lattice. In Figure 6.6(b), additionally, a spatially variable fill fraction map is applied.

In Figure 6.7, an example 3D lattice is shown. The result is raw, and parts of the lattice (for low fill fraction values) are 'floating' and are not printable. This problem can be avoided, e.g. by imposing constraints on the minimum fill fraction value.

6.1.3 Example application – flat Luneburg lens

A Luneburg lens is a classic example of a gradient-index lens. Its refractive index decreases radially from the centre to the outer surface. In the past, it was made, but not without technical difficulty, with discrete concentric shells of a different but homogenous refractive index. More design flexibility can be achieved by varying the effective electric permittivity using additive manufacturing techniques, such as 3D printing with dielectric filaments. In addition, the lens can be designed as isotropic by choosing an 'isotropic' unit-cell shape. Example permittivity distribution of a three-parameter oblate Luneburg-like lens is given by the following equation:

$$\left\{ \begin{array}{ll} \dfrac{(x_m - x_c)^2}{x_r^2} + \dfrac{(y_m - y_c)^2}{y_r^2} + \dfrac{(z_m - z_c)^2}{z_r^2} \leq 1 & \Rightarrow \varepsilon_{eff2} \\ \text{otherwise} & \Rightarrow \varepsilon_{eff1} \end{array} \right\} \tag{6.5}$$

where x_c, y_c, z_c are coordinates of the object (e.g. cuboid) geometric centre, x_m, y_m, z_m are running coordinates of the target object, x_r, y_r, z_r are radii in the x, y, z directions, ε_{eff2} and ε_{eff1} are the values of the effective permittivity.

To realise the gradient permittivity, the inhomogeneous lens is divided into a finite number of unit cells with different fill fraction values. The gradient permittivity can be achieved by changing the fill fraction of the unit. The unit-cell size must be adjusted to the wavelength. In general, the size of a single unit cell must be significantly smaller than the wavelength λ, say, $1/20\lambda \div 1/10\lambda$. It must be observed that this condition applies to wavelength reduced by the factor $\sqrt{\varepsilon_r}$. It would be safe to make sure that this condition is satisfied for $\varepsilon_r$of the filament. In the considered example, the frequency centre is 10 GHz (corresponding to $\lambda = 30$ mm in free space), so the unit cell size should be smaller than 3 mm. The material used in our design was PLA. The choice was dictated by the popularity and availability of PLA 3D printing filaments. The electric permittivity of PLA is $\varepsilon_r \approx 3.0$ [5] reducing the wavelength to $\lambda \approx 17$ mm, so the unit-cell size should be smaller than 1.7 mm. First, we design and analyse the desired Luneburg lens structure in the Remcom XFdtd – 3D finite-difference time-domain (FDTD) simulator. The simulated system geometry is presented in Figure 6.8. The excitation was the harmonic transversal EM wave with the frequency 10 GHz. Rectangular surface sensors collect electric and magnetic wave components in time (with time step $\Delta t = 9.62$ ps. The Yee cell size was $x = y = z \approx 0.3$ mm to satisfy the Courant condition.

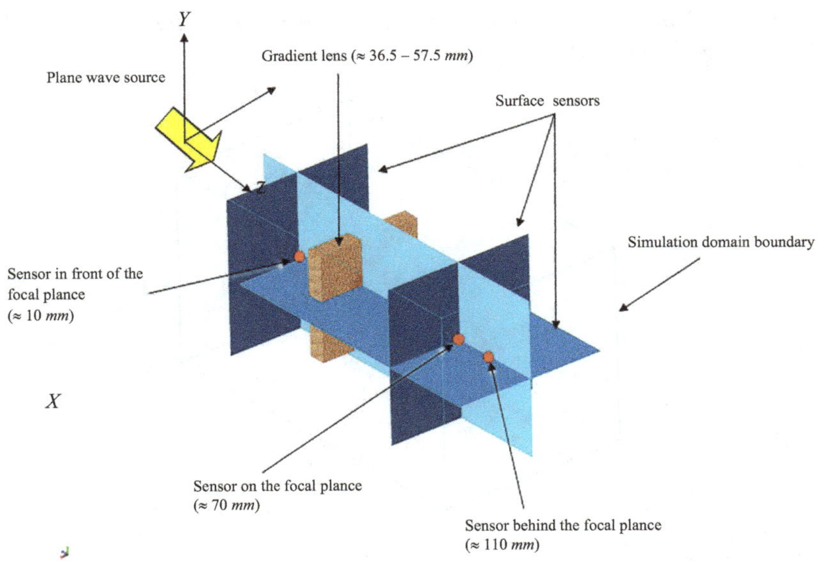

Figure 6.8 Geometry of the flat gradient lens simulation in Remcom XFdtd

Figures 6.9 (plane electric field sensors) and 6.10 (point electric field sensors) show simulation results and demonstrate the desired focusing effect. The lens was fabricated with a 3D printer using a PLA filament. The lens was tested in our microwave lab, which confirmed its performance.

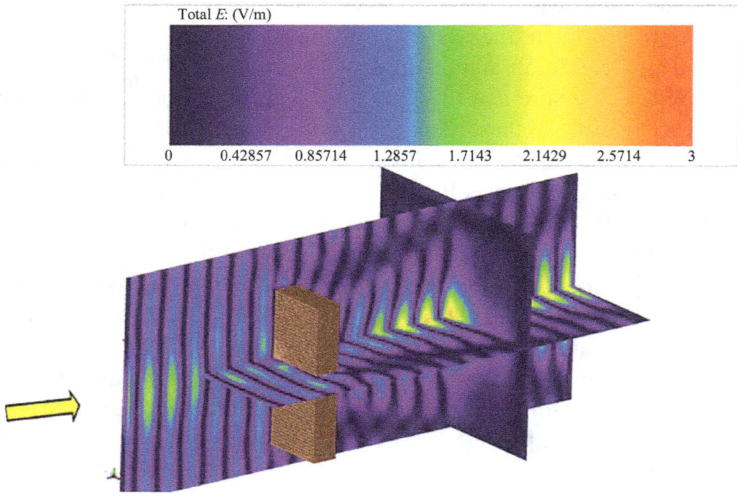

Figure 6.9 XFdtd simulation result for a flat gradient lens excited by a TEM plane wave shows the focusing effect

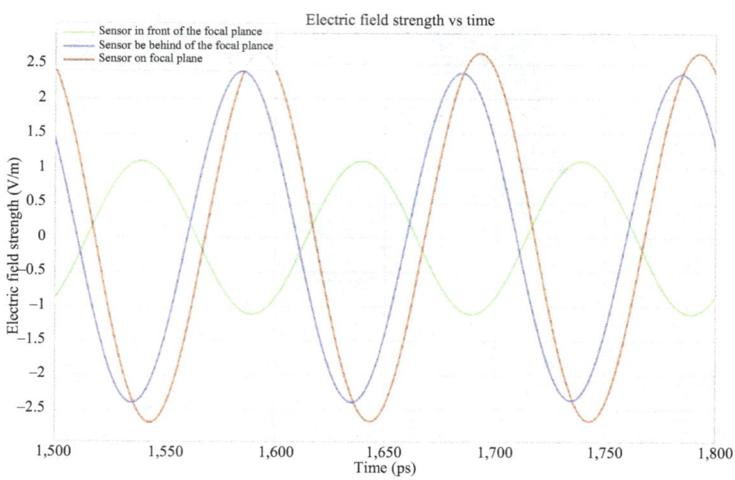

Figure 6.10 FDTD simulation of a flat gradient lens excited by a TEM plane wave. The focusing effect can be observed on three point-sensors

6.2 Resonators and resonator arrays

6.2.1 Introduction

In 1967, Russian physicist Victor Veselago hypothesised that there might exist materials that have negative permittivity and permeability. Today such materials are known under many different names such as left-handed (LH) media, media with negative refractive index, double-negative media, or backward wave media. Metamaterials make it possible to achieve effects that cannot be obtained with 'ordinary' materials, e.g. perfect lensing, EM cloaking (invisibility), reverse Doppler effect, or reverse Cherenkov effect.

Only after nearly 30 years since Veselago's paper was published, the first LH material was conceived and demonstrated experimentally by Pendry *et al.* in 1999 [6] using 'magnetic atoms' or 'meta-particles' very similar to the SRRs. SRRs are typical building blocks of metamaterials engineered to have negative permittivity and permeability. They are effective composite media, which use standard right-handed materials (dielectrics and conductors) at the constituent level. It should be noted that LH properties can only be observed in a narrow frequency band, typically just above the fundamental resonant frequency of the SRRs. Most generally, a magnetic resonator is a closed section of a conductor. Often, the resonator has a slot (to add capacitance) and then it is called an SRR (see Figure 6.11).

A single SRR can be regarded as a resonant circuit with an inductance (conductor loop geometry), capacitances (resulting from the ring gap and distributed capacitances), and a conductor resistance. Typically, dimensions of the structure are small compared to the resonant wavelength, which helps to achieve low radiative losses and high-quality factor (Q-factor) [7]. Metamaterial surfaces comprise many SRRs, most often arranged within (one or more) planes. Typically, each surface contains SRRs arranged periodically or quasi-periodically in rows and columns in the form of a 2D array. As a result, significant mutual coupling (mutual inductance and capacitance) between the SRRs occurs, leading to the modification of the fundamental resonant frequencies of the individual SRRs, the emergence of multiple

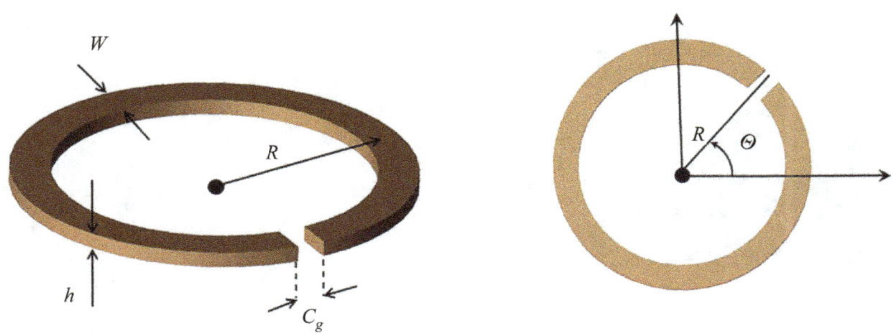

Figure 6.11 Geometry of the split ring resonator and gap orientation α

resonant frequencies, and the change of the Q-factor values. The mutual coupling strongly affects metasurface properties; therefore usually cannot be neglected.

Oscillating electric current is induced when a single SRR is positioned in a temporally varying magnetic field (see Figure 6.12). The induced electromotive force (EMF) follows the Faraday law [8,9] (6.6). The minus sign means that the induced EMF opposes the change of the flux, i.e. the direction of the EMF is given by the Lenz law [10]:

$$EMF = -\frac{d\Phi_B}{dt} \tag{6.6}$$

where EMF is the electromotive force, $\mathbf{\Phi}_\mathrm{B}$ is the magnetic flux.

When an electric current is induced in an SRR, it leads to the accumulation of charge across the gap of the structure (with the formation of an electric field there). The electric field opposes the induced current. Example-simulated instantaneous electric field strength distribution is shown in Figure 6.13.

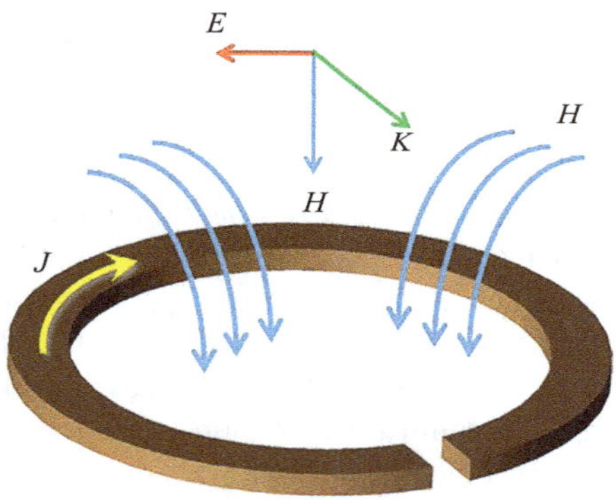

Figure 6.12 Current induced in an SRR conductor by the incident wave

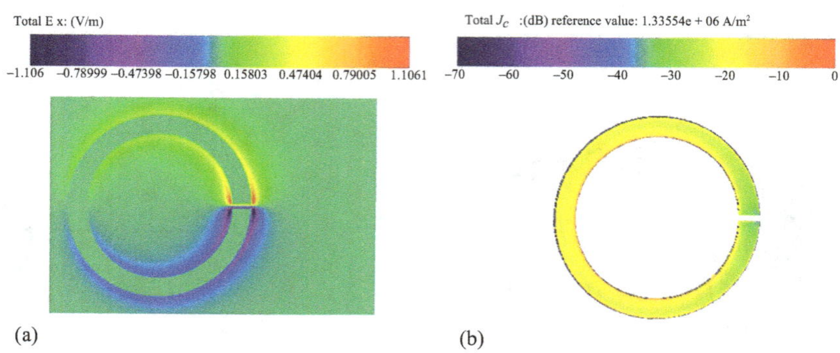

Figure 6.13 Example electric field distribution (a) and current density distribution (b) of SRR excited by an external harmonic TEM field

The coupling between the SRR and the EM field depends on the mutual orientation of the resonator geometry and the magnetic and electric components. The resonator is strongly coupled with the magnetic field component perpendicular to its plane, and the electric field component parallel to the plane.

6.3 Circuit-equivalent models of resonators

Although an SRR has a simple geometry, its interaction with the EM field is complex. It can be most accurately modelled using full-wave simulators, which directly implement Maxwell's equations. However, equivalent-circuit models are also helpful because of their simplicity, small computational burden, and low memory requirements. The simplest lumped-element SRR model is shown in Figure 6.14. It consists of an inductance L, capacitance C, and resistance R. This model can estimate the resonant frequency of the SRR and the quality factor Q.

For a conductor loop, the induced *EMF* will cause current I to flow in agreement with the following equation:

$$I = \frac{EMF}{R + jX} \tag{6.7}$$

where R is the conductor resistance of the resonator,

$X = X_L + X_C = \omega L + \frac{1}{\omega C}$

is the reactance of the resonator (resonant circuit), where X_L is the inductive reactance, X_C is the capacitive reactance, ω is the angular frequency, $\omega = 2\pi f$, L is the effective inductance, C is the effective capacitance.

The inductance L can be approximated by that of a closed (not split) ring. Many methods for the estimation of the inductance were proposed in the literature. One of the most accurate ones is Weinstein's equation as follows [68]:

$$L = \mu_0 \overline{R}(\lambda + \eta) \tag{6.8}$$

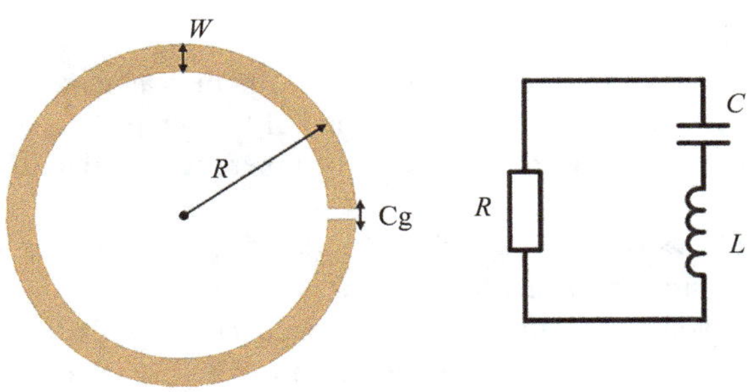

Figure 6.14 Single split-ring resonator and its equivalent-circuit model

where

$$\lambda = \ln\left(\frac{8\overline{R}}{w}\right) + \frac{1}{12} + \frac{\pi\chi}{3} - 0.5\ln\left(1+\chi^2\right) + \frac{1}{12\chi^2}\ln\left(1+\chi^2\right) + \frac{1}{12}\chi^2\ln\left(1+\frac{1}{\chi^2}\right) + \frac{3}{2}\left(\chi - \frac{1}{\chi}\right)\mathrm{arctg}\chi$$

$$\chi = \frac{h}{w}$$

$$\eta = \frac{w^2}{96R^2}\left[\begin{array}{c}\left(\ln\left(\frac{8\overline{R}}{w}\right) - \frac{1}{2}\ln\left(1+\frac{1}{\chi^2}\right)\right)\left(1+3\chi^2\right) + 3.45\chi^2 \\ -1.6\chi^3 + 3.2\chi^3\mathrm{arctg}\chi - \frac{1}{10\chi^2}\ln\left(1+\chi^2\right) + 0.5\chi^4\ln\left(1+\frac{1}{\chi^2}\right)\end{array}\right]$$

$$\overline{R} = R - \frac{w}{2}$$

where R is the radius (see Figure 6.11), w is the conductor width, h is the conductor depth, $\chi = t/w$, where t is loop thickness, and w is the width of the loop conductor, μ_0 is the magnetic permeability of the vacuum.

The inductance can be approximated using a simpler equation as follows [68]:

$$L = \mu_0\overline{R}\left(\ln\frac{8\overline{R}}{h+w} - 0.5\right) \tag{6.9}$$

The difference in the value of impedance L estimated from (6.8) and (6.9) is about 0.4%, so in practice, both methods can be used to determine the inductance of a single split resonator. The capacitive constituent has two major components: the gap capacitance and the surface (distributed) capacitance. Besides, in some cases, an additional lumped capacitor $C_L$is added across the gap. Assuming that all the capacitances are connected in parallel, the total capacitance C_{Total} is given by the following equation:

$$C_{Total} = C_g + C_{surf} + C_L \tag{6.10}$$

where C_{Total} is the total capacitance of SRR, C_g is the capacitance of gap, C_{surf} is the surface capacitance, C_L is the capacitance of the lumped capacitor (optional).

The lumped capacitor $C_L$can be used when there is a need to lower the resonant frequency, but it is not possible to increase the resonator size to increase its inductance. If $CL \gg \left(C_g + C_{surf}\right)$ then $C_{Total} \approx C_L$. Moreover, a varicap may be used instead of a lumped capacitor. Then, it is possible to tune the metasurface by applying a bias voltage to the varicaps, which controls their capacitance. The authors developed an experimental set-up to optimise the performance of a 5×5 matrix of resonators by tuning them in real time. The most commonly encountered formula for the gap capacity is as follows [9]:

$$C_g = \varepsilon_0 \left(\frac{hw}{g} + h + w + g \right) \tag{6.11}$$

where ε_0 is the electric permittivity of vacuum, g is the gap size, h, w, g are the resonator geometry (see Figure 6.11).

Another, more accurate, method of determining the gap capacity is based on Schwarz–Christoffel transformation as follows [12]:

$$C_g = \varepsilon_0 w \frac{h}{g} \left(1 + \frac{g}{\pi h} \left(1 + \ln \left(2\pi \frac{h}{g} \right) \right) \right) \tag{6.12}$$

The surface capacitance is the result of charge distribution on the surface of the ring. For infinitely long split cylinders with an infinitesimal gap, the surface charge density σ, and the voltage V, surface capacitance could be estimated using the following equations:

$$\sigma = \frac{\varepsilon_0 V_0}{2\pi R} \cot \frac{\theta}{2} \tag{6.13}$$

$$V_0 = \frac{C}{\pi} (\pi - \theta) \tag{6.14}$$

where V_0 is the voltage across the gap, and the θ is the angle defined in Figure 6.11. Finally, the surface capacitance is given by the following equation:

$$C_{surf} = \int_{\theta}^{\pi} \frac{\sigma R d\theta}{V} = \epsilon_0 h \int_{\theta}^{\pi} \frac{\cot(\theta/2)}{\pi - \theta} d\theta \approx \frac{2\varepsilon_0 h}{\pi} \ln \frac{4R}{g} \tag{6.15}$$

where

$$\theta = \arcsin \left(\frac{q}{2R} \right).$$

The total resistance R_{Total} has several components: DC loss resistance (R_{DC}) resulting directly from the conductor material properties and skin effect loss (R_{SE}). Furthermore, radiation resistance (R_{rad}) may also be taken into account. The 'radiation resistance' is a virtual resistance representing radiative loss:

$$R_{Total} = R_{DC} + R_{SE} + R_{rad} \tag{6.16}$$

$$R_{DC} = \rho \times \left(\frac{L}{A} \right) \tag{6.17}$$

where ρ is the resistivity of the conductor, L is the conductor length, A is the cross-sectional area of the conductor (for the resonator in Figure 6.11 it is equal to $w \cdot h$).

The skin effect may be calculated as follows:

$$R_{SE} = \sqrt{\frac{2\rho}{\omega\mu_r}}\sqrt{\sqrt{1+(\rho\omega\varepsilon_r)^2}+\rho\omega\varepsilon_r} \tag{6.18}$$

where ρ is the resistivity of the conductor, ω is the angular frequency, μ_r is the permeability of the conductor, ε_r is the relative permittivity of the conductor.

Radiation resistance may be calculated as follows:

$$R_{RAD} = 320\pi^4\left(\frac{A}{\lambda^2}\right)^2 \tag{6.19}$$

where A is the resonator area, λ is the wavelength.

The resonant behaviour of SRRs was described, e.g. in [13–16]. In simple terms, SRRs behave as *(R)LC* resonators excited by an external time-varying magnetic flux. Like any *(R)LC* circuit, equivalent-circuit of SRR can be characterised by its resonant frequency (6.20) and quality factor Q (6.21) [69]:

$$f_0 = \frac{1}{2\pi\sqrt{LC}} \tag{6.20}$$

$$Q = \frac{1}{\omega_0 RC} = \frac{1}{R}\sqrt{\frac{L}{C}} \tag{6.21}$$

where L is the total inductance, C is the total capacitance, R is the total resistance, B_f is the 3dB bandwidth.

Since we investigate metamaterials, it is helpful to enable one to estimate effective relative permeability as in the following equation [17]:

$$\mu_{eff} = 1 - \frac{\pi}{2}\frac{s^2\mu_0 rC}{1+sRC+s^2LC} \tag{6.22}$$

where r is the inner radius of the metal ring, s is the area of SRR, R is the total resistance, μ_0 is the magnetic permeability of vacuum.

The effective magnetic permeability and electric permittivity can be estimated from the *S*-parameters (scattering parameters) [17]. The unit cell is simulated by putting it in a waveguide with perfect electric conductor (PEC) and perfect magnetic conductor (PMC) boundaries. On both sides of the unit cell, there are ports (or field sensors). The first one is associated with the source of the wave, the second with the receiver or load. Effective magnetic permeability and dielectric permittivity can be calculated from the following equations:

$$\mu_{eff} = \frac{2}{jk_0 d}\frac{1-V_2}{1+V_2} \tag{6.23}$$

$$\varepsilon_{eff} = \frac{2}{jk_0 d}\frac{1-V_1}{1+V_1} \tag{6.24}$$

where

$$V_1 = S_{21} + S_{11}$$
$$V_2 = S_{21} - S_{11}$$

where S_{11} is the S-parameter for reflection, S_{21} is the S-parameter for transmission, k_0 is the wave number in free space ($k_0 = \omega/c$), d is the thickness of the substrate, ω is the angular frequency.

6.4 Distributed circuit-equivalent models of resonators

In the previous section, a simple *RLC*-lumped model of the SRR was presented. This approach may be sufficient for elementary geometries. However, for more complex geometries, e.g. a double-SRR (DSRRs)-distributed circuit model may be adequate [18,19]. As an example of this approach, a distributed model of a DSRR was presented in [18,19] (see Figure 6.15).

Each elementary sub-circuit sector is determined by an elementary angle $d\varphi$. Each element of the sub-circuit is described as the self-inductances L_1 and L_2 are self-inductances of external and internal ring, defined as per unit angle. Similarly, other parameters are defined as per the unit angle – the mutual inductance M between the two rings and inter-ring capacitance C. G_1 and G_2 are induced EMFs per unit angle. For the resultant lumped circuit, Kirchhoff's law is applied. The differential equations from there are formulated. It allows the determination of the resonant frequencies of the DSRR for specific geometric parameters. The equations can be discretised by dividing the circumference of the DSRR into N sectors

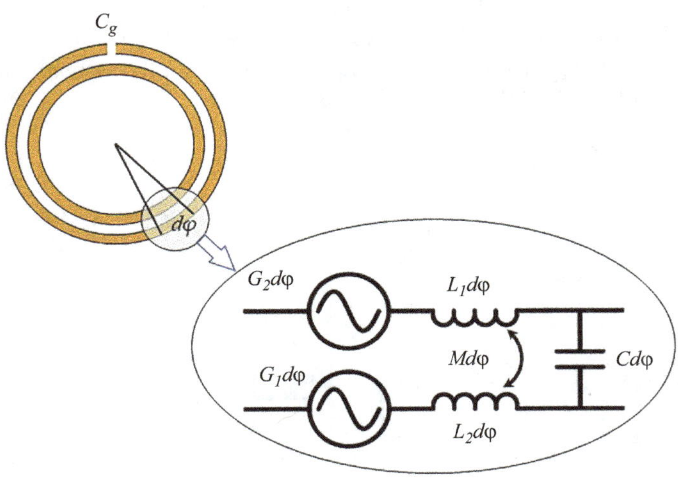

Figure 6.15 Equivalent distributed-circuit model for a double split-ring resonator (DSRR) sector $d\varphi$

$d\varphi = 2\pi/N$ corresponding to N discrete circuits. The circuits are identical. If N is sufficiently large, we may expect that this simulation will yield resonant frequencies close to those found by the differential equation method. For more details, readers are referred to [18,19].

6.4.1 Resonator arrays – mutual coupling

When resonators forming a set of SRRs are placed close to each other, the coupling between the resonators leads to complex frequency characteristics. When electromagnetically coupled resonators are considered, one important issue is their multi-peak frequency response [20]. Let us consider the system containing two magnetically, electrically, or mixed coupled resonators. Analytical determination of its resonant frequencies as a function of parameters for the lumped-element circuit model of the system, even in such an elementary case, is a complex task. In the case of identical SRRs, it can be shown that if the coupling coefficient $=|M|L_1L_2$ exceeds the critical value $= 1_{1\ 2}$, where ${}_{1}$, ${}_{2}$ denote the Q-factor of the respective resonant circuits, then the current amplitude characteristics in both resonators have two maxima, at frequencies different from the frequencies of the individual resonant circuit without coupling. Two types of models will be considered, i.e. a circuit-based model (Figure 6.16(b)) and a field-based model (Figure 6.16(a)).

For magnetically coupled resonators, the multi-peak frequency response must be accounted for. It can be observed when the coupling factor k (6.25) increases above a critical value [20]:

$$k = \frac{M}{\sqrt{L_1 L_2}} \tag{6.25}$$

where M is the mutual inductance factor that depends on the common part of magnetic flux, L_1, L_2 are the inductances of the resonators [68,70].

Applying Kirchhoff's voltage law equations for both the primary and secondary loops results in the following equations:

$$\left[R + j\omega L + \frac{1}{j\omega C}\right] I_1 + j\omega M \cdot I_2 = V_0 \tag{6.26}$$

$$j\omega M \cdot I_1 + \left[R + j\omega L + \frac{1}{j\omega C}\right] I_2 = 0 \tag{6.27}$$

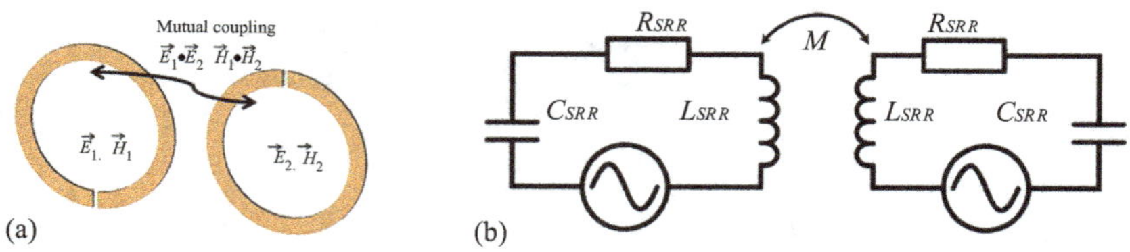

Figure 6.16 Equivalent circuit model for two coupled SRRs: coupled SRRs geometry (a) and equivalent circuit diagram (b)

It is assumed here that $L_1 = L_2$ and $R_1 = R_2$. Then these equations can be written in the matrix form as (6.28). In general, it can be written for N-coupled circuits [20]:

$$\begin{bmatrix} a & b \\ b & a \end{bmatrix} \cdot \begin{bmatrix} I_1 \\ I_2 \end{bmatrix} = \begin{bmatrix} V_0 \\ 0 \end{bmatrix} \tag{6.28}$$

where

$$a = R + j\omega L + \tfrac{1}{j\omega C}$$
$$b = j\omega M$$

The currents in the first and the second resonator circuit may be found following Cramer's rule as in the following equations:

$$I_1 = \frac{aV_0}{a^2 - b^2} \tag{6.29}$$

$$I_2 = -\frac{bV_0}{a^2 - b^2} \tag{6.30}$$

Resonance occurs at the two frequencies given by the following equation:

$$\omega_1^2 = \frac{1}{LC\left[1 - \frac{M}{L}\right]} \tag{6.31}$$

and

$$\omega_2^2 = \frac{1}{LC\left[1 + \frac{M}{L}\right]} \tag{6.32}$$

For natural resonance to occur, the condition is as follows:

$$\frac{1}{j\omega L_m} + \frac{j\omega C_1}{1 - \omega^2 C_1 (L_1 - L_m)} + \frac{j\omega C_2}{1 - \omega^2 C_2 (L_2 - L_m)} = 0 \tag{6.33}$$

After some manipulations may be reduced to the following equation:

$$\omega^4 \left(L_1 L_2 C_1 C_2 - L_1 L_2 L_m^2\right) + \omega^2 (L_1 L_2 + C_1 C_2) - 1 = 0 \tag{6.34}$$

This equation has four eigenvalues. However, only the two positive real solutions represent the resonant frequencies that are identifiable as follows:

$$\omega_{1,2} = \sqrt{\frac{(L_1 L_2 + C_1 C_2) \mp \sqrt{(L_1 L_2 - C_1 C_2)^2 + 4 L_1 L_2 L_M^2}}{2(L_1 L_2 C_1 C_2 - L_1 L_2 L_M^2)}} \tag{6.35}$$

Next, a new K_M parameter may be defined (we assume $\omega_2 > \omega_1$) as

$$K_M = \frac{\omega_2^2 - \omega_1^2}{\omega_2^2 + \omega_1^2} \tag{6.36}$$

Alternatively, the magnetic coupling coefficient may also be defined as the ratio of the coupled magnetic energy to the average stored energy as follows:

$$K_M = \frac{L_M}{\sqrt{CL_1 L_2}} \tag{6.37}$$

The electrical coupling can also be defined. For natural resonance with electrical coupling to occur, the condition is given by the following equation:

$$\frac{1}{j\omega C_M} + \frac{j\omega L_1}{1 - \omega^2 C_1 (L_1 - C_m)} + \frac{j\omega L_2}{1 - \omega^2 C_2 (L_2 - C_m)} = 0 \tag{6.38}$$

After some manipulations may be reduced to the following equation:

$$\omega^4 \left(L_1 L_2 C_1 C_2 - L_1 L_2 C_m^2\right) + \omega^2 (L_1 L_2 + C_1 C_2) - 1 = 0 \tag{6.39}$$

This equation has four eigenvalues. However, only the two positive real solutions represent the resonant frequencies that are identifiable as follows:

$$\omega_{1,2} = \sqrt{\frac{(L_1 L_2 + C_1 C_2) \mp \sqrt{(L_1 L_2 - C_1 C_2)^2 + 4 L_1 L_2 C_m^2}}{2\left(L_1 L_2 C_1 C_2 - L_1 L_2 C_m^2\right)}} \tag{6.40}$$

Like the magnetic coupling coefficient (6.37), the electric coupling coefficient can be defined as [20,71]

$$K_E = \frac{C_M}{\sqrt{C_1 C_2}} \tag{6.41}$$

Finally coupling coefficient as the ratio of the coupled magnetic energy to the average stored energy, we get

$$K = K_M + K_E \tag{6.42}$$

SRR operation is similar to that of the lumped-element series *RLC* resonant circuits. Generally, resonant frequencies depend on the coupling between the two resonators, despite whether they are synchronously or asynchronously tuned. The mutual inductive coupling between two coils (Figure 6.16) can be related to their self-inductance through the coupling constant K_M as follows [22]:

$$\frac{K_M = \iiint \mu \overrightarrow{H_1} \cdot \overrightarrow{H_2} dv}{\sqrt{\iiint \mu \left|\overrightarrow{H_1}\right|^2 dv \times \iiint \mu \left|\overrightarrow{H_2}\right|^2 dv}} \tag{6.43}$$

The coupling factor K_M depends only on the geometry of the SRRs (size of the rings, their mutual distance and orientation, and material parameters ε_r and μ_r of the SRR substrate). The coaxial orientation of the coils k_m factor will tend to be positive, while the parallel orientation k_m will tend to be negative. In the case of the orthogonal orientation of the coils, K_M will tend to zero.

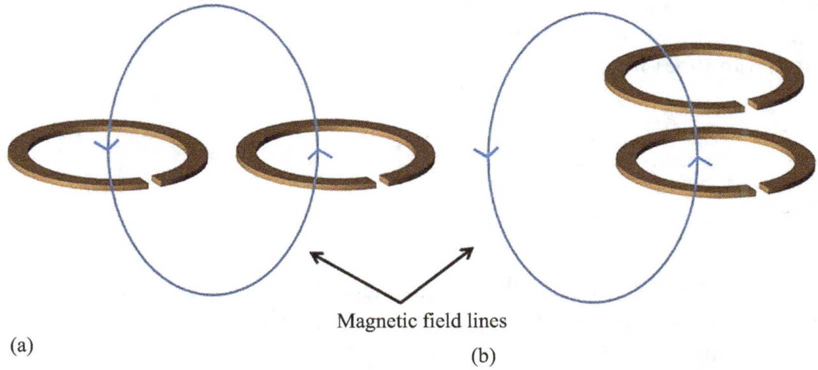

Figure 6.17 Mechanism of negative (a) and positive (b) magnetic coupling between two SRRs

The electric coupling K_E may also be defined based on the ratio of coupled energy to stored energy as follows:

$$K_E = \frac{\iiint \varepsilon \overrightarrow{E_1} \cdot \overrightarrow{E_2} dv}{\sqrt{\iiint \varepsilon \left|\overrightarrow{E_1}\right|^2 dv \times \iiint \varepsilon \left|\overrightarrow{E_2}\right|^2 dv}} \tag{6.44}$$

Finally, a complete, mixed magnetic and electric coupling may be defined as follows:

$$K = K_M + K_E = \frac{\iiint \mu \overrightarrow{H_1} \cdot \overrightarrow{H_2} dv}{\sqrt{\iiint \mu \left|\overrightarrow{H_1}\right|^2 dv \times \iiint \mu \left|\overrightarrow{H_2}\right|^2 dv}} + \frac{\iiint \varepsilon \overrightarrow{E_1} \cdot \overrightarrow{E_2} dv}{\sqrt{\iiint \varepsilon \left|\overrightarrow{E_1}\right|^2 dv \times \iiint \varepsilon \left|\overrightarrow{E_2}\right|^2 dv}} \tag{6.45}$$

The magnetic coupling factor value may be negative or positive [20,21]. Depending on the direction of the magnetic field vector, the result is either the sum (Figure 6.17(b)) or the difference (Figure 6.17(a)) of both fields [17,21]. The negative magnetic coupling factor decreases total inductance and increases the resonant frequency, while the positive coupling factor increases total inductance and decreases the resonant frequency.

Opposite to magnetic coupling, the electrical coupling can only be positive, which means that the electrical coupling can only increase the total capacitance of the resonator [21].

6.5 Full-wave FDTD numerical models of resonators

Maxwell's set of equations constitute the complete classical theory of EM phenomena. The four equations (in Heaviside's formulation) describe electric and magnetic fields as a function of charges and currents [24] in the vacuum and other media. Ampere's (6.46) and Faraday's (6.47) laws are two dynamic evolutionary (evolution in time) laws:

$$\nabla \times \vec{H} = \frac{\partial}{\partial t}\vec{D} + \vec{J} \tag{6.46}$$

$$\nabla \times \vec{E} = -\frac{\partial}{\partial d}\vec{B} \tag{6.47}$$

where $\nabla\times$ is the curl operator, $\vec{B}$ is the magnetic flux density (induction) (Wb/m^2), $\vec{D}$ is the electric flux density (displacement) (C/m^2), $\vec{H}$ is the magnetic field strength (A/m), $\vec{E}$ is the electric field strength (V/m), $\vec{J}$ is the electric current density (A/m^2).

To fully describe the EM fields, source defining relationships (6.48), (6.49), and material equations (6.50), (6.51) are also needed:

$$\nabla \cdot \vec{D} = \rho_v \tag{6.48}$$

$$\nabla \cdot \vec{B} = 0 \tag{6.49}$$

$$\vec{D} = \varepsilon_0 \varepsilon_r \vec{E} \tag{6.50}$$

$$\vec{B} = \mu_0 \mu_r \vec{H} \tag{6.51}$$

where $\nabla\cdot$ is the divergence operator, ρ_v is the electric charge density (C/m^3), $\varepsilon_0 \approx 8.85418 \times 10 - 12$ (F/m) is the vacuum permittivity, $\mu_0 \approx 1.2566 \times 10 - 6$ (H/m) is the vacuum permeability, μ_r is the relative permeability, ε_r is the relative permittivity.

Analytical solution of Maxell's equations is possible only for some canonical problems. In more complex cases, computational EMs must be employed to obtain approximate (but often sufficiently accurate) solutions to Maxwell's equations. Discrete methods such as FD (finite difference) or FEM (finite-element method) are often used. In further discussion, the FDTD method will be exploited based on the discretisation of the computational domain and the replacement of differential operators with difference quotients [25]. FDTD belongs to grid-based differential numerical modelling techniques. It was initially described in 1966 by Yee but fully developed by Taflove [24,26]. In its most condensed version, the algorithm could be defined as follows:

1. Replace all the derivatives in Ampere's and Faraday's laws with FDs. Discretise space and time so that the electric and magnetic fields are staggered in both space and time (see Figure 6.18).
2. Solve the resulting difference equations to obtain 'update equations' that express the (unknown) future fields in terms of (known) past fields.
3. Evaluate the magnetic fields one time-step into the future, so they are now known (effectively, they become past fields).

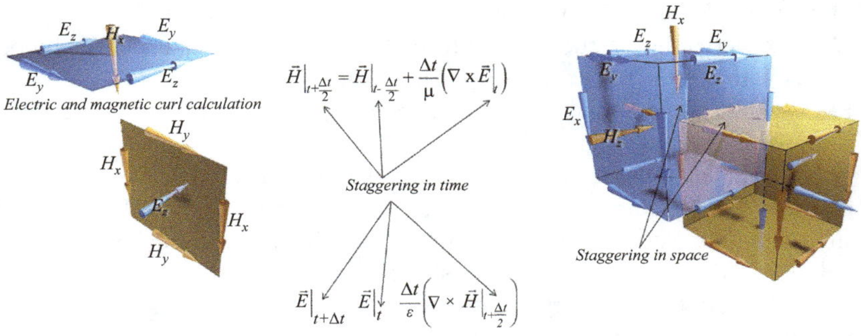

Figure 6.18 In finite-difference time-domain (FDTD) method, electric and magnetic fields must be staggered in both space and time to obtain the necessary update equations

4. Evaluate the electric fields one time-step into the future, so they are now known (effectively, they become past fields).
5. Repeat the previous two steps until the fields have been obtained over the desired duration of time.

The choice of the discretisation step (the size of Yee cells) is crucial for the algorithm's stability and accuracy. It is assumed that the results can be considered valid when condition, as in the following equation, is satisfied:

$$\Delta x, \Delta y, \Delta z < \frac{\lambda_{\min}}{10 \div 20} \Rightarrow \Delta x, \Delta y, \Delta z < \frac{c/f_{\max}}{10 \div 20} \tag{6.52}$$

where $\lambda_{\min}$ is the minimum wavelength propagating in a given medium, $\Delta x, \Delta y, \Delta z$ are the size of the grid cell, c is the speed of light in free space, and $f_{\max}$ is the maximum frequency component of the EM field in a given simulated system. Additionally, the grid size must allow for the placement of at least a few grid cells inside each object (in the direction of the EM wave propagation). Finally, the minimum grid size ($\Delta x, \Delta y, \Delta z$) determines the maximum time-step value to guarantee numerical stability, which is known as the Courant condition:

$$\Delta t \leq \frac{1}{c\sqrt{\frac{1}{(\Delta x_{\min})^2 + (\Delta y_{\min})^2 + (\Delta z_{\min})^2}}} \tag{6.53}$$

where c is the speed of propagation of the EM wave in free space.

Some boundary conditions must be defined to truncate the simulation domain (computational region) correctly. There are four main types of boundary conditions:

1. Absorbing – A simple absorbing boundary condition establishes a lossy layer at the edge of the computational domain, which has only minimal reflections regardless of the incidence angle or frequency. Owing to that, the grid appears

to extend infinitely. The two most well-known boundary conditions are Liao [27] and perfectly matched layer [28–30]. The boundary can absorb outgoing waves with a refection coefficient of less than −60 dB.
2. PEC – The computational domain is limited by a PEC. PEC may be used, for example, to provide a ground plane.
3. PMC – The computational domain is limited by a PMC.
4. Periodic – The computational domain between two opposite sides of the simulation domain is reproduced on the other side of the boundary. With periodic conditions, both the physical structure and the EM fields must be periodic. The periodic boundary helps reduce the size of the FDTD mesh, memory requirements, and calculation time.
5. Mixed – a combination of the above boundary conditions.

Correct spatial discretisation of the simulated system is critical for modelling accuracy. Typically, defining the simulated system begins with abstract (not discretised) geometrical primitives, such as circles, rectangles, annuli. Then they are assigned material parameters (μ_r, ε_r, and σ). The definition of material parameters may be more intricate, e.g. describing dispersive media. The second step is gridding which determines how the 3D geometry and surrounding simulation space is discretised into Yee cells. Some FDTD simulation packages offer variable grid size, which can be refined at geometry edges or in media with $\varepsilon_r \gg 1$ and/or $\mu_r \gg 1$, where the wavelength is reduced by the factor $1/\sqrt{\varepsilon_r \mu_r}$. Each Yee cell edge requires that a single material be assigned to it. The process of assigning material properties (μ_r, ε_r, and σ) is known as meshing.

The boundary conditions (absorbing, PEC, PMC, periodic or mixed) must also be defined and excitation as a lumped feed, external excitation, waveguide, etc., depending on the simulation software. Since the method operates in the time domain, the excitation must be a waveform, e.g. a harmonic or broadband wave. Finally, we define sensors that can store field values such as E, H, B, J_c in selected Yee cells for each simulation step. The user can set the time and space boundaries of the computational domain. Eventually, the core of the FDTD simulation is run.

Let us consider a model of a single SRR in free space. There are at least two practical methods to excite the SRR – using an external excitation wave (far field) (Figure 6.19(a)) or using a loop antenna (near field) (Figure 6.19(b)).

A point sensor placed on the conductor surface collects the current density J_c values over time. Absorbing boundary conditions were employed in the first case. The point sensor is located on the resonator surface (see Figure 6.19(b)).

The geometrical sizes of the SRR (the same for the exciting loop antenna in the case when it is used) are $R = 25$ μm, $w = 5$ μm, $h = 2$ μm, $C_g = 1.5$ μm.

It should be pointed out that magnetic coupling will arise between the SRR and the loop antenna. In our example, for the distance between the loop antenna and the SRR of 50 μm, we estimated (using the Babic method [22,31]) the mutual impedance as $M = 3.5465\text{e}^{-12}$ (H) (3.5 pH). The magnetic coupling affects the resonant frequencies of the SRR, as shown in Table 6.1. Namely, the resonant frequency is increased because the value of M is positive.

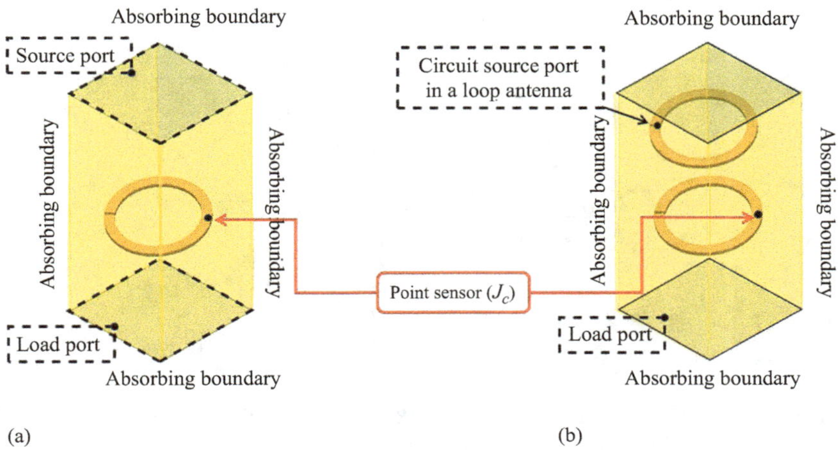

Figure 6.19 Set-up for testing a single SRR: (a) far-field case with external plane wave, (b) near-field case with a loop antenna as a wave source. The dimensions of the SRR are R = 25 μm, w = 5 μm, h = 2 μm, and C_g = 1.5 μm

Table 6.1 Resonant frequency for near- and far-field excitation of the same SRR (obtained using FDTD)

Excitation	**Resonant frequency (GHz)**
Near field	875.1
Far field	875.65

Full-wave simulators also offer the possibility of simulating an infinite array of SRRs using periodic boundaries (or an appropriately selected combination of PMC and PEC boundaries) (see Figure 6.20).

Simulations were performed with the same resonator geometry as in the previous case. Two cases were considered: absorbing boundaries (single resonator) (see Figure 6.20(a)) and periodic boundaries (corresponding to an infinite array of identical resonators (see Figure 6.20(b)). We located a point current density sensor on the resonator, as shown in Figure 6.20(c). The EM excitation was broadband (700 GHz to 1 THz). In Figure 6.20(d), the Fourier transform magnitude of the current (on the point sensor) is plotted. The resonant frequency is higher for the periodic array, which can be attributed to the negative value of the mutual coupling between the adjacent resonators.

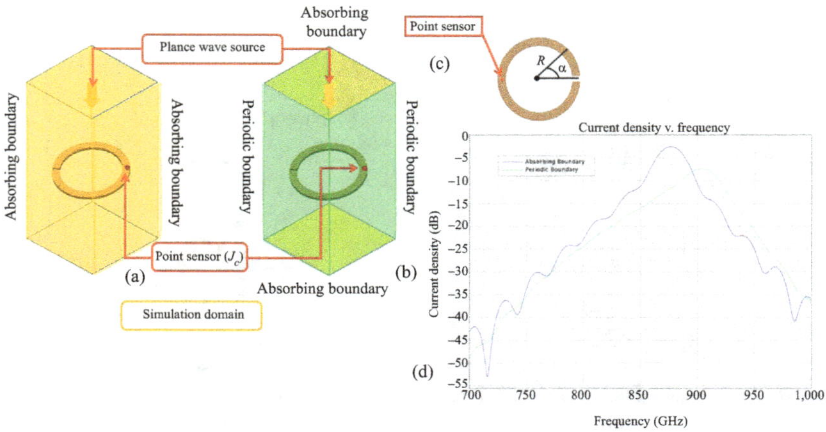

Figure 6.20 FDTD model for a single SRR (a), model of an infinite array of SSRs (b), location of point current sensor (c), current vs frequency for absorbing (single resonator) and periodic (infinite resonator array) boundary conditions (d)

6.6 Flat metasurface lens for MRI applications

Since its development in the 1970s and 1980s, MRI [32] has turned out to be a highly versatile medical imaging technique [33] used to construct images of various anatomical structures (e.g. normal and pathological tissues) and physiological processes (e.g. blood flow in blood vessels) of the human or animal body. The images are constructed from radio frequency (RF) currents induced in the receiver coils by the EM fields generated by hydrogen nuclei precession in the direction transversal to the main longitudinal axis of the MRI scanner. The precession arises at the so-called Larmor frequency, which is $f_L = B{\cdot}42.58$ (MHz) for hydrogen nuclei, where B (T) is the static induction in a given MRI scanner. The amplitude of the signal received by RF coils depends on the concentration of protons in the body tissues, RF energy absorption and spatial spreading. The quality (e.g. resolution) of MRI images depends on the signal-to-noise ratio (SNR), which can be defined as the ratio of the amplitude of the RF magnetic resonance signal received to the amplitude of the noise. The noise has many sources – physiological (such as thermal fluctuations of electrolytes in the subject), electronic (internal amplifier noise), and interference (external sources of EM radiation) which are difficult to reduce and control. Therefore, the maximisation of the SNR has been an essential issue for the improvement of MRI diagnostic value [33]. It should also be stated that imaging time can be traded off for image quality.

For several decades, one of the methods of increasing SNR in MRI has been developing MRI systems with ever stronger magnetostatic fields (1, 1.5, 3.0, 7.0 T, and experimentally even 13 T). However, this implies unfavourable compromises, such as increased hardware costs and concerns about hazards involved in imaging

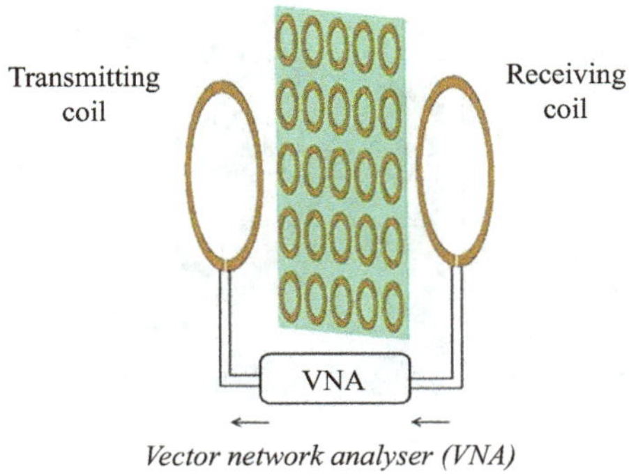

Figure 6.21 Test set-up for full-wave computer simulations or measurements of S_{21} vs frequency

patients with implanted medical devices. SNR can also be increased by oral or intravenous administration of contrast agents, usually gadolinium-based, with all the involved toxicity concerns. Developments have also been made in RF coil technology, including multiple coils, coil arrays, and specialised coil set-ups for particular body parts. Advanced MRI pulse sequences and image-processing techniques yield further improvements in SNR and overall diagnostic power.

Beyond the techniques mentioned earlier, the potential of metamaterial structures has been investigated to improve SNR in MRI scanners. In particular, metamaterial lenses were considered [34]. When applied to MRI RF fields at Larmor frequency, the resonant mode of the metamaterial focusing lens may be exploited, leading to an increase in RF field strength and consequently improving the SNR. The lens is located between the subject and the RF coil.

Let us consider a flat EM lens in the form of the array of capacitor-loaded metallic ring resonators adopted from [34]. The lens can potentially improve the SNR because of its focusing power. The lens in its basic fixed-parameter version is shown in Figure 6.21. We have also developed a novel feature of tunability of the individual SRRs (see Figure 6.22). It can be tuned to the B_1 field at the Larmor frequency range from *ca* 120 to 170 MHz, which is suitable, e.g. for 3 T MRI scanners.

Our full-wave simulation set-up is shown in Figure 6.21. It consists of the lens (SRR array), two near-field coils representing the operation of MRI RF coils and S_{21} (two-port network scattering-matrix element) calculator, simulating a vector network analyser. Near-field test coils correspond to the real MRI scenario where the magnetic component normal to the RF coil is considered. In our study, we also estimate the resonant frequency of each resonator using its circuit-equivalent model as follows:

$$f_r = \frac{1}{2\pi\sqrt{LC}} \tag{6.54}$$

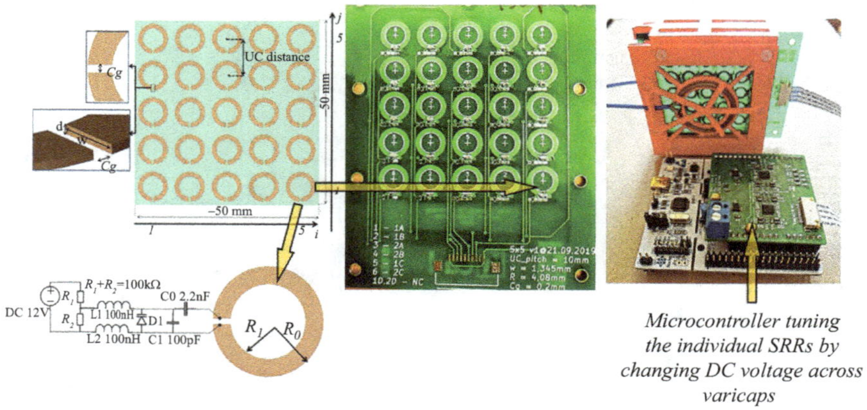

Figure 6.22 5×5 capacitor-loaded ring resonator array with varicap tuning; $R = 4.09$ mm, $w = 1.33$ mm, $C_g = 0.4$ mm, $d = 0.2$ mm, substrate electric constant $\varepsilon = 3$; varicap voltage is 12 $VR_2/(R_1+R_2)$

where

$$\begin{aligned} L &= L_{RR} + L_{MUT} \\ C &= C_{cap} + C_{surf} + C_{capacitors} + C_{MUT} \end{aligned}$$

where L_{RR} is the ring resonator inductance, L_{MUT} is the mutual inductance between resonators, C_{cap} is the gap capacitance, C_{surf} is the surface capacitance, $C_{capacitors}$ is the capacitance of the connected lumped capacitor(s),

$$L_{RR} = \mu_0 \times R_O \times \left(\log\left(\frac{8R_O}{w}\right) - 2 \right)$$

$$L_{MUT} = \frac{1}{(MN-1)} \sum_{\substack{i=n \\ x \neq i \wedge y \neq j}}^{N} \sum_{j=m}^{M} L_m([x,y],[i,j]),$$

It should be observed that the mutual inductance L_{MUT} between the resonators, calculated using the methodology proposed in [31], significantly depends on their neighbourhood, i.e. their position in the array (see Figure 6.23 for the 5×5 case). The negative sign of the mutual inductance can be interpreted so that current in one SRR creates a negative magnetic flux in the second SRR.

If all the resonators have the same geometry and consequently the same value of self-inductance L_{RR} but different mutual inductance, those result in a different tuning f_r of the resonators assuming equal values of the effective capacitance C.

As previously noted, for a finite resonator array, the mutual impedance L_{MUT} depends on the position i,j of an individual resonator in the array. To accomplish equal values of the resonant frequency $f_r(i,j)$ for each resonator, it is necessary to guarantee the same value of LC product in (6.54). When the array is tuned with varicaps, it is easy to tune each SRR independently by adjusting its individual

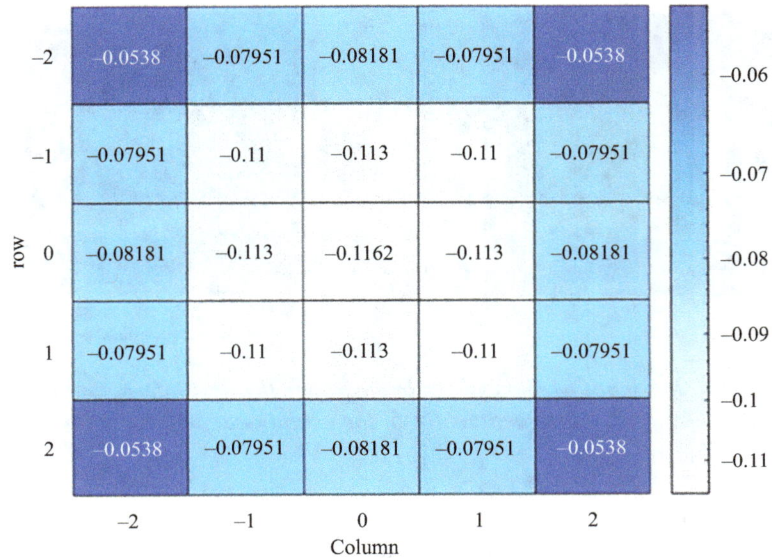

Figure 6.23 Resultant mutual magnetic coupling coefficient values for each SRR in the 5 × 5 array, i.e. the sum of all K_M values for a given SRR

varicap DC voltage. The differences in the values of L_{MUT} can also be compensated by adjusting the geometry of the resonators. It should be noted that the modification of any resonator will influence the effective inductance of all other resonators in the array because of the mutual inductances. The adjacent SRRs will be influenced the most.

Let us define the array geometry optimisation problem as follows. Given an initial solution (e.g. all resonators of identical geometry), find the minimiser with respect to the design vector $g = (R_I)$ of the following objective function $\Psi(g)$:

$$\Psi(g) = \sum_{i=1}^{N} \sum_{j=1}^{N} \left(f_{i,j} - f_r\right)^2 \tag{6.55}$$

where R_I is the set of inner radii of the resonators (subject to design constraints), $f_{i,j}$ is the resonant frequency of the i,jth resonator; f_r is the desired resonant frequency; N is the resonator array size (in the considered case $N = 5$).

We used a derivative-free Estra optimisation algorithm that exploits an evolutionary approach. A cost-effective version of the lowest order algorithm (i.e. a single parent generates a single offspring) was chosen. A detailed description of the algorithm can be found, e.g. in [35], while its successful applications in EMs are presented, e.g. in [36,37]. In Chapter 3 an overview of different methods is proposed. The optimisation process is illustrated in Figure 6.24 which corresponds to the state before optimisation – it is evident that the resonant frequencies $f_r(i,j)$ differ

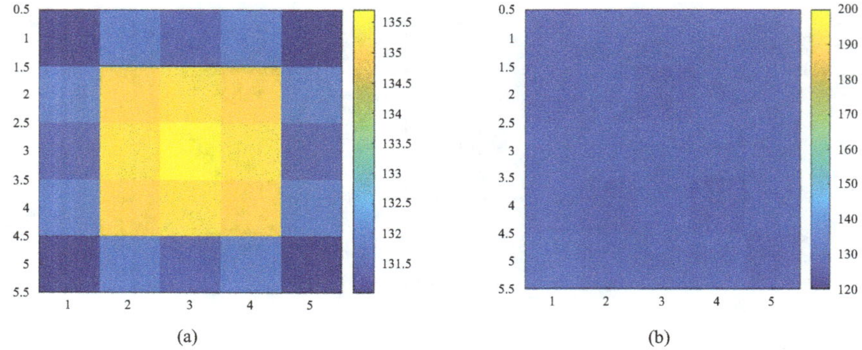

Figure 6.24 Resonant frequencies estimated after the first interaction (a) and after 800 interactions (b) of the evolutionary optimisation algorithm

significantly, i.e. approximately by 4 MHz (about 3% of the centre frequency). To test the performance of the optimisation algorithm, we assumed no symmetries in the array. Therefore, we optimised all the 25 $R_f(i,j)$ values independently, while it was also possible to consider the symmetries and reduce the parameter space dimensionality.

Figure 6.24(a) illustrates the resonant frequencies after the first iteration of the optimisation algorithm. Random values of the resonant frequencies in the range of approximately 120–200 MHz can be observed. After 800 iterations, the resonant frequencies have converged with a range of fewer than 1 MHz (Figure 6.24(b)). This illustrates the significant improvement of the lens tuning we achieved using our approach [23].

6.7 Artificial magnetic conductor

AMCs were proposed for the first time in 1999 and since then have been extensively explored and widely used in antenna and EMs applications [38]. These structures consist of a periodic grid of metallic patches mounted on a thin grounded dielectric substrate. The unique property of AMCs is their in-phase reflection coefficient at a particular frequency which makes them (frequency selective) feasible alternative for PMCs. AMC structures [38], also known as high impedance surfaces, are especially interesting in the design of efficient and low-profile antennas [39,40] owing to their inherent in-phase reflection properties. Additionally, the same AMC characteristic reduces the radiation towards the user's body in wearable applications [24], such as wireless body area networks. AMC surfaces [38–42] with boresight radiation antennas reduce their radiation towards the human body (without deteriorating propagation conditions for other directions) and make the antennas tolerant to positioning on the human body for single and multiband wearable applications.

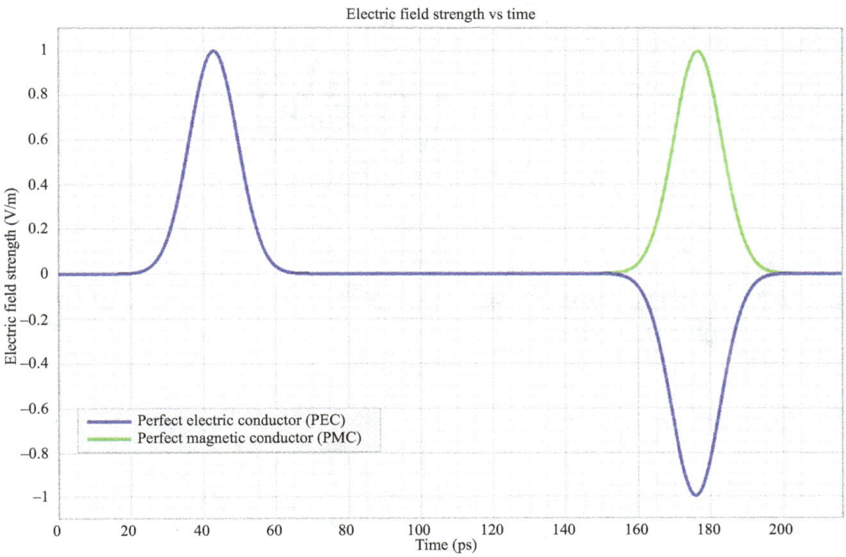

Figure 6.25 FDTD simulation of a wideband wave reflected from a perfect electric conductor (PEC) – blue curve and a perfect magnetic conductor (PMC) – green curve

Currently known AMCs are intrinsically frequency-selective and suffer from narrow-band performance, limiting their wideband or multi-band antenna applications. That is why AMC surfaces are also known as frequency-selective surfaces.

If a plane wave is reflected from a flat PEC surface, the reflected component's phase will be shifted by 180°, as shown in Figure 6.25 (blue curve). However, if the reflecting surface is a PMC, the phase of the reflected wave will not change (Figure 6.25, green curve).

In practice, the reflection phase of AMC is 0° at only one frequency (or several distinct frequencies), but its useful bandwidth can often be assumed for the ±90° range on either side of the 0° frequency.

A typical cross-section of an AMC metasurface structure is shown in Figure 6.26. It consists of two main parts: the AMC main resonator layer and the dielectric with a metal plate behind it. The AMC layer consists of periodically spaced resonators (see the example in Figure 6.27).

6.7.1 Reflection phase estimation method

Figure 6.28 shows the simulation scenario used to compute the AMC reflection phase vs frequency, which was used to estimate the 0° centre frequency for the AMC and its ±90° bandwidth.

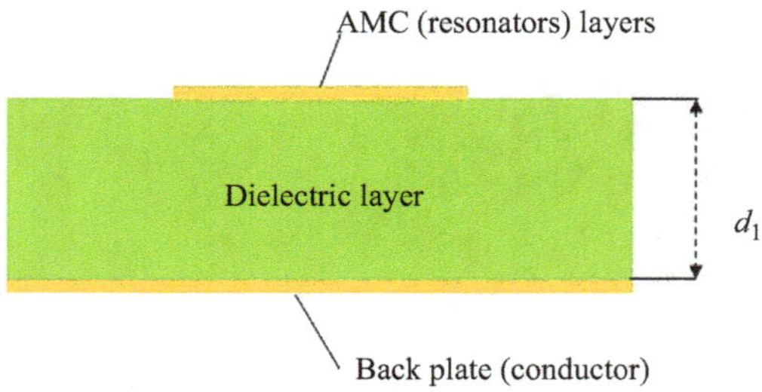

Figure 6.26 FDTD geometry set-up cross-section for the estimation of the AMC reflection phase

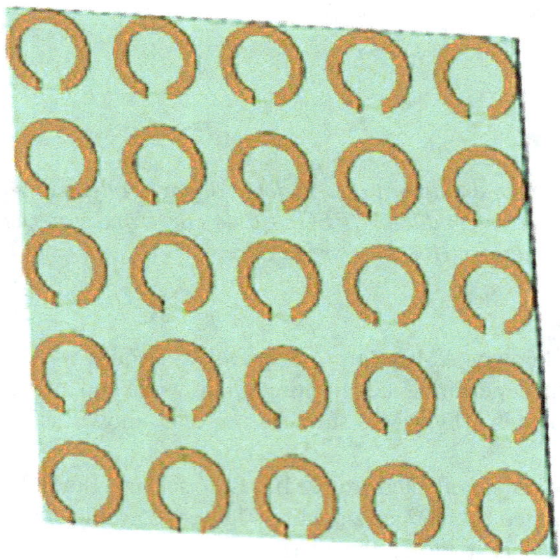

Figure 6.27 An example of AMC (artificial magnetic conductors) meta-surface. In this case, the resonators in the AMC layer are visible as rings with a slit (SRR – split ring resonator)

We used Remcom XFdtd (implementing FDTD full-wave EM simulation method) to compute the reflection phase by exciting the AMC surface with a Gaussian broadband wave [43]. Distance d between the sensor and the AMC surface is chosen so that the incident and reflected pulses do not overlap in time and can be separated. Therefore, using time windowing of the waveform received by the near-field sensor, we can isolate the incident wave from the reflected wave. Next, we transform the signal from the time domain into the frequency domain using the Fourier transform. The last step is estimating phase reflection as a

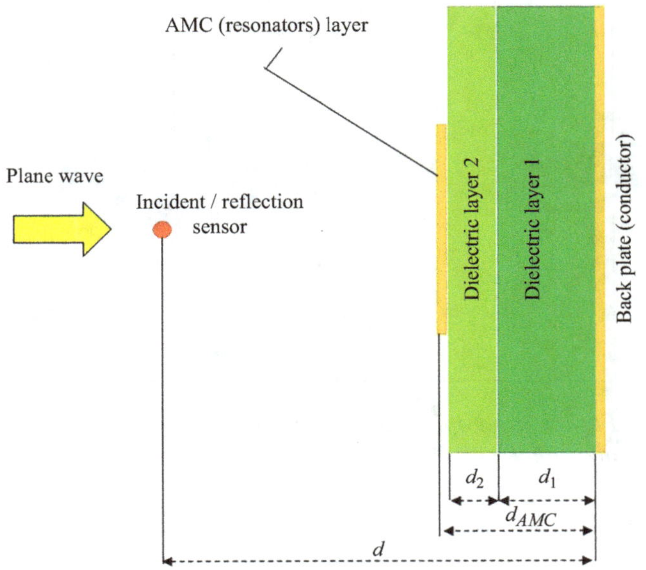

Figure 6.28 FDTD geometry set-up cross-section for the estimation of the AMC reflection phase

function of frequency. The general equation for AMC reflection phase as a function of frequency is given by the following equation [23]:

$$\varphi_{AMC}(f) = (\varphi_R - \varphi_I) + \left(\frac{2\pi f 2d}{c}\right) \tag{6.56}$$

where f is the frequency in Hz, φ_{AMC} is the AMC reflection phase shift, φ_R is the reflected wave phase obtained from the near-field sensor, φ_I is the incident wave phase obtained from the near-field sensor, c is the speed of light in vacuum, is the distance between the near-field sensor and the AMC surface in meters.

6.7.2 AMC structure optimisation

EM parameters of SRRs depend on their geometry and their mutual position in space (e.g. on the dielectric substrate surface). One of the methods of designing the AMC is the optimisation of the parameter of the resonator network itself. The Pareto-Estra (P-Estra) [36,37] optimisation algorithm was used to optimise the AMC geometry. The optimisation goal was to improve the structure in terms of two objectives: the bandwidth and the centre frequency f_c deviation f_{dev} from the desired value f_0. The two components were evaluated from the numerical simulations performed using Remcom XFdtd (see Figure 6.29). The AMC geometry model was created automatically in each iteration of the optimisation loop using the geometry parameter values for each new offspring of the evolutionary algorithm. The values

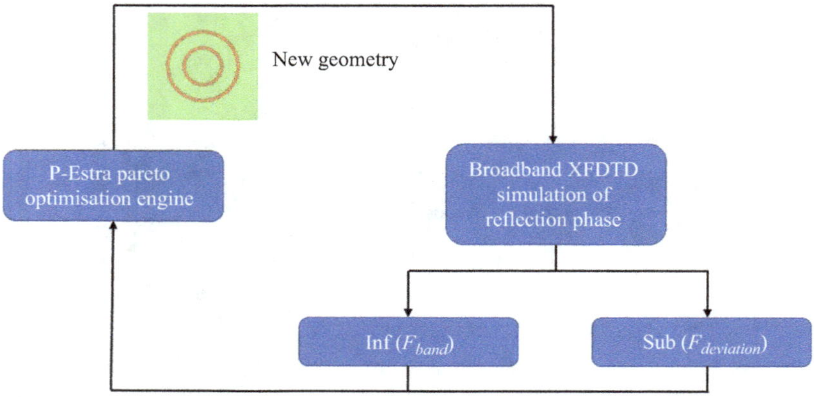

Figure 6.29 Flowchart of optimisation loop combining the P-EStra optimisation algorithm and XFdtd EM full wave simulation software

Table 6.2 AMC size for the double ring resonator

No	Symbol	Physical meaning	Initial (mm)	Constrains		Final (mm)
				Min (min)	Max (mm)	
1	R_1	The radius of the inner ring	3	2	10	3.1
2	R_2	The radius of the outer ring	5	4	15	4.43
3	W_1	Width of path of inner ring	1	0.1	2.5	0.78
4	W_2	Width of path of outer ring	1	0.1	2.5	2.1
5	T_S	Total size with half-distance between unit cells	6	8	18	18

(parameters listed in Tables 6.2 and 6.3) were generated for each iteration made by the P-Estra.

A simplified flowchart of the optimisation loop is presented in Figure 6.30. One objective was to improve the bandwidth B of the AMC, when the phase angle for the reflected wave ranged from $-90°$ to $+90°$, which is a maximisation task. The other objective was to minimise the difference f_{dev} between a target centre frequency and the centre frequency of the solution, which is a minimisation task. Therefore, we solve a bi-objective optimisation problem with two objective functions: (1) the frequency deviation from the desired 0° reflection phase frequency f_{dev}(6.57) and (2) $\pm 90°$ reflection phase bandwidth B (6.58):

$$f_{dev} = f_c - f_0 \tag{6.57}$$

$$B = f_{\max} - f_{\min} \tag{6.58}$$

Table 6.3 AMC size for the double T resonator

No	Symbol	Physical meaning	Initial (mm)	Constrains		Final (mm)
				Min (min)	Max (mm)	
1	UC_S	Size of unit cell total size with half-distance between unit cells	11	7	50	13.35
2	T_S	Total size with half-distance between unit cells	15	11	70	15
3	T_L	Thickness of structure	5	0.1	5	0.98262
4	P_A	Width of external arm of cross	9.09	6	20	10.576
5	P_{B1}	Internal length of external arm of cross	5.44	4	10	4.5573
6	P_{B2}	External length of external arm of cross	6.44	5	11	5.140
7	P_C	Width of internal arm of cross (depth of the cross)	1.6	1	3	1.0533

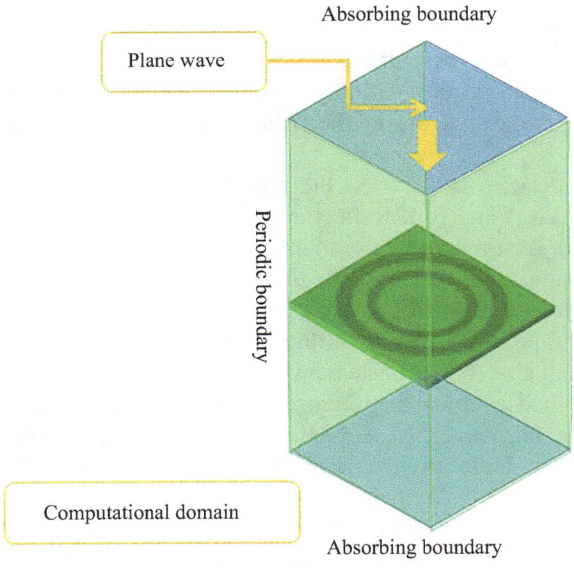

Figure 6.30 Setup of AMC full EM simulation in Remcom XFdtd

where f_c is the desired 0° reflection phase frequency, f_0 is the real 0° reflection phase frequency, $f_{\min}$ is the bandwidth minimum frequency, $f_{\max}$ is the bandwidth maximum frequency.

As an example, let us select the 5.8-GHz (Wi-Fi) frequency band and two different geometries of resonators: a basic double ring (Figure 6.31(a)) and a double

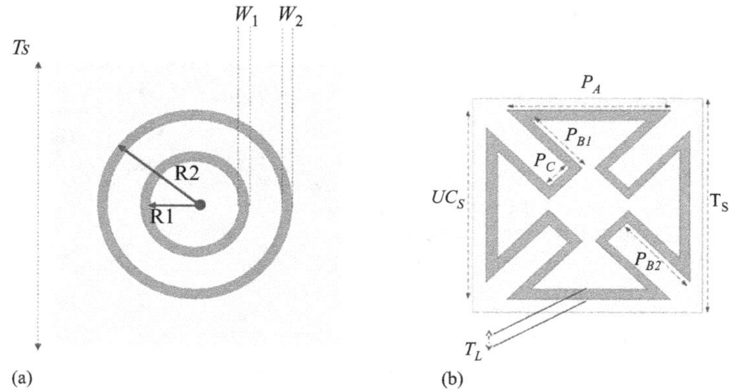

Figure 6.31 Single resonator (a) double ring resonator, (b) based on double T-Ring, and its geometrical parameters, used as a design variables in optimisation process

T-ring-based resonator [43] (Figure 6.31(b)). The main goal of the experiment was to check the possibility of optimising different versions of the metasurfaces for a given frequency band(s). For simplicity, infinite resonator arrays were assumed; consequently, all the resonators in the metasurface have the same geometrical parameter values.

For each of the two geometries, the resonator's initial and final shapes are shown (Figure 6.32(a) and (c)). Reflection phase vs frequency is also shown in Figure 6.32(a)–(c). As a first model, a double ring resonator as a unit cell was used. There are five optimisation variables: the thickness of conductor paths (w_1, w_2), the radii of the conducting rings (R_1, R_2), and the unit cell size T_S – equal for all the resonators. The w_1, w_2, R_1, R_2 mainly influence the resonance frequency and the Q-factor. The unit cell size (the distance between adjacent resonator centres) strongly influences the mutual coupling between resonators. To reduce the computational complexity of the full-wave EM simulation necessary to evaluate the objective functions, we used periodic boundary conditions surrounding a single unit cell (see Figure 6.29). This simulates an infinite array of identical resonators. As a second model, we use a resonator based on a double T-Ring (see Figure 6.31(b)) [43] with eight design variables (seven geometry parameters of the resonator and the unit cell size T_S).

We prepared a fully automatic optimisation software set-up with direct communication between the Remcom XFdtd simulator and P-Estra optimiser implemented in the MATLAB® programming environment. As a result of the evolutionary process, we can observe a clear improvement in both objective functions, i.e. the $\pm 90°$bandwidth B and centre frequency deviation f_{dev}. Moreover, the broadest bandwidth with AMC behaviour (reflection phase less or equal to $\pm 90°$) is even several GHz. Likewise, for each of the two cases, the final value of f_c is closer to the target frequency.

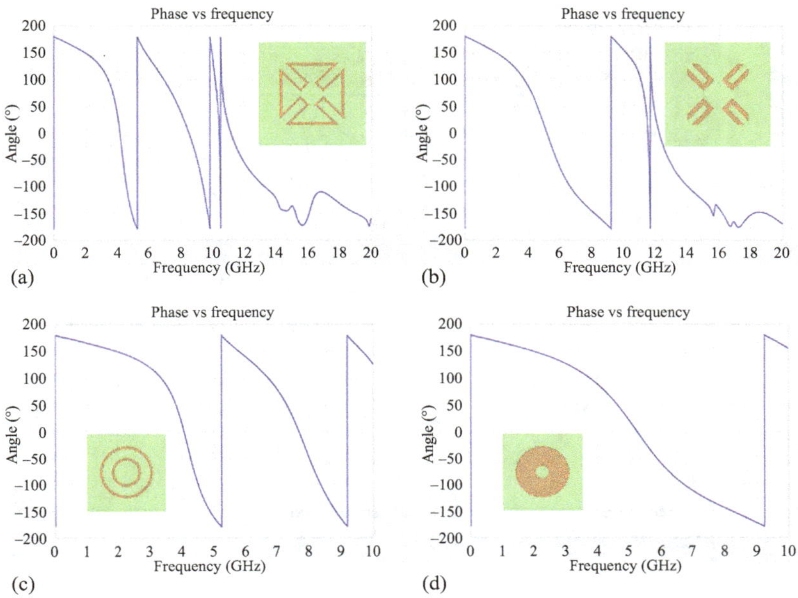

Figure 6.32 The results of optimisation of AMC surface: initial shape and characteristic of phase reflection for initial shape (a and c), (b and d) optimisation result (final shape) for f_c=5.8 GHz

It can be observed that in our Pareto-like optimisation approach, the two initial geometries evolved significantly (see Tables 6.2 and 6.3) with clear performance improvement in terms of both the objective functions B and f_{dev}.

Possible future research directions can be 'free-form' geometries (see Chapter 3) and the use of multi-fidelity optimisation. The multi-fidelity approach with circuit-equivalent models (low-fidelity) and full-wave models (high-fidelity) could significantly reduce the computational complexity of the optimisation loop.

6.8 Frequency-selective metasurfaces in the THz band

The terahertz gap is a frequency band between radio waves and infrared. It is defined as 0.1–10 THz (corresponding to wavelength from 3 mm to 30 μm), although the upper boundary is somewhat arbitrary and is considered by some sources as 30 THz (a wavelength of 10 μm) (Figure 6.33(a)). Presently, at frequencies within this range, practical power generation and sensor devices are challenging to construct. Nonetheless, terahertz frequencies could be exploited in several attractive emerging applications. EM radiation at terahertz frequencies is non-ionising as opposed to, for example, X-rays. Consequently, THz radiation does not damage biological tissues or electronic devices with apparent benefits for medical or diagnostic applications. In general, the level of detail or image spatial

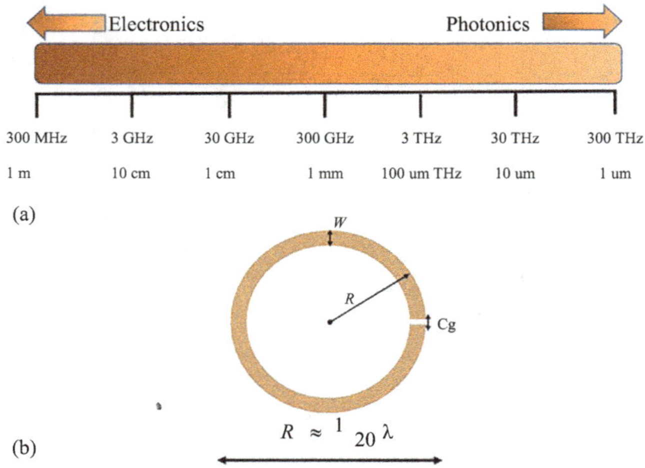

Figure 6.33 Relationship between wavelength and size of split rings resonators as elements of meta-surfaces

resolution depends on the wavelength of the EM radiation being used. The smaller the wavelength, the better the resolution. Image resolution for terahertz waves may be between X-rays and microwaves. The most important of applications of terahertz band radiation include terahertz pulsed imaging and terahertz pulsed spectroscopy [44–46], terahertz-biosensing [47–49], polarisation conversion, wave front conversion or reflection [50–52], identification tags [53–55], and smart textiles applications [56].

As for other frequency ranges, the size of the resonator should not be larger than $1/10\lambda$ (Figure 6.33(b)). This determines the sizes of a single resonator between 100 and 10 μm. Special techniques of printing are needed because of that. Currently, inkjet printing is increasingly used to produce metamaterials sensitive in the field of terahertz radiation. It has been shown that it is possible to print metamaterials on flexible substrates [57,58]. The resolution with which geometrical structures are printed reaches tens of microns [59,60]. Other THz-metasurface fabrication techniques are as follows:

Photolithography. This is a conventional microfabrication technique that can be employed to synthesise metamaterials that operate at terahertz frequencies. It enables the fabrication of subwavelength structures of micrometre size necessary at terahertz wavelengths (30 μm to 3 mm). This makes this technique well suited for terahertz metamaterials. This technique is extensively used to fabricate single-layer and multilayer (3D) metamaterials [61].

Shadow mask lithography is an acid-free fabrication technique for planar and multiple layer micro- and nano-scale structures. It involves direct deposition materials (metal films, oxides, or dielectric layers) onto a substrate through a stencil which is analogous to the solder paste or conductive ink screen printing approaches [61,62].

Soft lithography. It is named 'soft' because it uses elastomeric materials, most often polydimethylsiloxane. Soft lithography allows multi-scale patterning of micro- and nano-scale patterns on polymers or other substrates [61,62].

Electron beam lithography uses the small wavelength of a beam of electrons at high accelerating voltages to obtain nano-scale patterns: analogous to conventional photolithography [61].

6.9 2D and 3D printing techniques

6.9.1 3D printing complex internal structures

Remarkable progress in the design methods for dielectric 3D metamaterials has opened the way to their exploitation in various applications. However, after designing and testing a metamaterial model with a numerical simulation, the next step is to fabricate a prototype. An overview of 2D/3D fabrication technologies can be found in Chapter 3. One of the most effective, widely known, and (in principle) easy to use 3D printing technologies is filament-based 3D printing. The flexibility of that type of 3D printing process makes it possible to fabricate objects of almost any shape. This unique feature can tremendously help the realisation of the 3D dielectric metamaterial prototyping, where a surface could exhibit geometric features hardly implementable employing former technologies. The methods of free-form printing in 3D appear to be particularly suited for prototyping 3D dielectric metamaterials. However, typical 3D printing software is focused on the best mapping of the shape of the (outer) surface of the printed object. The internal surfaces, especially when they are small in size in relation to the rest of the surrounding structure, can be mapped incorrectly. It is essential to point out that the problem is not a limitation of the printer hardware but its software. This section introduces a proprietary algorithm to generate G-code directly from the voxel representation of metamaterials. The major advantage of our algorithm is to eliminate (lattice) defects in the internal structure of printed metamaterials.

G-code is a widely used Computer Numerical Control programming language containing a sequence of commands sent to the printer to control the three-axis movement of the printed object and the parameters of the extruder. It is the native language of most 3D printers.

In the case of metamaterials, unlike in most popular applications of 3D printing, we must also consider their complex internal geometry, i.e. the structure of the unit cells. This is like a new 'dimension', and therefore the number of G-commands may grow exponentially. The more complex and detailed the object's internal structure, the more printer commands are needed; therefore, the bigger G-code files are.

The standard procedure for generating G-code from a three-dimensional raster image is shown in Figure 6.34. First, the voxel image is converted into an STL (which stands for Standard Triangle Language or Standard Tessellation Language)

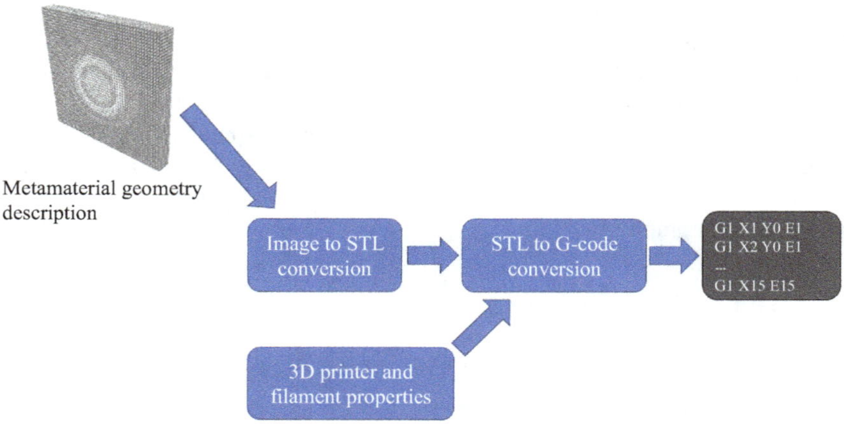

Figure 6.34 A block diagram of the typical algorithm for creating G-code from raster images

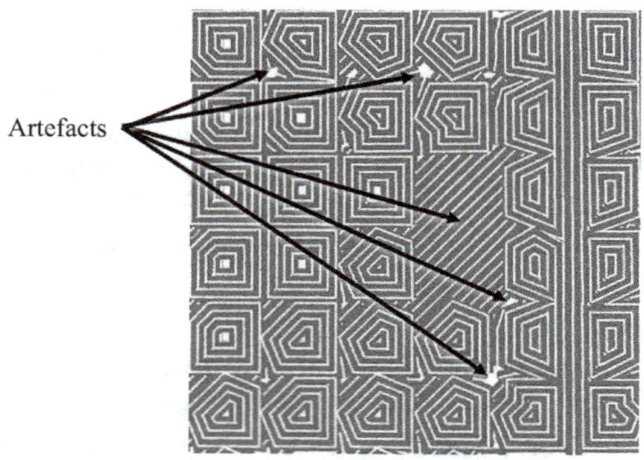

Figure 6.35 Example artefacts on the internal surfaces

file. STL files are in the form of looping instructions that define a polygonal mesh. Next, the STL files are converted into G-code.

At these two conversion stages, significant distortion (artefacts) of the structure geometry occurs (Figure 6.35). Additionally, the generation of STL files requires a lot of RAM. For example, for a voxel file with a spatial resolution of $600 \times 600 \times 70$ voxels, and with unit-cell size *ca* $60 \times 60 \times 60$ voxels, the time needed to create an STL file, using typical software and a typical personal computer is around 6–7 h and about 16-GB RAM. Then, to generate G-code (Repeat host^{TM}) another 8–10 h are needed. These two problems (artefacts and computational/memory requirements) motivated and inspired us to develop a more efficient approach.

6.9.2 Transforming 3D voxel images to G-code directly

We have developed our proprietary algorithm for creating G-code [63] directly from the raster image. The schematic diagram of the algorithm is presented in Figure 6.36. The input file is a raster 2D/3D image with a given spatial resolution. If the input image is a 3D image, then it is split into 2D raster images in the z-axis. The first layer (raster) is the bottom layer.

The first step is known as connected-component labelling [64–66]. The outcome is a set of different (in the sense of a discrete topology) sets of pixels – objects. The next step is the extraction of a single path. In Figure 6.37, this step is implemented in the blocks '*Extract boundary object*', '*Generate path*', and '*Line simplification*'.

The path becomes the 'way' for the printer head (extruder) – these are points with the (x, y) coordinates in the printing plane. The relationship between the pixel position and its actual position for the printer is defined by pixel spacing and global reference coordinates. The pixel spacing determines the actual size of the printed object. Path extraction has two stages: iterative binary dilation and image subtraction and line simplification (Figure 6.38). The final result of this step is a set of (x,y) coordinates marking successive points on the path of the printer extruder. For the line simplification, the Ramer–Douglas–Peucker algorithm [67] can be applied (Figure 6.38).

The last step is G-code generation. Subsequent commands in the G-code file are interpreted and executed separately. The current command or its result may

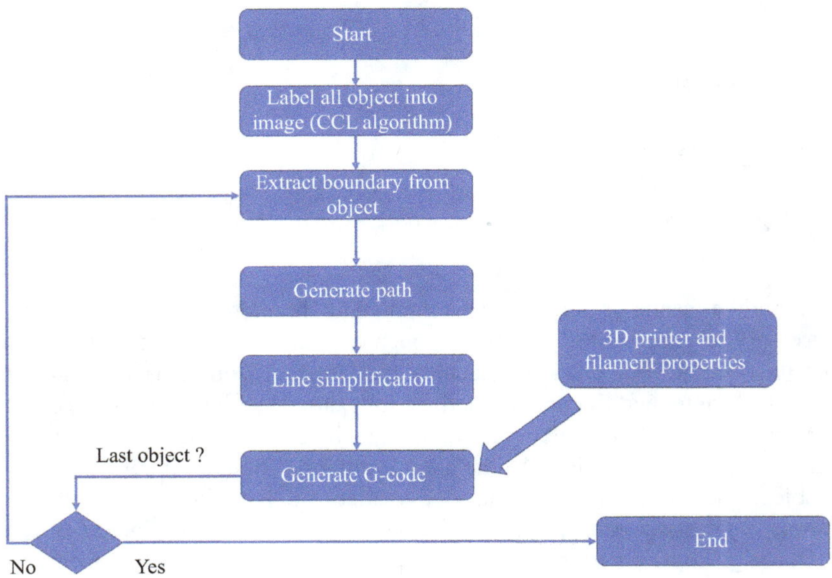

Figure 6.36 A novel G-code generation method (without the intermediate STL format)

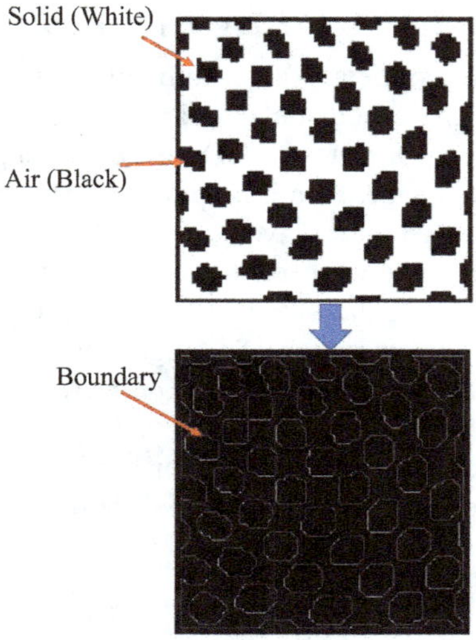

Figure 6.37 Extracting object boundaries

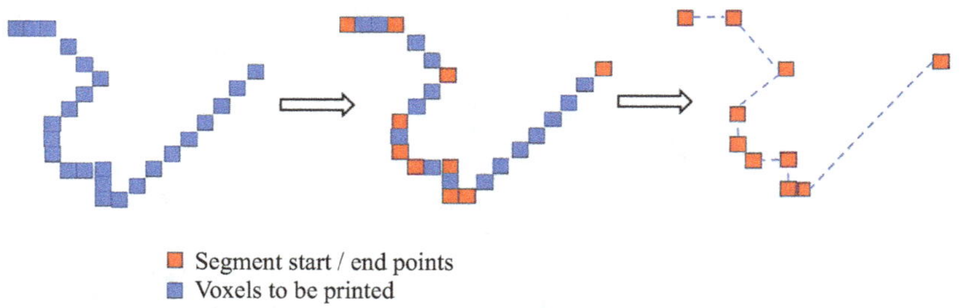

Figure 6.38 Generation of a G-code: example of line simplification

depend on the results of the previous commands. Therefore, a finite state machine is necessary to generate the complete sequence of commands. Except for the previous (last) command, the state machine (or language compiler) must take into account the printer and filament properties, for example path thickness, speed (for a given printing mode), temperatures, and many others. These parameter values must be set before the algorithm starts the actual printing. The loop iterates through all the objects identified by the connected component labelling (CLL) algorithm. Example objects and corresponding printing paths are shown in Figure 6.39.

Our algorithm converts raster images into the G-code with much higher accuracy (fewer artefacts/lattice defects). Moreover, it does it in a few minutes compared to many hours for the standard algorithm (for the same flat Lüneberg lens example) and with much lower memory requirements.

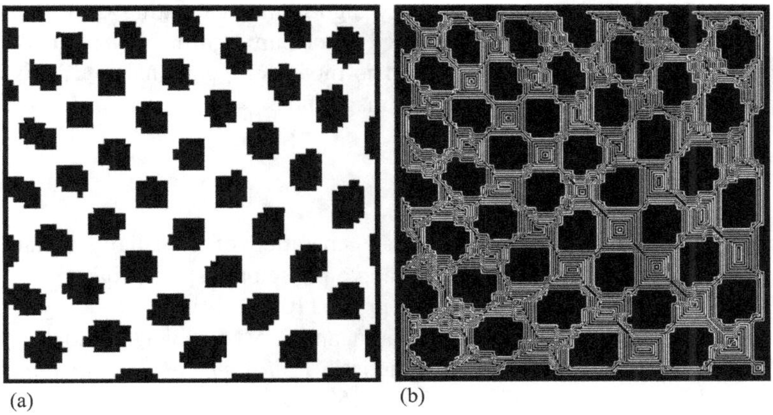

Figure 6.39 Results of converting single 2D raster image (from a 3D file) (a) to a set of 'paths' for printer input image (b)

6.10 Conclusion

Section 6.1 presented the principles of gradient dielectric metamaterials (a subclass of effective media) that exploit local, quasi-periodic changes in the (effective) permittivity values. We explained an algorithm for synthesising this type of metastructures. It is based on a crystal-like lattice with quasi-periodically distributed unit cells. The lattice is spatially modulated using parameter maps such as unit cell size, directional orientation, and fill fraction.

Section 6.2 discusses split resonators, which are the fundamental building blocks (metaparticles) of metasurfaces. We described how the resonators are coupled to the electric and magnetic EM wave components. Metasurfaces consist of arras of resonators usually positioned on a dielectric substrate. The resonators in the array are mutually coupled inductively and capacitively, which modifies the performance (e.g. the resonant frequency and quality factor) of the individual resonators. In Sections 6.3 and 6.4, we presented some circuit-equivalent models which can take this coupling into account. The models are significantly less computationally complex than full-wave models, such as FDTD. This complexity reduction is particularly advantageous in optimisation loops if many iterations are required to improve/optimise the solution. On the other hand, full-wave models (Section 6.5) are irreplaceable for cases where circuit-equivalent models do not exist or are inaccurate.

Section 6.6 is devoted to an example design and optimisation of an AMC metasurface with potential application as an MRI lens.

An overview of applications of metamaterials in the terahertz-gap band is presented in Section 6.8. In addition, the most important methods of printing micrometre size structures are also briefly presented.

Section 6.9 describes our proprietary algorithm for creating G-code directly from 2D/3D raster images. The presented algorithm significantly reduces lattice defects in 3D-printed metamaterials and is more computationally effective than standard algorithms by several orders of magnitude.

References

[1] M. Kumar and J. Joseph, "Optical generation of a spatially variant two-dimensional lattice structure by using a phase only spatial light modulator", *Applied Physics Letters*, vol. 105, pp. 051102, 2014.

[2] R. C. Rumpf, J. J. Pazos, J. L. Digaum, and S. M. Kuebler, "Spatially variant periodic structures in electromagnetics", *Philosophical Transactions of the Royal Society A*, vol. 373, pp. 20140359, 2015.

[3] R. C. Rumpf, C. R. Garcia, H. H. Tsang, J. E. Padilla, and M. D. Irwin, "Electromagnetic isolation of a microstrip by embedding in a spatially variant anisotropic metamaterial", *Progress In Electromagnetics Research*, vol. 142, pp. 243–260, 2013.

[4] R. C. Rumpf and J. Pazos, "Synthesis of spatially variant lattices", *Optics Express*, vol. 20, no 14, pp. 15263–15274, 2012.

[5] J. Felicio and C. Fernandes, "Complex permittivity and anisotropy measurement of 3D-printed PLA at microwaves and millimeter-waves", *22nd International Conference on Applied Electromagnetics and Communications (ICECOM)*, pp. 1–6, 2016.

[6] J. B. Pendry, A. J. Holden, D. J. Robbins, and W. J. Stewart, "Magnetism from conductors and enhanced nonlinear phenomena", *IEEE Transactions on Microwave Theory Techniques*, vol. 47, pp. 2075–2084, 1999.

[7] V. G. Veselago, "The electrodynamics of substances with simultaneously negative values of and μ", *Soviet Physics Uspekhi*, vol. 10, no 4, pp. 509, 1968.

[8] M. N. O. Sadiku, *Elements of Electromagnetics*, New York: Oxford University Press, 2014.

[9] O. Sydoruk, E. Tatartschuk, E. Shamonina, and L. Solymar, "Analytical formulation for the resonant frequency of split rings", *Journal of Applied Physics*, vol. 105, no 1, pp. 014903-1–014903-4, 2009.

[10] U. Ehret and M. Neuper, "Applying the Ramer-Douglas-Peucker algorithm to compress and characterize time-series and spatial fields of precipitation", EGU General Assembly, Wien, Österreich, 2014.

[11] S. Lee, B. Kang, H. Keum *et al.*, "Heterogeneously assembled metamaterials and metadevices via 3D modular transfer printing", *Scientific Reports*, vol. 6, 2016.

[12] H. B. Palmer, "The capacitance of a parallel-plate capacitor by the Schwarz-Christoffel transformation", *Electrical Engineering*, vol. 56, no 3, pp. 363–368, 1937.

[13] R. Marques, F. Mesa, J. Martel, and F. Medina, "Comparative analysis of edge- and broadside- coupled split ring resonators for metamaterial

design – Theory and experiments", *IEEE Transactions on Antennas and Propagation*, vol. 51, no. 10, pp. 2572–2581, 2003.

[14] F. Bilotti, A. Toscano, L. Vegni, K. Aydin, K. B. Alici, and E. Ozbay, "Equivalent-circuit models for the design of metamaterials based on artificial magnetic inclusions", *IEEE Transactions on Microwave Theory and Techniques*, vol. 55, no 12, pp. 2865–2873, 2007.

[15] J. D. Baena, J. Bonache, F. Martin *et al.*, "Equivalent-circuit models for split-ring resonators and complementary split-ring resonators coupled to planar transmission lines", *IEEE Transactions on Microwave Theory and Techniques*, vol. 53, no 4, pp. 1451–1461, 2005.

[16] R. Marques and M. Francisco, "Role of bianisotropy in negative permeability and left-handed metamaterials", *Physical Review B*, vol. 65, 2002.

[17] A. Sethi and R. Rajni, "Determination of electromagnetic parameters of a new metasurface comprising of square loop", *Journal of electrical science and technology, Malaysia*, vol. 13, pp. 48–57, 2018.

[18] L. Solymar and E. Shamonina, *Waves in Metamaterials*, Illustrated edition. Oxford; New York: Oxford University Press, 2009.

[19] M. Shamonin, E. Shamonina, V. Kalinin, and L. Solymar, "Properties of a metamaterial element: Analytical solutions and numerical simulations for a singly split double ring", *Journal of Applied Physics*, 2004.

[20] "Theory of Two Magnetically Coupled RLC Circuits". ukessays.com. 11 2018; UKEssays. 09 2021. Available: https://www.ukessays.com/essays/physics/theory-magnetically-coupled-rlc-6772.php?vref=1.

[21] L. T. Wood, R. M. Rottmann, and R. Barrera, 'Faraday's law, Lenz's law, and conservation of energy', *American Journal of Physics*, vol. 72, no. 3, pp. 376–380, 2004.

[22] S. Babic, C. Akyel, J. Martinez, and B. Babic, "A new formula for calculating the magnetic force between two coaxial thick circular coils with rectangular cross-section", *Journal of Electromagnetic Waves and Applications*, vol. 29, pp. 1181–1193, 2015.

[23] Ł. Jopek, S. Hausman, and P. Di Barba, "Optimization of an artificial magnetic conductor geometry using a Paretian approach", *EuCAP*, 2019.

[24] A. Taflove and S. C. Hagness. Computational Electrodynamics: The Finite-Difference Time-Domain Method, 3rd ed. Norwood, MA: Artech House, 2005.

[25] J. M. Dohlus, P. Hahne, X. Du, B. Wagner, T. Weiland, and S. G. Wipf, "Using the Maxwell grid equations to solve large problems", *IEEE Transactions on Magnetics*, vol. 29, no 2, pp. 1914–1917, 1993.

[26] K. S. Yee, "Numerical solution of initial boundary value problems involving Maxwell's equations in isotropic media", *IEEE Transactions on Antennas and Propagation*, pp. 302–307, 1966.

[27] R. L. Wagner and W. C. Chew, "An analysis of Liao's absorbing boundary condition", *Journal of Electromagnetic Waves and Applications*, vol. 9, pp. 993–1009, 2012.

[28] E. Bécache, S. Fauqueux, and P. Joly, "Stability of perfectly matched layers, group velocities and anisotropic waves", *Journal of Computational Physics*, vol. 188, no 2, pp. 399–433, 2003.

[29] S. G. Johnson, "Notes on Perfectly Matched Layers (PMLs)". arXiv:2108.05348, 2021.

[30] F. L. Teixeira and W. C. Chew, "Complex space approach to perfectly matched layers: A review and some new developments", *International Journal of Numerical Modelling: Electronic Networks, Devices and Fields*, vol. 13, no 5, pp. 441–455, 2000.

[31] S. Babic, F. Sirois, C. Akyel, and C. Girardi, "Mutual inductance calculation between circular filaments arbitrarily positioned in space: Alternative to Grover's formula", *IEEE Transactions on Magnetics*, vol. 46, no 9, pp. 3591–3600, 2010.

[32] P. C. Lauterbur, "Image formation by induced local interactions: Examples employing nuclear magnetic resonance", *Nature*, vol. 242, no 5394, Art. no 5394, 1973.

[33] W. Catherine, *MRI in Practice 4e*. Chicester, UK: Wiley, John & Sons, 2012.

[34] J. Wu and B. Li, "Development of a magnetoinductive lens for magnetic resonance imaging", *IEEE Instrumentation & Measurement Magazine*, vol. 20, pp. 56–60, 2017.

[35] P. D. Barba, M. E. Mognaschi, R. Palka, and A. Savini, "Optimization of the MIT field exciter by a multiobjective design", *IEEE Transactions on Magnetics*, vol. 45, no 3, pp. 1530–1533, 2009.

[36] P. di Barba, M. E. Mognaschi, R. Palka, and A. Savini, "Optimization of the MIT field exciter by a multiobjective design", *IEEE Transactions on Magnetics*, 2009.

[37] Ł. Januszkiewicz, P. Di Barba, and S. Hausman, "Field-based optimal placement of antennas for Body-Worn wireless sensors", *Sensors*, vol. 16, no 5, 2016.

[38] D. Sievenpiper, L. Zhang, R. F. J. Broas, N. G. Alexopolous, and E. Yablonovitch, "High-impedance electromagnetic surfaces with a forbidden frequency band", *IEEE Transactions on Microwave Theory and Techniques*, vol. 47, no 11, pp. 2059–2074, 1999.

[39] R. Kuse, T. Hori, M. Fujimoto, T. Seki, K. Sato, and I. Oshima, "Equivalent circuit analysis for AMC with double layer patch type FSS", *IEEE Antennas and Propagation Society International Symposium (APSURSI)*, 2013, pp. 1360–1361.

[40] J. H. Yoon, E. Y. Kim, Y. Lim, and Y. J. Yoon, "Equivalent circuit model and reflection phase control methods for dual-band AMC", *5th European Conference on Antennas and Propagation, EUCAP*, 2011, pp. 1222–1226.

[41] P. D. Barba, *Multiobjective Shape Design in Electricity and Magnetism*. Lecture Notes in Electrical Engineering, Springer, Dordrecht Heidelberg, London, New York, 2010.

[42] J. Chaires, D. Schumerth, C. Drawdy, and W. Yang, "Gain incorporated split-ring resonator structures for active metamaterials", *Advances in OptoElectronics*, vol. 2015, pp. 1–9, 2015.

[43] K. Agarwal, Y.-X. Guo, and B. Salam, "Wearable AMC backed near-endfire antenna for on-body communications on latex substrate", *IEEE Transactions on Components, Packaging and Manufacturing Technology*, vol. 6, no 3, pp. 346–358, 2016.

[44] W. Yang, K.-W. Tam, W.-W. Choi, W. Che, and H. T. Hui, "Novel polarization rotation technique based on an artificial magnetic conductor and its application in a low-profile circular polarization antenna", *IEEE Transactions on Antennas and Propagation*, vol. 62, no 12, pp. 6206–6216, 2014.

[45] L. Yu-Sheng, L. Shaoquan, L. Xiaoyan *et al.*, "Tunable terahertz metamaterial by using three-dimensional double split-ring resonators", *Optics & Laser Technology*, vol. 112, pp. 215–221, 2019.

[46] H.-T. Chen, W. J. Padilla, J. M. O. Zide, A. C. Gossard, A. J. Taylor, and R. D. Averitt, "Active terahertz metamaterial devices", *Nature*, vol. 444, no 7119, pp. 597–600, 2006.

[47] A. J. Fitzgerald, V. P. Wallace, J.L. Mercedes *et al.*, "Terahertz pulsed imaging of human breast tumors", *Radiology*, vol. 239, no 2, pp. 533–540, 2006.

[48] A. Menikh, R. MacColl, C. A. Mannella, and X-Ch. Zhang, "Terahertz-biosensing technology: Progress, limitations, and future outlook", *Optical Guided-wave Chemical and Biosensors II*, vol. 8, pp. 283–295, Berlin, Heidelberg, Springer-Verlag, 2010.

[49] S. Yan, L. Xia, D. Wei, H.-L. Cui, and C. Du, "Terahertz biosensing of protein based on a metamaterial", *IEEE International Conference on Manipulation, Manufacturing and Measurement on the Nano-scale (3M-NANO)*, 2016, pp. 327–330.

[50] H. Zhou, Ch. Yang, D. Hu *et al.*, "Terahertz biosensing based on bi-layer metamaterial absorbers toward ultra-high sensitivity and simple fabrication", *Applied Physics Letters*, vol. 115, pp. 143507, 2019.

[51] J. Fan and Y. Cheng, "Broadband high-efficiency cross-polarization conversion and multi-functional wavefront manipulation based on chiral structure metasurface for terahertz wave", *Journal of Physics D: Applied Physics*, vol. 53, pp. 025109, 2019.

[52] W. Yang, K.-W. Tam, W.-W. Choi, W. Che, and H. T. Hui, "Novel polarization rotation technique based on an artificial magnetic conductor and its application in a low-profile circular polarization antenna", *IEEE Transactions on Antennas and Propagation*, vol. 62, pp. 6206–6216, 2014.

[53] X.-F. Zang, S.-J. Liu, H.-H. Gong, Y. Wang, and Y.-M. Zhu, "Dual-band superposition induced broadband terahertz linear-to-circular polarization converter", *Journal of the Optical Society of America*, vol. 35, 2018, pp. 950–957.

[54] M. Hamdi, F. Bonnefoy, M. Bernier *et al.*, "Identification in the terahertz domain using low cost tags with a fast spectrometer", *12th IEEE International Conference on Anti-counterfeiting, Security, and Identification (ASID)*, 2018, pp. 111–115.

[55] R. Mitsuhashi, K. Murate, S. Niijima, T. Horiuchi, and K. Kawase, "Terahertz tag identifiable through shielding materials using machine learning", *Optics Express*, vol. 28, pp. 3517, 2020.

[56] R. Mitsuhashi, T. Horiuchi, K. Murate, and K. Kawase, "Terahertz tag identifiable through shielding material", *44th International Conference on Infrared, Millimeter, and Terahertz Waves (IRMMW-THz)*, 2019, pp. 1–3.

[57] D. Yang, "Applications of terahertz wave technology in smart textiles" Tao X. (eds) *Handbook of Smart Textiles*. Singapore: Springer, pp. 809–842, 2015.

[58] D. Lee, H. K. Sung, and S. Lim, "Flexible subterahertz metamaterial absorber fabrication using inkjet printing technology", *Applied Physics B*, 2016.

[59] K. Kashiwagi, L. Xie, X. Li *et al.* "Inkjet-printed silver-nanoparticle THz metamaterial", *41st International Conference on Infrared, Millimeter, and Terahertz waves (IRMMW-THz)*, 2016, pp. 1–2.

[60] P. M. Grubb, H. Subbaraman, S. Park, D. Akinwande, and R. T. Chen, "Inkjet printing of high performance transistors with micron order chemically set gaps", *Scientific Reports*, vol. 7, 2017.

[61] W. Voit, W. Zapka, L. Belova, and K. V. Rao, "Application of inkjet technology for the deposition of magnetic nanoparticles to form micron-scale structures", *Science, Measurement and Technology, IEE Proceedings*, vol. 150, pp. 252–256, 2003.

[62] S. Walia, M. Ch. Shah, P. Gutruf *et al.*, "Flexible metasurfaces and metamaterials: A review of materials and fabrication processes at micro- and nano-scales", *Applied Physics Reviews*, vol. 2, no 1, 2015.

[63] T. R. Kramer, F. M. Proctor, and E. R. Messina, "The NIST RS274NGC Interpreter – Version 3", 2000.

[64] L. He, X. Ren, Q. Gao, X. Zhao, B. Yao, and Y. Chao, "The connected-component labeling problem: A review of state-of-the-art algorithms", *Pattern Recognition*, vol. 70, pp. 25–43, 2017.

[65] L. G. Shapiro, "Connected component labeling and adjacency graph construction", *Machine Intelligence and Pattern Recognition*, vol. 19, pp. 1–30, 1996.

[66] H. Samet and M. Tamminen, "Efficient component labeling of images of arbitrary dimension represented by linear bintrees", *IEEE Transactions on Pattern Analysis and Machine Intelligence*, vol. 10, no 4, pp. 579–586, 1988.

[67] P. Cebrian and J. C. Moure, "GPU-accelerated RDP algorithm for data segmentation", *Computational Science – ICCS 2020*, Cham, 2020, pp. 234–247.

[68] E. B. Rosa and W. G. Frederick, *Formulas and Tables for the Calculation of Mutual and Self-Inductance*. Washington: Govt. Print. Off, 1916.

[69] C. Alexander and M. Sadiku, *Fundamentals of Electric Circuits*, 6th edition. New York, NY: McGraw-Hill Education, 2016.

[70] S. Babic and C. Akyel, "Calculation of some electromagnetic quantities for circular thick coil of rectangular cross-section and pancake with inverse radial currents", *IET Electric Power Applications*, pp. 1306–1310, 2018.

[71] 'Positive coupling, negative coupling, and all that'. Available: http://aaronscher.com/Circuit_a_Day/Impedance_matching/positive_negative_coupling/positive_negative_coupling.html. [Accessed 16 June 2021].

Chapter 7
Industrial design perspectives

Xose M. Lopez-Fernandez[1]

The fields of both additive manufacturing (AM), also known as three-dimensional (3D) printing, and metamaterials have gained great popularity over the last two decades. The emerging field of metamaterials promises to drive artificial materials engineering that will make it possible to tailor the properties of material. That, in turn, will create extraordinary new possibilities, transcending what can currently be done by industry. Without doubt, the rapid advance of AM has acted as an enabling technology for metamaterials, contributing as a valuable tool for prototyping. The AM process is very flexible both for designing parts and joining materials. 3D printing is potentially capable of placing different materials arbitrarily in three dimensions. With the increasing capability of computational simulation and topology optimisation (TO), metamaterials with extremely complex structures can be conceived under premises of optimal design and exploit 3D printing techniques.

With the exotic properties they exhibit, metamaterials are considered to have many potential industrial applications. At the same time, AM has emerged and promises to stimulate the acceleration of industrial growth by applying the emerging metamaterials.

In general, metamaterials promise a wide range of applications in areas like medicine, architecture, textiles, defence, entertainment, sports, automobiles and aerospace. However, the industry is always reluctant to take on new business initiatives without sufficient market demand. Moreover, researchers still need to develop feedstock to devise prototypes and justify pilot-scale experimentation that will ultimately lead to market demand. With this in mind, it is crucial that the unique ability of AM to create superior complex shapes and structures is also turned into functional products. At the same time, costs should become competitive once AM technologies are fully integrated into manufacturing processes over the entire product lifecycle. Aligning application opportunities with emerging technologies could accelerate innovative manufacturing of metamaterials. Such a

[1]Department of Electrical Engineering, Escola de Enxeñeiros Industriais, Universidade de Vigo, Vigo, Spain

scenario will require collaboration and understanding between manufacturing researchers, development leaders, end-users from industry, academia and governments.

About 63 per cent of enterprises currently using AM deploy the technology for prototyping, while only 21 per cent use it for items that cannot be built with any other manufacturing technology.

The US Department of Energy (DOE) estimates that AM might slash waste and materials costs by nearly 90 per cent and reduce manufacturing energy use in half, if compared to traditional manufacturing. The AM industry is expected to grow to $23 billion globally by 2022. This figure could be increased if metamaterial manufacturing and applications are also included.

The purpose of this chapter is to provide an overview of current and potential industrial design combining metamaterials and 3D printing methods. Practical considerations concerning how adaptable 3D printing is for metamaterial manufacturing and applications are explained, and some promising application fields in industry are described. In particular, there is a review of the limitations when integrating AM and metamaterials. That is followed by an overview of market segments and the different key business strategies adopted by important players operating in the global AM market, concentrating on the end users, applications and geography. The chapter ends with some current examples and promising developments for the future that will enable AM and metamaterials to reach their full potential in industry.

7.1 Current and potential trends

AM is defined by the International Committee F42 on AM technologies as the 'process of joining materials to make objects from 3D model data, layer by layer, as opposed to subtractive manufacturing methodologies' [1].

In the 1980s, techniques to create objects layer by layer using computer-aided design (CAD) started to appear. The emergence of stereolithography (SLA) meant they could be used for rapid prototyping [2]. Pieces were made with epoxy resin in 0.10-mm layers and with a minimum thickness of 0.6 mm.

AM is capable of integrating functionalities into the designs. Examples include embedded objects, wiring conducts [3], complex noise mitigation devices [4] and optimised fluid channels [1].

The first AM technology was patented in the United States in 1979 by Housholder [5]. In the early 1990s, Kruth [6] categorised various AM processes from three perspectives: liquid-, powder- and solid-based systems according to different material creation and direct-3D and -2D layers techniques according to different shape building [7]. There are currently more than 20 different recognised AM technologies [8]. Most recently, The American Society for Testing and Standards (ASTM) has classified AM technologies into seven categories based on the specific manufacturing processes used [7]:

1. material extrusion;
2. powder bed fusion;
3. vat photopolymerisation;
4. material jetting;
5. binder jetting;
6. sheet lamination;
7. directed energy deposition.

The most widely used AM technologies for polymeric materials are fused deposition modelling for consumer applications, SLA for models and prototypes of high complex geometries and selective laser sintering for industrial applications. The two main AM technologies for metallic materials are direct metal laser sintering and electron-beam melting.

A summarised overview of AM technology is shown in [7,9]. It uses the ASTM's seven discrete groups and concentrates on AM of plastic polymers, metal materials and ceramic materials.

7.1.1 Design of complex geometries for manufacturability

When they adopt AM, businesses can look forward to increased manufacturing potential and greater freedom in design. Parts can be designed for function rather than manufacturability, and difficult features like undercuts, conformal structures and living hinges can easily be built. 3D printing offers a new opportunity to design and produce geometries and assemblies previously thought impossible [10].

AM's limited geometric constraints are an advantage because they allow complexity in part designs through strategies based on in-design optimisation. Designs can therefore integrate features, use fewer materials or include TO [1].

AM systems combined with design for additive manufacturing (DfAM), which allows adaptation, modification and optimisation, can lead to weight reduction in a part such as aircraft components [11]. The well-known example of the LEAP (Leading Edge Aviation Propulsion) jet engine fuel nozzle by GE Aviation reduced costs by 75 per cent and weight by 25 per cent.

Current possibilities for reducing material usage in parts produced by AM processes are found in several manufacturing strategies. These include direct modification of the design itself by reducing non-necessary material, by eliminating sections on the basis of topological optimisation, integrating components or controlling the infill density for designs that are conceived as not completely solid [1].

The possibilities of manufacturing free-standing complex 3D geometries, obtaining low manufacturing costs and fabricating directly from CAD design, enhance the design process. Complex structures can be made with predefined electromagnetic, optical, acoustic, thermal and mechanical properties [12].

For over a third of a century, AM resolution and material availability have been improved, and so complex structures can be designed and manufactured in robust engineering materials. That was something that could not be done using conventional subtractive manufacturing techniques [13].

As mentioned, the behaviour of metamaterials is dependent on their internal structure, whose topology is generally a complex, small repeating one. AM has the potential to overcome structural design constraints and manufacturing defects by managing such aspects as accuracy, structural connectivity, additional support structure, surface roughness or material properties. If it is to take full advantage of AM to maximise product performance, DfAM requires close integration of product design and manufacturing that takes into account AM process constraints and real material properties during the AM process [14].

7.1.2 3D printing in manufacturing metamaterials

3D printing, also called AM, refers to a set of techniques able to create objects or parts by a process of joining materials from 3D model data, usually adding layer upon layer. This is in contrast to classical subtractive manufacturing and formative manufacturing methodologies [15].

3D printing is inherently a computationally driven technique, where a computer program is controlling how a specific material is applied in layers to create a 3D object. This object is based on a previous 3D digital model, given in STL (Standard Tessellation Language), OBJ or other file types, for a 3D printer to follow [16]. Chapter 3 explains some critical aspects due to the complex geometry of the model which influence the time needed to convert it into GCODE formats. Once those critical aspects have been taken into account, the real capability of 3D printing, therefore, lies in the way it facilitates the manufacture of specific objects on demand.

It is the digital model that tells the printer what constituent material or materials are layered onto the substrate and how they are scaled. The tools used to layer the material are based on procedures controlled and operated digitally by software.

AM technologies promise to effectively overcome many challenges in metamaterials manufacturing as they have intrinsic advantages when dealing with complex structures. Metamaterials are artificially engineered materials based on a structural design which is constructed to incorporate unprecedented properties that are only found in strange and expensive materials and not seen in nature [17]. A typical feature of metamaterials is that their design is often achieved by repeating small-scale unit patterns (unit cell) to large scale.

Usually, metamaterials are built by assembling several elements based on metal or plastic. The scale of the repeating pattern is related to the physical phenomena that the metamaterials have to influence, e.g. wavelengths. Fundamental metamaterial properties, such as the collapse mechanism, wavelength of incident light, force concentration or thermal expansion, can be governed by manipulating their internal structures [18]. Meanwhile, the unit cell shape, scale and orientation can be further optimised to increase the external properties. This makes it possible to create and control the properties of metamaterials by manipulating their internal geometry structure.

Advances in metamaterials and AM offer new opportunities for intra- or interdisciplinary research to create new metamaterial structures. New developments using AM in electromagnetic, optical and thermal metamaterials have not been widely examined, likely due to the submicron resolutions required in those metamaterial structures [15].

It is important to mention that the selected manufacturing method/technology is a critical choice that dictates the resolution, material and working performance of the resulting metamaterial. Moreover, most metamaterial applications continue to be underexploited because there are currently no manufacturing processes that can fabricate their complex geometries on a large scale [12]. Therefore, AM can greatly contribute to increasing production capability and decreasing manufacturing costs in metamaterials design and fabrication.

Focusing on optimisation of metamaterials – and their high-precision, multi-structural, multi-material nature and topology – and combining them with AM fabrication techniques will make it possible to overcome the challenges in research and make the most of the opportunities in industry.

7.1.3 Opportunities and future directions

AM allows the creation of complex shapes which are not possible by any other manufacturing means [19]. 3D printing on demand is essential in research on new metamaterials, with properties tailored on demand [16].

CAD software has in part driven the way in which AM has expanded and been adapted. A CAD file containing a solid 3D model starts the AM process, which slices it into the layers that become the input for the AM part building machine [15]. The process is cost efficient as there is no need for assembly, and there is very little waste material [20,21]. AM processes can also be scaled to print parts that range in size from nanometres to metres.

Opportunities and future directions in the field of metamaterials include finding ways to bind structural and functional properties more intimately and overcoming the physical constraints when engineering metamaterials with acceptable physical performance and environmental properties [22]. However, even though metamaterials are increasingly becoming commercially available, there are often doubts about how dependable they are in terms of quality or suitability [23].

There is a need to develop and improve critical feedstocks, create prototypes and justify pilot-scale experimentation that will ultimately generate opportunities and drive market demand. The involvement of public–private partnerships will help as industry is reluctant to act without sufficient market demand.

Most metamaterials are presently applied in the development of devices for controlling electromagnetic (and also acoustic) waves in the fields of optoelectronics and electronic engineering and, to some extent, mechanical and thermal engineering [1]. Specific technologies include photonic crystals, plasmonic structures and sensors, light cloaks, superlenses and frequency selective surfaces [24].

Metamaterials are being introduced in more traditional fields, such as mechanical and aeronautical engineering, and also architecture. They improve mechanical characteristics by adding features during the design and manufacturing processes. Porous materials, for example, are used in product design and lattice structures are used in architecture. They normally provide greatly reduced weight, with remarkable relationships between Young's modulus and overall density and between resistance and weight. Some materials have a porous structure like trabecular bone and natural foams, or a lattice structure like supporting skeletons or exoskeletons, which can be ideal for specific applications [1].

The metamaterial concept does not only apply to classical mechanics but also to thermodynamics (including elastostatics, elastodynamics, acoustics and fluid dynamics) [25]. Fluid dynamics is a field with rapid growth potential into new areas of research and application. For example, controllable metamaterials that lead to smart structures and smart fabrics are a key trend [22].

There are many exciting possibilities that can lead to highly imaginative ideas and futuristic concepts of advanced science and technology. A group of metamaterials has emerged that can be called smart materials, whose properties change when triggered by stimuli such as pressure, heat. Personalised metamaterials – microstructures engineered in order to find certain properties that differ from those of the base material – can also be created [19]. The challenge is to print at a microscopic scale and obtain the expected and desired macroscopic results. Being able to predict the final performance will again be crucial. Effective use of both smart materials and metamaterials will depend on how the manufacturing process is able to add complexity. The following issues will be important [19]:

- CAD-CAE systems will need improved characteristics that can enable these complex structures to be designed with agility. This is currently feasible but not in the agile way demanded by market deadlines;
- Enriched G-code files – containing the information on how the machine will produce the part according to some given parameters – will have to be able to admit parameters that support microstructure and multi-physics phenomena;
- Capabilities for predicting operating behaviour in both steady and transient situations.

A promising area of development in metamaterials is in 4D printing [15]. This is a subset of 3D printing involving smart materials that can modify their behaviour over time based on an environmental stimulus, which could be thermal, mechanical, electrical, magnetic, chemical or many combinations of those and other stimuli. 4D metamaterials are a very recent development and have only been demonstrated practically in the lab [15].

Fully extending the opportunities of AM requires that it be promoted among all stakeholders to create a joint understanding of the benefits, the changing processes involved and what role each stakeholder should play [19].

7.1.4 Disruptive design perspectives

AM can be called disruptive because it is able to control the geometry, material and functionalities of manufactured products in a way that no other method has so far been able to [1]. However, its most outstanding capability is the ability to produce complex designs in a direct manufacturing process that are not feasible by conventional methods. This advantage is so relevant that it is even changing the way in new products could be conceived [1]; fabrication with AM now means that the functionality of a product takes centre stage in the design process; it is a disruptive approach.

3D printing is disruptive and spreading fast, which means research is needed to look into fundamental fabrication concepts so that engineers can take full advantage of the technology [26]. The US Assistant Secretary of Defense for Research and Engineering had already identified metamaterials as one of the disruptive basic research areas in 2012 [27].

The emergence of AM processes goes way beyond conventional concepts of design, and its main advantage is the great freedom it gives designers. An approach combining TO and additive production procedures promises optimal solutions [28].

Once all the parameters have been calibrated, the cost of AM processes will depend directly on the mass of the material to be fused rather than on the final part complexity, contrary to what occurs in subtractive manufacturing [28].

TO methods [29] are, in theory, an ideal numerical tool for creating optimum shapes automatically with no need to consider the manufacturing process. They specify the functionality criteria that need to be minimised in a given design space and are potentially much better at optimising size or shape than classical methods [28]. Changes can now not only be made in the geometry of the structure, but particularly in its topology.

In recent years, TO modules have been integrated into industrial design and calculation software, and the main commercial software packages already include them.

Potentially, AM makes it possible to embed into the printed part such things as cooling channels, any sort of join, small metal parts or memory alloys, sensors, batteries or conductive track. This means that the completed object can perform better and have improved functionality [20].

Therefore, AM is one of the potential manufacturing paradigm changers that could disrupt the complete design and manufacturing value chain, shifting the emphasis from mass production to individually tailored products, and from centralised to distributed production [19].

AM's most exciting disruptive perspective is the convergence of design and manufacturing. Traditionally, those functions have been largely separate. 3D printing can bring both disciplines together at all points of product development. So, when a product is 3D printed, designers do not have to consider production separately because they start their prototyping work on the same platform that will be used to produce the item [30].

7.2 Applicability and adaptability

Metamaterials, originally theorised and developed for electromagnetic applications, have now spread to optical, thermal, acoustic and mechanical applications [15] where internal structures can be complex and difficult to design and fabricate.

The main reason AM has been used in metamaterial fabrication has not always been to add new properties. Instead, it has operated as an enabling technology for the applicability and adaptability of metamaterials. Metamaterial properties can be obtained, in theory, without the need for AM. Many, however, would never be able to prove their theoretically desired properties practically without AM [15].

In order to achieve the unprecedented behaviours seen in metamaterials, their internal structures must be manipulated at the scale where the natural laws governing their properties operate. As a consequence, depending on the application, using the internal structure of metamaterials means dealing with complex topology structures that can be anywhere from millimetres to nanometres in scale. Complex topology structures, including porous structures (holes or voids), lattice structures, cellular structures, etc., and combined different materials, enhance the metamaterial's properties. The specific properties of metamaterials have to be manipulated at the unit cell level, typically at the scale of wavelengths as when dealing with electromagnetism. In that case, when the wave enters the metamaterial, it interacts with the cell structure and can be managed in ways that can often seem counter-intuitive [22]. The applicability and adaptability of these 3D printing technology capabilities represent new opportunities and new challenges to the design methodologies of complex topology structures.

Classical design for manufacturing and assembly, on the other hand, sees designers typically customising their designs to overcome manufacturing difficulties and reduce manufacturing, assembly and logistics costs to a minimum. DfAM offers a set of wider flexible concepts [31] as follows:

- AM enables the use of complex geometries and provides a freedom inherent to the design.
- Integration of components to consolidate parts, avoiding assembly issues.
- Customisation and functionalisation of consolidated products.
- Combination of materials within a unique geometry.

A new concept of product design is emerging thanks to the possibility of designing and additively manufacturing items with porous or lattice structures, inner details, intricate surface topographies, cantilevers or other extremely complex (or previously impossible) features [1]. This helps to incorporate new optimised functionalities able to adjust the mechanical, electromagnetic, thermal and fluidic performances, to drive physical aspects of the environment and to personalise products. There is also clear potential for promoting a gradual shift from mass-production to mass-customisation and revealing new horizons for applicability and adaptability in fundamental fields including [1]: the transport and aerospace industries, energy and environmental engineering, automation and production technologies, biomedical engineering and health, architecture and art or conventional product development.

7.2.1 Rapid prototype

Until 2014, the main use of AM was in setting-up prototypes. However, AM will be much more widely used in the coming years, spurred on and supported by several factors: its relevance for lightweight parts, availability of results at the point of use, instantaneous addressing of customisation requirements, reduced assembly cost, significantly reduced time to market, additional functionalities, reduced utilisation of source material and its positive impact on the supply chain [19].

All the 3D printing technologies mentioned above have revolutionised many scientific fields thanks to their inherent ability to make rapid prototype designs feasible [32].

The fast construction of prototypes allows developers to readily improve and change their designs based on 3D prints. In addition, the AM cost for rapid prototyping is about 50 times less than classical prototype machining.

For many service bureaus, prototyping, particularly based on polymer 3D printing, represents their key business area. Furthermore, there are also exciting opportunities in other emerging areas of industry that are in the early stages of adopting 3D printing, such as metal 3D printing [33].

Among the advantages offered by 3D printing for rapid prototyping developments [34–36] are

- fabrication of novel, complex structures based on lattices, internal geometries, etc.;
- direct manufacturing of 3D CAD models;
- easy customisation of products;
- economical solutions for small batches;
- material saving in terms of recyclability, waste material;
- light-weighted parts and optimisation for downsizing;
- functionalisation of products;
- integration of components in design;
- energy efficient production systems;
- improved durability of products;
- simplified supply chain;
- inventory reduction and part consolidation;
- lower transportation related expenses;
- sustainability and environmental impacts.

A Rapid Prototyping and 3D Printing Service Bureau Directory is available online (https://www.rapidtoday.com/index.html).

7.2.2 Industrial design challenges

Metamaterials have become increasingly easier to fabricate with the development of AM technologies (see an overview of technologies in Chapter 3). There are still limitations that are mostly related to the precision of the build, which is generally at the sub-millimetre range of what is desired in metamaterial fabrication. A number of high-precision AM methods are thus being proposed, even to achieve submicron and nanometre-scale resolutions. The most prominent of these methods is projection micro-SLA, which is based on SLA, a relatively simple and well-understood process, able to achieve micron-scale precision [15]. The high-precision AM methods developed so far have primarily been intended for non-metallic materials but adapted processes now also print metallic and ceramic materials. These metamaterials have the ability to offer structural stiffness in addition to the other more challenging properties. Nevertheless, this area needs further development, especially the development of high-precision metallic AM methods [15].

The first metamaterials were experimentally shown in 2001. Since then, enhanced 2D and 3D fabrication techniques for metamaterials have heightened interest. One of the largest, current challenges is to build parts at a macroscale while still retaining sufficient resolution of the unit cell when it is scaled from 2D to 3D [15].

Progress in 3D print technology now allows the printing of structured acoustic absorbent objects at microscopic scale. The pattern of repeatability of the basic cell unit of these metamaterials is a gateway to developing viable macromodels to simulate constructed structures based on detailed models of the individual cell units. However, verification of such models in actual manufactured structures still represents a challenge [37].

AM technologies are unquestionably beneficial; however, many limitations and challenges still exist such as a lack of standardisation for materials and processes to facilitate repeatability, intellectual property protection and recognition, high capital and material cost, exposure to potential hazards for polymeric AM systems (laser and materials in the case of desktop printers) [38], increased recyclability, segmented services, etc. A major limitation is the lack of training and education that focuses on using AM in design. Such specific training would boost the benefits of applicability and adaptability of AM technology. Furthermore, there is a need for novel design tools that better represent AM capabilities and material integrations. Traditional CAD software resources should be reconceived. Designers have to explore new areas of design space based on spontaneous ideas, freedom from conventionalisms imposed by weight and by the constraints of traditional training.

DfAM requires different conceptual rules for the design process to those usually considered [39]. The new process will design and optimise a product together with its production system to increase efficiency, quality and performance and reduce lead time and cost [1]. This way of designing implies that designers and engineers must overcome the barriers imposed on them by the conventional education they have received: they have more freedom to be creative.

However, incorporating a new design approach is not easy for conventionally trained minds given the time a designer has to spend on the design process.

In fact, during the design of a new product, designers and engineers traditionally invest 5 per cent of their time choosing the appropriate software tool, 20 per cent thinking of a solution and 75 per cent modelling the 3D CAD object [40]. Specific DfAM, on the other hand, requires that the maximum amount of human resources be invested in the 'thinking' step, supported by optimal design strategies.

Current design strategies for the AM process can be classed as [1]

- manufacturing-driven design strategy;
- functionality-driven design strategy;
- simulation-driven design strategy.

A summary of challenges for 3D printing (design) [35,41,42] could be listed as follows:

- changing the designer's conception of product design;
- expanding the utilisation of AM for direct component and product manufacture;
- production of larger parts and profitable process for mass production;
- inability to recycle support structures;
- intellectual property issues;
- continuous entry of competitors;
- education of traditional manufactures to understand the potential uses of AM;
- lack of standardisation and certification of materials and processes;
- reliability and repeatability issues;
- lack of services more advanced than lower-end consumer 3D printers;
- limited integration to other manufacturing techniques;
- potential hazards for materials;
- high investment rate for the metal AM systems.

Standardisation and software resources are two explicit issues. They have the clear potential to contribute to the consolidation of the AM sector in the sphere of industrial design. For that reason, they are commented on individually in the following subsections.

7.2.2.1 Standardisation in the AM sector

Currently, only a few materials can be processed within the required quality specifications, and standardisation is still necessary for those having the potential to do so [43]. Challenges and priority ranking for the aspects involving AM standardisation are summarised in [44].

7.2.2.2 Software resources

Software resources can be classified into three groups, depending on the stage of the manufacturing process [1]: design, those for the creation of the 3D model; simulation, the software for the optimisation and validation of the design; slicing, involving the requested tools to prepare the printing job.

7.2.3 Limitations in integration AM with metamaterials

Many limitations still need to be overcome before AM techniques see widespread application in industry. These include production efficiency, high costs, inconsistency and material inflexibility. The challenges, and therefore future opportunities, for fabricating metamaterials using AM in high precision, lie in the possibility of developing multi-structures and combining multi-materials, which would be easier by TO. Other limitations include support structure, anisotropy and heterogeneity, and residual stresses. Therefore, the main limitations are

- high-precision AM;
- multi-structural and multi-material metamaterials;
- TO of metamaterials combining with AM.

Additional details about the listed limitations are in [45].

7.2.4 Topology optimisation and additive manufacturing

AM builds parts by joining material layer by layer. The current proliferation of AM technologies opens up a wide variety of design possibilities that can be linked to optimisation tools to manufacture complex structures.

Just like many design techniques, optimisation in AM is considered to be a process of parameter optimisation [46]. Some of the AM aspects are related to design optimisation, and more particularly to TO, especially for topologically optimised structures.

The combination of TO and 3D printing leads not only to lighter weights, but also enhanced structural performance, while reducing the time from design to manufacturing. Moreover, there are designs with geometric complexity restrictions, which might not be addressed by traditional manufacturing techniques supported by TO. However, AM could greatly reduce those restrictions.

An advanced structural design method, TO obtains the optimal structure configuration by distributing material reasonably while meeting the defined load conditions, performance and constraints [14]. Compared to sizing and shape optimisation, TO does not depend on the initial configuration, and so there is a broader design space. Examples of applications where TO has been applied to multifunctional structure design techniques can be found in the aerospace and automotive industries, and architecture.

TO is concerned with such things as design simplicity, material choice efficiency, improved multifunctionality at multiscale and integrated technical solutions [41]. This has a direct consequence in the AM process, which is based on point-by-point material control, from which emerge the concepts of multiphase materials and functionally graded materials [36,47]. This equips AM with the capability to create freeform designs from TO [46].

Several important industrial considerations should be taken into account regarding TO as a support for robust AM solutions. These are related to the following aspects:

- geometry of the CAD model;
- path generation;
- technology selection;
- technology resolution;
- feed material properties;
- support materials.

More information about these points can be found in [46].

7.3 Research progress and industry application

It is likely that further innovative solutions will be developed to fill the gaps in research and the industries that can benefit will gain confidence and adopt metamaterials for critical applications [31]. In recent years, research relating to metamaterials has boomed, as has the development of AM technologies. Both technologies have experienced a parallel rise in popularity in the past 20 years in terms of published works, and there is a very wide range of research and applications involving them [15].

The evolution of 3D printing is driving increased adoption of AM. The early 1980s and 1990s saw 3D printing emerge and develop beyond SLA, the first widely used additive technology with the potential to produce fast prototypes and first iterations commercially [10].

During the first decade of the twenty-first century, commercial 3D printing reached a new maturity promoted by the idea of 'a printer in every home'. The idea of 3D printing grew beyond manufacturing activity to be adopted by 3D designers and the general public. During this time, very few AM processes were adopted for production. Now, 3D printing is going through a new stage, focusing on its technological capabilities and their place in industrial production and applications.

AM experienced exponential growth, which began as 2010 approached and can be attributed to projects like the RepRap Project. The goal of this project was to develop a 3D printer that could, eventually, become a fully self-replicating machine able to fabricate all its own parts. All of the software and hardware related to the RepRap Project was released under an open-source license, and so, anyone could use and modify the technology [48].

Thanks to this widely accessible free information and the availability of low-cost mechanical components and printed circuit boards, numerous derivations of the RepRap Project appeared, leading to the current ubiquity of 3D printing technology [49]. The cost of a 3D printer has dropped an order of magnitude over the past 10 years.

AM and metamaterials overlap. The fabrication of metamaterials is in many ways limited by AM technology, which makes new designs of internal metamaterial structures possible and advances the state-of-the-art [15].

It was not until recent years, however, that the potential of AM became broadly realised [10]. However, the real benefits of AM will not be fully realised until companies alter the way they approach product design. They must stop viewing AM as a replacement manufacturing method for traditionally designed components and start embracing new designs with the features and capabilities of 3D printing in mind to disrupt traditional value chains.

Professionals use 3D printing for two main purposes [30]:

- a prototyping solution to accelerate product development (rapid design iterations; low-cost, functional prototypes; widely accessible solution; effective design communications);
- a manufacturing technology for low-run production, and one-off custom parts (few design restrictions; on-demand production; mass customisation; distributed manufacturing).

Modern AM can turn a wide range of 3D complex geometry models into real-world products in a way that is efficient and economic, which has led to it being adopted in many industrial applications [50].

Industrial-scale mass production of metamaterials is still a major challenge [27]. When 3D printing can impact operations on a broader scale, that is, once such challenges as equipment and material costs have been overcome, more firms will see the competitive advantages of 3D printing [10].

The most pressing issue now is how to realise the application potentials of manufactured metamaterials. The throughput required for many metamaterials applications in commercial production is still beyond the capabilities of the techniques and equipment that have so far been developed. Furthermore, if high-quality metamaterials are to be applied practically, innovative methodologies in metrology and modelling are called for along with new standards [31]. In addition, industry appears to be slow or even reluctant to adopt metamaterials that have been developed, because little research has been done in the largely uncharted field of reliable metamaterial structures.

7.3.1 Manufacturing and prototyping

Progress in AM has been swift over the last 10 years, serving as a valuable tool for prototyping metamaterials [23]. The technology is increasingly being adopted outside the areas of prototyping and tooling to produce end and spare parts [43]. However, even though there is increased use of 3D printing for production, the benefits of AM have not been considered much beyond the prototyping stage or for small product runs [50]. Regrettably, moving on from the prototyping business model and into a model based purely on production is still a critical challenge [33]. Industry has to be educated in order for it to change its perception and demonstrate that there is a real added value in production with 3D printing at scale [51].

Current users of 3D printing worldwide are shown in Figure 7.1. It is predicted that a number of significant 3D printing supply chain developments on not only the design side but also the production side will be a reality within the next 5 years according to [52].

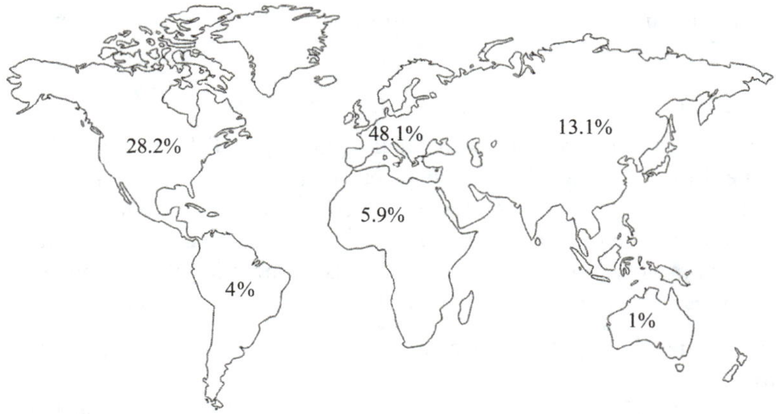

Figure 7.1 3D printing worldwide users [52]

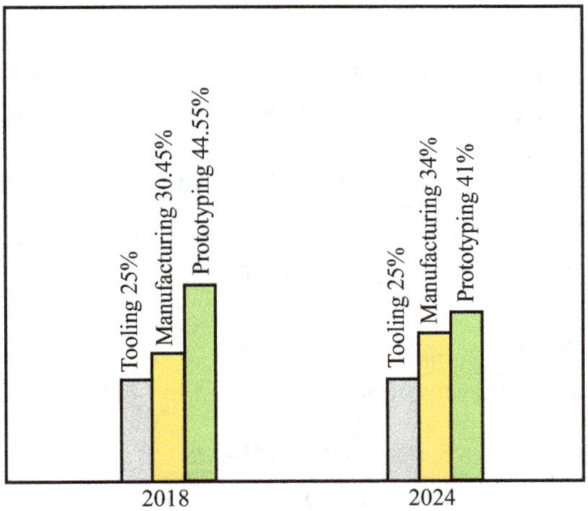

Figure 7.2 3D Printing market size by application, 2018 and 2024 [53]

Rapid prototyping continues to be the main application for AM; however, end part production has taken on increased relevance in recent years. As industry moves towards production applications, service bureaus will gain a more prominent role with applications providing the greatest opportunity for the services market [33]. It is estimated that rapid prototyping continues to generate over half of company revenue [52].

The prototyping application is expected to hold the largest 3D printing market share through the next 5 years as shown in Figure 7.2.

A large number of firms should soon offer CAD–CAM designs for downloading by final consumers and retailers. Individuals will be able to make tailored items at home, and firms will be able to order replacement parts whenever they require [52].

7.3.2 AM integration challenges to scaled production

AM has clear benefits, but the proof of the pudding will be whether the printed objects perform as expected when they reach the market. This unique ability to create superior shapes and structures must also create useful products.

AM first emerged commercially as a way of producing concept models, design or functional prototypes and visualisation tools in the mid-1980s. However, more recent progress in printer and material technology has led to the following [43]:

- AM to expand to applications such as factory tooling, spare parts and end-use products.
- AM is currently used primarily for low-volume parts production but is also slowly becoming a valued part of production processes more generally.
- Sixty-three per cent of enterprise AM users still deploy the technology for prototyping, while only 21 per cent use AM for items that cannot be manufactured any other way.

The last point clearly indicates that something is still preventing manufacturers from integrating AM technologies more generally into their manufacturing processes for scaled production. There are still challenges that must be addressed first if AM is to achieve widespread adoption. These include the following issues:

- technical challenges: materials and procedural manufacturing;
- digital challenges: 3D printing interfaces;
- design challenges;
- financial challenges.

More details about these points can be found in [43].

7.3.3 Industrial sectors and applications

Research into metamaterials involves a wide range of fields, including energy, biomedical, optics and communication. The concept originated in electromagnetic metamaterials [31] but was followed by a blossoming of various other types of metamaterials such as mechanical metamaterials [54], acoustic metamaterials [55] and thermal metamaterials [56]. However, with their exotic properties, metamaterials also have the potential to be applied in a boarder range of areas including medical, architectural, defence, entertainment, sports, automobile and aviation applications.

As pointed out in the previous sections, metamaterials go hand-in-hand with AM technologies, which have been enabled and adopted in the fabrication of metamaterials. But AM technologies also cover a wider range of application in industrial sectors like architecture, defence, electronics, education, research, fashion, jewellery or aerospace. However, the lion's share of AM activity is concentrated in the automotive, consumer goods and medical industries.

7.4 Market segments

AM processes must satisfy market demands and ensure that products live up to the same performance levels as those made traditionally. At the same time, they must comply with current regulations [19].

It is already assumed that there will be vast reductions in AM technology costs, and the sector is already providing affordable solutions for the personal use of individuals [1]. That has happened because low-cost AM has brought the technology into the public domain, and so the possibilities of AM technology have increased exponentially in recent years, providing a new portfolio of potential solutions and applications that can use them, including those related to metamaterials.

With ubiquitous applications across multiple market segments, 3D printing is changing new product development and aftermarket supply chains globally.

Both AM and metamaterials are identified as opportunities that can have a significant impact in different market segments with the potential to offer substantial benefits in terms of savings, efficiencies and new capabilities [22].

There is no doubt that metamaterial applications offer major technical and economic opportunities. Their market segment looks set to expand rapidly over the next decade and offer advances in novel products.

Global 3D printing market segmentation by applications can be classified as follows:

- prototyping;
- tooling;
- functional parts.

Global 3D printing market segmentation by the following components:

- hardware;
- materials;
- software: Support for the design, control, printing and scanning;
- services: AM systems by providers, manufacturers, maintenances, consulting, training, seminars, conferences, etc.

Global AM market segmentation by the following services:

- automotive;
- aerospace;
- machinery;
- consumers;
- medical;
- defence;
- architectural;
- power and energy;
- others.

Global metamaterials market segmentation by the following end-users:

- healthcare;
- telecommunication;
- aerospace;
- defence;
- electronics;
- optics;
- energy and power;
- other.

Despite this wide range of opportunities, metamaterials currently lack the wide-scale adoption needed to deliver value to society or the economy through practical applications [23].

7.4.1 An overview of current size and growth trends

An overview of the current size and growth trends of the 3D printing market reveals that the landscape has changed dramatically in the last decade (2010–2020). What began as a prototyping solution has burgeoned into a technology for manufacturing end-use parts for industrial applications. The year 2019 can be regarded as a milestone in this evolution [30]. Three times more professionals were using 3D printing in that year than in 2017 according to [30]. The market grew annually by 47 per cent from 2012 to 2017 [43]. Senior analysts agree that, as adoption steadily increases, the total 3D printing market will continue to approximately double in size between 2020 and 2023. Important patents for AM will expire, and this will no doubt stimulate the expansion of AM with cheaper consumer-oriented printers.

Manufacturing services have always been a mainstay of the AM industry from the outset [33]. In many ways, the fact that enterprises can outsource 3D printing production has helped spread AM technology into different industries and enlarge the range of applications.

In 2019, the market for service bureaus offering 3D printing was more competitive than ever. Independent bureaus grew in number, particularly helped by reduced cost of machines. Additionally, some hardware manufacturers now provide services as part of their business models [33].

The AM market over the past 13 years for segments by services is shown in Figure 7.3.

The estimated global market size for AM technologies for the period 2014–2027 for market segments by components, in a professional environment, is shown in Figure 7.4. The per cent of 3D printing investments by components for the period 2021–2025 as follows:

- hardware 25 per cent (3D printing manufacturers);
- materials 9 per cent (materials producers);
- software 13 per cent;
- services 30 per cent applied 3D printing, 19 per cent manufacturing platforms;
- others 4 per cent.

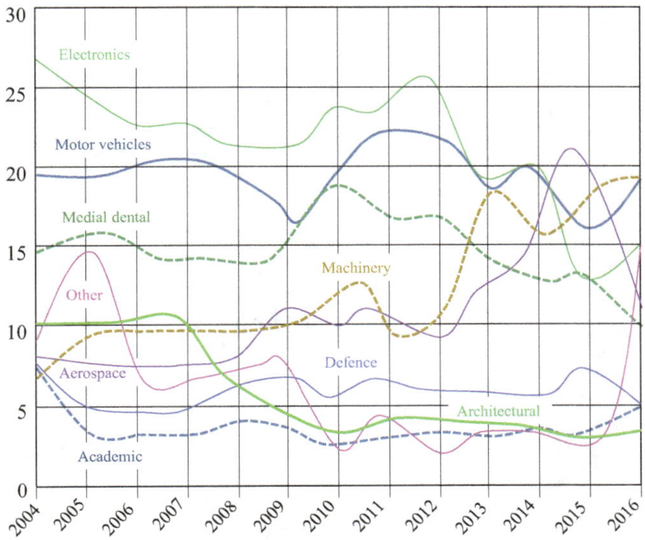

Figure 7.3 AM services industries during the past 13 years in per cent of sales (interpreted from [57])

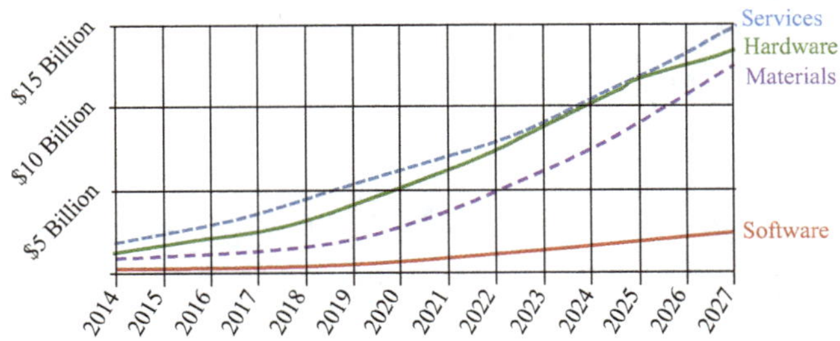

Figure 7.4 Estimated global market size for AM technologies for the period 2014–27 by segment (interpreted from [58])

The estimated AM market size by technologies for the period 2018–2024 is shown in Figure 7.5.

The global market revenue of AM technologies estimated by region for the period 2015–2025 is shown in Figure 7.6. According to this estimation, the AM industry will grow by 15 per cent, with China having more than 70 per cent of the business in the Asia-Pacific region.

Finally, the global metamaterials market size by year for the period 2017–2025 is shown in Figure 7.7, which estimates an average growth of 36 per cent in that period.

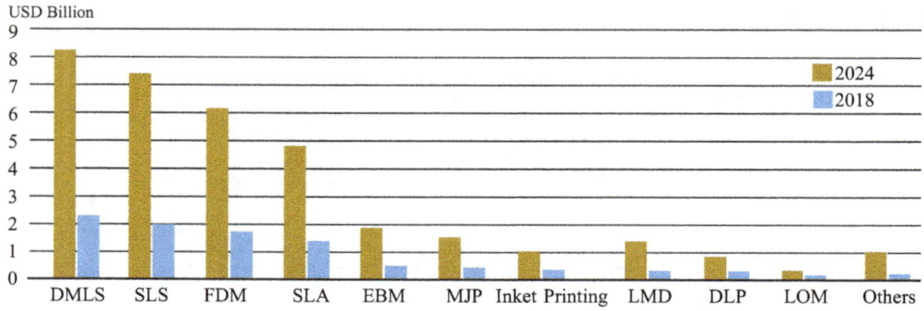

Figure 7.5 *AM market size by technologies for the period 2018–2024 (analysed from [53])*

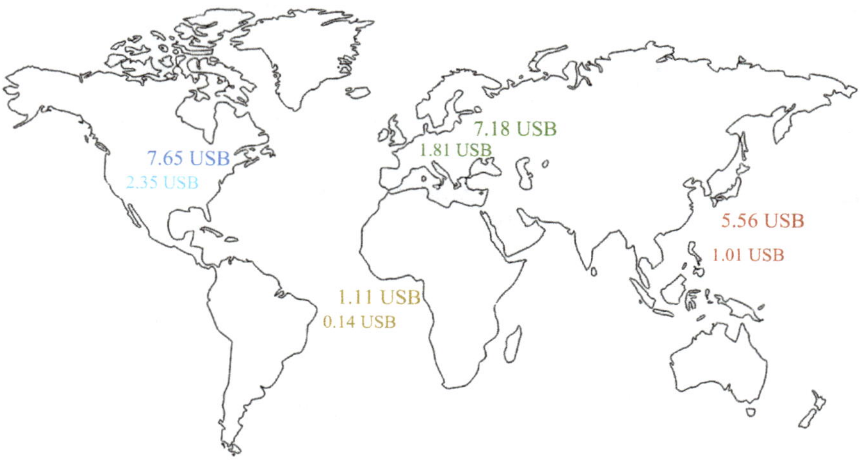

Figure 7.6 *AM market revenue-estimation by regions for the period 2015–2025 (extracted from [59])*

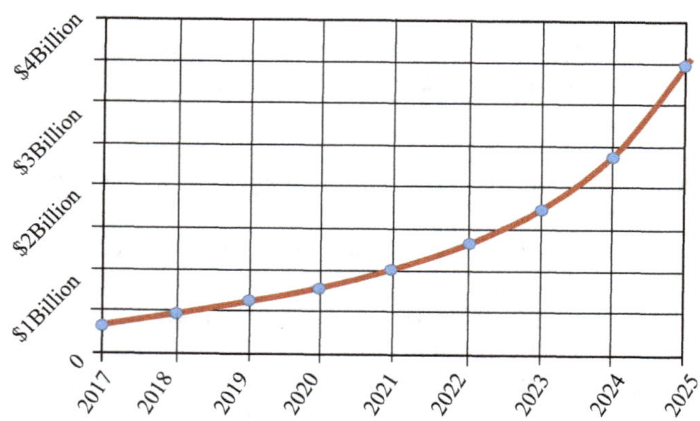

Figure 7.7 *Metamaterials market size by year for the period 2017–2025 (interpreted from [23])*

7.4.2 Online 3D printing and services providers

3D printing services are companies that will print out objects on demand. One of the most important benefits of such a service is that an individual can have access to very expensive, industrial-grade 3D printers.

When an online platform automates procurement and sourcing in a manufacturing industry, it is called 'online manufacturing' or 'manufacturing-as-a-service' [60].

An online 3D printing platform is simple for users, who only have to upload a design to find out immediately how much it will cost to produce. Production is then just a click away [30]. The steps are as follows:

1. upload parts and explore capabilities;
2. specify technology, material and tolerance requirements. The price is updated in real time;
3. select lead times that fit the project planning;
4. go straight into production.

North America and Europe together accounted for more than 95 per cent of the online 3D printing global demand in 2019 [30].

Figure 7.8 shows the distribution of online 3D printing demand by industry based on a representative sample of professional users. The manufacturing of industrial, electrical or consumer goods represent over 65 per cent. The aerospace, automotive and medical industries prefer to produce parts in-house or outsource by means of conventional supply chains.

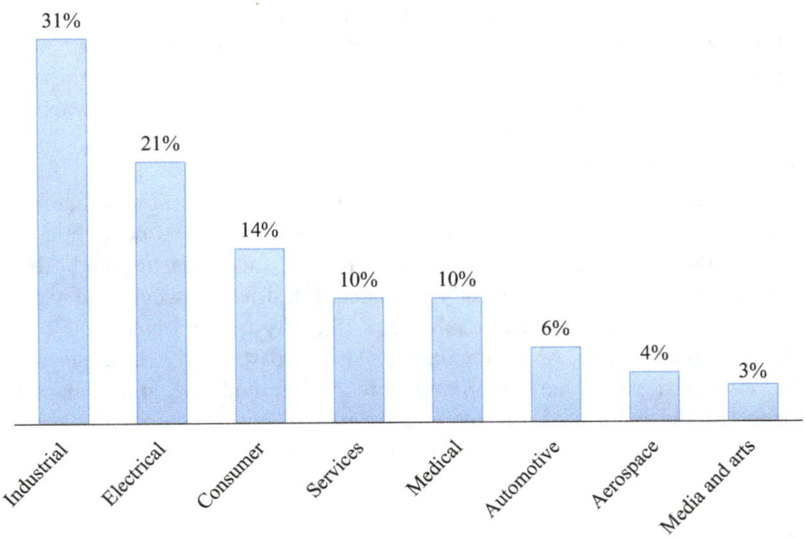

Figure 7.8 Distribution of online 3D printing demand by industry (interpreted from [30])

Distributed manufacturing has always appealed because of its low transport costs, while centralised manufacturing has predominated because of economies of scale. This is the reason why online manufacturing platforms based on a manufacturing-as-a-service model are also changing the services market by offering companies on-demand access to a network of suppliers all over the world [33].

There are three major benefits for engineering and manufacturing companies when they use online manufacturing platforms:

- Online manufacturing automates and expedites the order placement process.
- Online manufacturing provides price transparency to a previously closed marketplace.
- Online manufacturing increases the usage of installed manufacturing systems, reducing their overhead costs and allowing them to innovate faster.

More details about these points can be found in [30].

Online manufacturing platforms are helping engineering companies worldwide to bring their products to market faster in practice. Forty per cent of online 3D-printed parts in 2019 were for serial production [30].

7.5 Promising directions and examples

AM as a 3D printing technology has developed rapidly since 1990. It shows great promise, but it is still unclear what effect it will have on traditional industry [61].

AM optimises structures and uses material only where needed to make products lighter. It gives functions to structures, such as a complex integrated cooling channel [19]. It allows simplification of structures, reducing several parts into one, thus shorting assembly time. It represents a new paradigm, which must change the concept of 'installation' and consider integrated designs. It is expected that the next industrial revolution – an explosion of new materials – is ready to break out, stimulated by the customised and tailor-made fabrication AM offers.

Eventually, parts made using AM could meet or exceed existing functional requirements and also be much lighter. This would also be done in way that is energy efficient [62]. Current AM technologies already benefit critical economic sectors, from buildings to aerospace and defence part repairs, mould casting and other consumer and industrial products in tightly regulated industries. However, the inability to validate and verify the structural quality and integrity of the objects built by AM restricts the wider use of AM components. This is particularly the case in tightly regulated industries like aerospace and defence, biomedicine and transportation, where one of the biggest challenges is certification. High material costs also will mean a need for improvements in process controls, surface finishes and throughput, so that AM techniques and products can be adopted on a wider scale [62].

There will probably be a combination of different fabrication techniques used in industry. Additive, equivalent and subtractive processes will be used together in one piece of equipment or combined in the production chain to form an integrated process capable of making complex products [61].

The following section lists prospective and current success stories in industry which could identify possible development directions for 3D printing in the near future. There is also a subsection dealing with forecasts for how AM could improve performance and manufacturability of future electrical machines.

As demonstrated in previous sections, metamaterials have also opened up the possibility of unprecedented and fascinating concepts and applications [63]. In general, the realisation of 3D metamaterials is a major challenge in the research and development fields.

7.5.1 Successful examples in industry

Famous success stories at large industry leaders have led to increased demand for AM systems, and considerable amounts have been invested in AM facilities at R&D centres worldwide, where they are looking for the key application that, when adopted, will break open the market and increase margins.

The aerospace industry is a pioneer and in the vanguard of the potential industrial application of AM. Aircraft manufacturers and their suppliers require geometrically complex manufacturing to create reliable high-tech aircraft parts. AM can drastically cut their production costs and weight. Moreover, AM significantly contributes in defence, which is a particular sector where many applications also often require miniaturised, custom-designed units in relatively small numbers.

AM is also widespread in the medical sector. The healthcare industry is investing in tailored prosthetics, dental implants, hearing aids and other types of medical devices and tools [64]. Advanced solutions provide physical models based on patient scans [1]. Examples of consumer good in this area are orthopaedics, prosthetics, implants, tissue engineering solutions and drug-screening models.

The automotive industry is where AM shows great promise and has the potential to accelerate innovation processes. Fewer parts will be needed in vehicle bodies and engines, and rapid redesigning will keep failures at a minimum [64].

Attention should also be paid to 3D printing for the energy industry. More specifically, the impact of AM on the renewable energy sector is likely to be of great interest. Green energies are currently one of the biggest challenges. Solar panels, for example, are still quite expensive and could be improved. 3D printing could be a good method for creating them, and it is believed that this could reduce production costs of the panels by 50 per cent. Furthermore, they could even be more efficient than traditional solar panels.

7.5.1.1 3D-printed fuel nozzles

One of the most illustrative cases is from General Electric Aviation, which started using metal AM technologies to produce the 19 fuel nozzles [65] for its LEAP engines as that shown in Figure 7.9. GE has been able to realise a relative weight reduction of 25 per cent. The single-piece new nozzle replaces 20 conventional parts in one part that is five times more durable and 30 per cent more cost efficient. Traditional methods could not have made it. A 3D printer functions like a laser pen,

Figure 7.9 The LEAP's fuel nozzles

following a computer drawing and fusing layer upon layer of fine metal powder into the final shape. 3D printers can build complex, dense parts like the fuel nozzle while generating a fraction of the waste produced by conventional manufacturing. The structure of the LEAP's fuel nozzles is so intricate that 3D is the only practical way to mass-produce them. A fuel nozzle requires instrumenting over 3,000 layers of powdered metal that are about the thickness of a human hair. Eight thousand such fuel nozzles were produced in 2017.

Beyond the LEAP engine, GE Aviation uses AM to make sensors, blades, heat exchangers and other parts for engines. GE Engineers have replaced 855 components with just a dozen using 3D printing.

More details can be found at: https://www.ge.com/news/reports/transformation-3d-walnut-sized-part-changed-way-ge-aviation-builds-jet-engines.

7.5.1.2 3D-printed solar panels

The global demand for solar installations has been increasing over recent years. In the United States, solar installations are expected to quadruple by 2030. This is an opportunity for metamaterial-based solar solutions. Metamaterials can absorb a huge amount of light with exceptionally high efficiency, thereby increasing the amount solar cells can use. For instance, Metamaterial Technologies, Inc. has developed the Lamda Solar technology, which helps to increase the solar cell efficiency by up to 100 per cent by collecting light from all angles and absorbing it.

Figure 7.10 3D-printed thin-film solar cells

Figure 7.10 illustrates the new printable solar cells shown by the Victorian Organic Solar Cell Consortium [66]. These cells are flexible, light weight and are so thin that they can cover most surfaces.

More details can be found at: https://www.sculpteo.com/blog/2018/01/24/3d-printed-solar-panels-meet-the-renewable-energy-revolution.

7.5.1.3 3D-printed windings for electric cars

The Additive Drives engineering team [67] has been able to realise greater efficiency for electric motors through 're-thinking' in design and engineering and applying new 3D-printed windings and coils.

While conventional manufacturing yields typical round wire winding, when compared to 3D-printed coils, the Additive Drives team found that the new design reduces the source of heat and increases dissipation.

With the creation of novel procedures based on AM, they are able to prevent voids and issues with any improper fit for grooves. The amount of copper is increased when filling in grooves, resulting in less electrical resistance. The higher copper content in the electric motor reduces losses and improves the thermal coupling of the winding.

Figure 7.11 shows the printable winding reconsidered by Additive Drives.

More details can be found at: https://3dprint.com/270322/additive-drives-3d-printed-windings-electric-cars.

Figure 7.11 Winding with assembled 3D-printed coils

7.5.1.4 3D-printed electric drive housing

Porsche has unveiled its first-ever electric drive housing created entirely using AM as that shown in Figure 7.12. The 3D-printed drive housing is 10 per cent lighter than a similar part created using traditional casting techniques. It benefits from unique lattice structuring in high-stress areas that only became possible due to 3D printing, so this new component is also 100 per cent stiffer than before.

This housing can incorporate the two-speed gearbox and its heat exchanger as well as the electric motor, which means that it is also much easier to cool the whole drive unit. This is a basic requirement for increasing performance, integrating as many functions and parts as possible, saving weight and optimising the structure.

Utilising the properties of AM, Porsche could integrate the bearings, heat exchangers and oil supply in the drive unit. Additionally, the electric motor and two-speed gearbox rest inside the same housing.

The drive housing is 3D printed, layer by layer, using premium aluminium alloy powder. Each layer is melted and fused via a laser metal fusion process using high purity metal powder. It took engineers several days and two build processes to complete its first drive unit prototype. The firm estimates, however, that future machinery could reduce the build time by 90 per cent.

More details can be found at: https://www.topgear.com.ph/news/technology-news/porsche-ev-parts-3d-printing-tguk-a3832-20210110.

7.5.1.5 Largest 3D-printed rocket engine

Orbex has manufactured the world's largest 3D-printed rocket engine. It was printed in a single piece and will be part of Orbex's Prime rocket. The rocket was

Figure 7.12 3D-printed electric drive housing

3D printed using metal AM, which reduces its turnaround time by 90 per cent and lowers costs by 50 per cent compared to more traditional production processes.

The issues solved in this 3D-printed rocket engine concern the peculiar pattern of the cooling system. This consists of a complex network of tubing that takes fuel from the tanks and routes it though pumps, the engine bell and then to the engine injector, where it is vaporised in another series of complex fluid paths for ignition. With 3D printing, all the tiny embedded channels are built together at once.

Cellcore and SLM Solutions are also involved in rocket propulsion engine AM. Their experience shows that due to rocket propulsion engine complexity, the traditional manufacturing process is cost-intensive and requires at least half a year. A structural cooling of a 3D-printed rocket propulsion engine conceived by Cellcore and SLM Solution is shown in Figure 7.13. In the 3D-printed engine, they demonstrated the possibilities SLM technology can offer for the aerospace industry, as the AM process takes fewer than 5 days to build, while creating a functionally optimised component.

Additional details can be found at: https://orbex.space/vehicle, https://www.etmm-online.com/3d-printing-a-rocket-engine-a-886960.

7.5.1.6 Largest 3D-printed internal combustion engine

In the LeiMot research project, a consortium from science and industry has produced an internal combustion engine using a laser melting process, saving almost

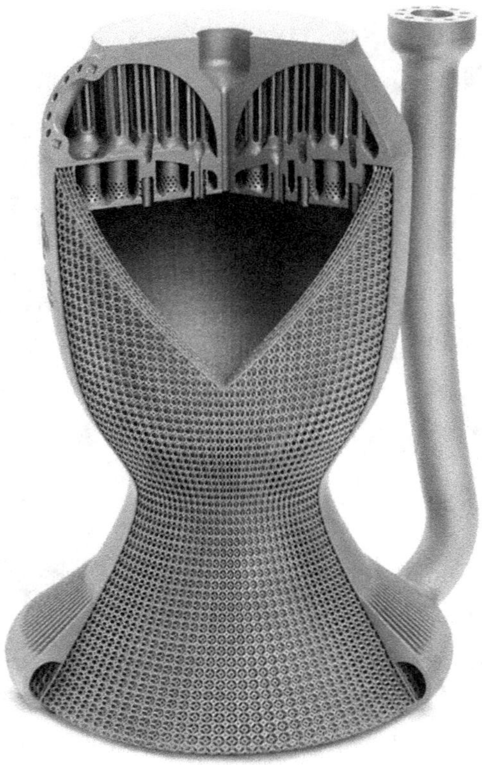

Figure 7.13 Structural cooling of a 3D-printed rocket propulsion engine conceived by Cellcore and SLM solutions

21 per cent in weight. In addition, the efficiency of the cooling system and the oil circuit was increased. The focus was on the cylinder head and crankcase of a modern diesel engine. Instead of being produced by aluminium casting, as was previously the case, the two components were manufactured using the AM process shown in Figure 7.14. The aluminium alloy was used as the powdered starting material.

The new cross-flow cooling system makes it possible to lower cylinder temperatures in a targeted manner and at the same time reduce the amount of water required. One main design difference is the use of individual cooling lines in the cylinder head to replace the large-volume water jacket. This reduces temperatures in the combustion chamber by up to 40 per cent [68].

Additional details can be found at: https://www.boellinger-group.com/en/latest-news/detailed-view/news/leichtbau-motor-1.

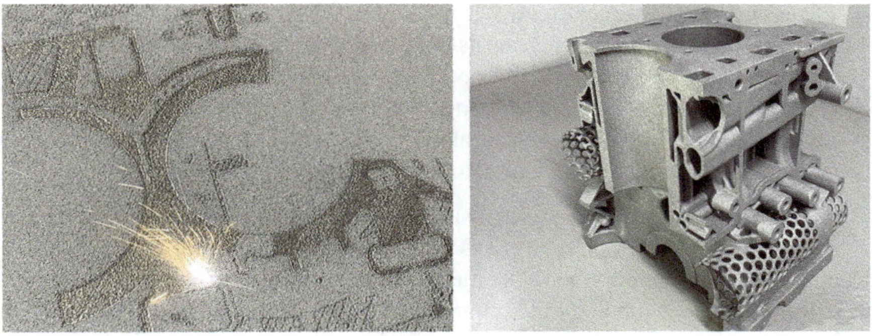

Figure 7.14 LeiMot project. Weight reduction and functional improvement in the combustion engine through additive manufacturing and the use of hybrid metal and plastic components. (Left) One-cylinder crankcase during the 3D printing process. (Right) Additively manufactured test one-cylinder made of aluminium

7.5.2 How AM could innovate future electrical machines

AM ticks many of the boxes for modern 'design for application' of electrical machines. In particular, tailored-made machines with specific functionalities.

AM has potential in the construction of all integral components of an electrical machine assembly. These include active parts like coils/windings, electrical insulation, stator/rotor magnetic core packs, permanent magnets, motor housing/structural parts and elements for machine thermal, magnetic, electric and mechanical strength management [69]. Additively manufactured electric machines not only enable complex geometry and high utilisation of materials, but there is also change in the microstructure and material properties [70]. Theoretically, material properties of additively manufactured electrical machine components can be purposely controlled by changing as built orientation and postprocessing conditions [71].

Management of current conductor distribution can be addressed in electrical machines by using an additively manufactured customised conductor fill factor and leakage flux control, which also improves heat dissipation.

There have been many attempts to additively manufacture structural parts of rotating electrical machines, including stator housings, end covers, rotor shafts, ball bearings, winding frames [69,70]. However, such work is scarce and mainly focused on the individual machine components and materials.

Areas where AM promises to be successfully used are:

- Reducing the mass of structural parts in the motor assembly. A lightweight rotor/shaft with a lattice structure was built, reducing the overall rotor mass by 25 per cent and decreasing the moment of inertia by 23 per cent compared with a more conventional design [69].

- Increasing heat dissipation. AM can be used to build cooling structures with porosity, complex passages/channels/ducts and extended surfaces [70]. It can also reduce weight and lead time by optimising geometries and minimising tooling constraints [70]. Components of thermal management systems include a variety of heat exchangers, cooling jackets, heat pipes, etc. [70]. There are reports of up to 20 per cent improvement in total heat conduction and cooling system efficiency [69].

AM clearly offers a virtually unrestricted way of designing and manufacturing new components. Nevertheless, there is still a need to understand how this can be successfully implemented when it comes to in-volume manufacturing of electrical machines [69]. The technology of soft and hard magnetic materials and electric conductive materials is still relatively immature. These show poorer physical properties compared with the existing solutions [69]. For example, tests show that the 3D-printed rotor is mechanically vulnerable under high-speed operation as the complex microstructure between layers might lead to crack initiation and propagation [70].

In the particular case of electrical transformers [72], the first experiences applying 3D printing technology were focused on printing components with non-complex geometry and no critical requirements, such as toroidal transformers [73] and high-voltage pulse transformers [74].

The appeal of AM in power transformers, which are key and costly components in networks [72], is that it can achieve full fill factors, high power densities and also reduce the time and cost of manufacturing, transportation and assembly [75]. However, power transformers are complex machines, where the integration of all components demands in-depth knowledge of how to harmonise all coupled operation restrictions (electrical, magnetic, thermal and mechanical). The mechanical constraints placed on power transformer components are a limiting factor and therefore the number one priority when selecting their materials. For example, they must ensure the mechanical integrity of the machine under short-circuit operation.

Nevertheless, first steps when applying 3D printing to power transformers could be addressed by innovating and optimising the cooling system, in the way already implemented in aerospace jet engines, adopting cooling microchannels, ducts, etc., integrated within insulating parts, for example. Such a level of integration required a new concept of graded sticks, spacers, angel rings, angel washers, block washers, shield rings, etc., with fashioned shapes to point to point electric and magnetic field control. This philosophy could also be extended to electrical conductors. These and other innovative solutions might become reality once the appropriate materials for making them have reached maturity.

It can be concluded that more comprehensive optimal design approaches based on the availability of appropriate materials will be the natural step forward in the development of AM in the next generation of electrical machines [69].

7.6 Conclusions

AM has an important role to play in a range of industrial design and manufacturing methods. As the technology continues to improve, AM will shift from being a disruptive technology used only by innovators to become a common method for core production. The aerospace and automotive industries are in the vanguard of promising application fields for AM, while metamaterials are expected to have an impact on energy harvesting (both solar electric and solar thermal) high-power long-distance power transmissions and wireless charging. Several of these are early commercial-stage technologies which are undergoing in-the-field proof-of-concept prototyping. Furthermore, AM is also a likely to be used in the production of innovative electrical machines.

AM holds a vast potential for promoting science, technology, engineering, education and industrial applications through its inherent impact on the very design of complex and optimised parts, simplifying assembling and processing.

Both AM and metamaterials encompass a very broad range of research and applications. Massive adoption of AM in the design and fabrication of metamaterials is still in its early stage. Numerous challenges remain to be addressed, requiring innovations and disruptive ideas. The pressing issues include, but are not limited to, the improvement in precision for current AM techniques, the integration of multi-structure and multi-material AM with metamaterials, the combination of AM with TO for the design of metamaterial structures, the AM support structure, the anisotropy and heterogeneity and residual stresses of AM products.

The close integration of TO and AM is and will continue to be a hot and attractive issue. Further research needs to be carried out to design and optimise structures for application that are industrially practical.

If AM is to be adopted for the fabrication of metamaterials, both at an industrial production scale and at the scale of users, it will need to adopt suitable design and simulation tools. From the earliest design stages, these tools will provide designers and engineers with the capability to deal with 3D models, multiple structure patterns and all the specific features that are inherent to fabricating and modelling a manufacturing process.

References

[1] De Blas Romero A. *Design strategies for additive manufacturing using vat photopolymerization systems*. PhD thesis [online]. 2018. Available: http://oa.upm.es/51156/. [Accessed 9 February 2021].

[2] Hull. C. W. *Apparatus for production of three-dimensional objects by stereolithography*, US Patent 4,575,330 [online]. 1986. Available: https://www.google.com/patents/US4575330. [Accessed 9 February 2021].

[3] Materialise. *Keeping intrion robots up to speed* [online], 2015. Available: http://www.materialise.com/en/cases/keeping-intrion-robots-up-to-speed. [Accessed 9 February 2021].

[4] Setaki F., Tenpierik M., Turrin M., and Van Timmeren A. 'Acoustic absorbers by additive manufacturing'. *Building and Environment*. 2014,

vol. 72C, pp. 188–200. Available: https://doi.org/10.1016/j.buildenv.2013.10.010.

[5] Housholder R. F. *U.S. molding process*, US Patent 4247508 [online]. 1979. Available: https://patentimages.storage.googleapis.com/56/71/3d/121c1360a03b69/US4247508.pdf. [Accessed 9 February 2021].

[6] Kruth J. 'Material incress manufacturing by rapid prototyping techniques'. *CIRP Annals*. 1991, vol. 40(2), pp. 603–14.

[7] Gao W., Zhang Y., Ramanujan D., *et al.* 'The status, challenges, and future of additive manufacturing in engineering'. *Computer-Aided Design*. 2015, vol. 69, pp. 65–9.

[8] Maddison A. *Automotive Applications*. John Wiley & Sons; 2005 https://doi.org/10.1002/0470014229.ch1.

[9] ASTM standard F2792. *Standard terminology for additive manufacturing technologies*; 2013 [Online]. Available: http://www.astm.org/Standards/F2792.htm. [Accessed 9 February 2021].

[10] Stratasys Direct Manufacturing. *3D Printing's impact on the value chain* [online]. 2016. Available: https://www.stratasysdirect.com/-/media/files/direct/white-papers/value-chain-white-paper_web.pdf. [Accessed 9 February 2021].

[11] Runze H., Riddle M., Graziano D., *et al.* 'Energy and emissions saving potential of additive manufacturing: The case of lightweight aircraft components'. *Journal of Cleaner Production*. 2016, vol. 135, pp. 1559–70.

[12] Askari M., Hutchins D. A., Thomas P. J., *et al.* 'Additive manufacturing of metamaterials: A review'. *Additive Manufacturing*. 2020, vol. 36, pp. 1–36.

[13] Garcia C. R., Correa J., Espalin D., *et al.* '3D printing of anisotropic metamaterials'. *Progress in Electromagnetics Research Letters*. 2012, vol. 34, pp. 75–82 https://doi.org/10.2528/PIERL12070311.

[14] Zhu J., Zhou H., Wang C., Zhou L., Yuan S., and Zhang W. 'A review of topology optimisation for additive manufacturing: Status and challenges'. *Chinese Journal of Aeronautics*. 2021, vol. 34(1), pp. 92–110.

[15] Saunders R. *Metamaterials using additive manufacturing*. Naval Research Laboratory. Washington DC [online]. 2020. Available: https://apps.dtic.mil/sti/pdfs/AD1103255.pdf. [Accessed 9 February 2021].

[16] Sparavigna A. C. *Metamaterials on demand. The use of 3D printing to create new metamaterials* [online]. 2017. Available: https://ssrn.com/abstract=3031417. [Accessed 9 February 2021].

[17] Koschny T., Soukoulis C. M., and Wegener M. 'Metamaterials in microwaves, optics, mechanics, thermodynamics, and transport'. *Journal of Optics*. 2017, vol. 19(8), pp. 1–16.

[18] Wu X., Su Y., and Shi J. 'Perspective of additive manufacturing for metamaterials development'. *Smart Materials and Structures*. 2019, vol. 28, pp. 1–24.

[19] Atos *Unlocking the potential of Additive Manufacturing* [online]. 2014. Available: https://atos.net/wp-content/uploads/2016/11/White-Paper-Additive-Manufacturing-web.pdf. [Accessed 9 February 2021].

[20] Thompson M. K., Moroni G., Vaneker T., *et al.* 'Design for Additive Manufacturing: Trends, opportunities, considerations, and constraints'. *CIRP Annals*. 2016, vol. 65(2), pp. 737–60.

[21] Gao W., Zhang Y, Ramanujan D., *et al.* 'The status, challenges, and future of additive manufacturing in engineering'. *CAD Computer Aided Design*. 2015, vol. 69, pp. 65–9.

[22] Burnett M., Ashton P., Hart A., *et al. Advanced materials and manufacturing – implications for defence to 2040*. Defence Science and Technology Group DST Edinburgh, Commonwealth of Australia [online]. 2018. Available: https://www.dst.defence.gov.au/sites/default/files/publications/documents/DST-Group-GD-1022_0.pdf. [Accessed 9 February 2021].

[23] Bishop-Moser J., Spadaccini C., and Andres C. *Metamaterials manufacturing. Pathway to industrial competitiveness*. Alliance for Manufacturing Foresight [online]. 2018. Available: https://medium.com/@MForesight/metamaterials-manufacturing-pathway-to-industrial-competitiveness-2f2791f9fdfb. [Accessed 9 February 2021].

[24] Capolino F. *Metamaterials Handbook. Vol. I: Theory and Phenomena of Metamaterials*. Boca Ratón, Florida: CRC Press Taylor & Francis Group; 2009.

[25] Kadic M., Buckmann T., Schittny R. and Wegener M. 'Metamaterials beyond electromagnetism'. *Reports on Progress Physics*. 2013, vol. 76(12), pp. 1–34.

[26] Dogana E. '3D Printing metamaterials towards tissue engineering'. *Applied Materials Today*. 2020, vol. 20, pp. 1–13.

[27] Liu R., Ji C., Zhao Z., and Zhou T. 'Metamaterials: Reshape and rethink'. *Engineering*. 2015, vol. 1(2), pp. 179–84.

[28] Saadlaoui Y., Milanb J. L., Rossib J. M., and Chabrand P. 'Topology optimisation and additive manufacturing: Comparison of conception methods using industrial codes'. 2017, vol. 43(1), pp. 178–86.

[29] Bendsøe M. P. and Sigmund O. *Topology Optimisation: Theory, Methods and Applications*. Berlin: Springer; 2003.

[30] Bukholt N., Juárez J., Lobovsky M., Bruner J., and Lüders C. *3D printing trends 2020. Industry highlights and market trends*. Hubs Manufacturing LLC [online]. 2020. Available: https://downloads.hubs.com/3D_printing_trends_report_2020.pdf. [Accessed 9 February 2021].

[31] Gibson I., Rosen D., and Stucker B. *Additive Manufacturing Technologies*. 2nd edn. New York, NY: Springer New York; 2015.

[32] Sadeqi A., Nejad H. R., Owyeung R. E. and Sonkusale S. '3D Printing of metamaterial embedded geometrical optics (MEGO)'. *Microsystems & Nanoengineering*. 2019, vol. 5(16), pp. 1–10.

[33] Dunham S., Warbrick J., Van Es N. and Perla C. *State of the 3D printing industry survey 2019*. AMFG, AM Service Providers [online]. 2019. Available: https://amfg.ai/wp-content/uploads/2019/07/AMFG-State-of-the-Industry-Report_-AM-Service-Providers.
pdf. [Accessed 9 February 2021].

[34] Huang S., Liu P., Mokasdar A., and Liang H. 'Additive manufacturing and its societal impact: A literature review'. *The International Journal of Advanced Manufacturing Technology*. 2013, vol. 67, pp. 1191–203.

[35] Ford S. and Despeisse M. 'Additive manufacturing and sustainability: An exploratory study of the advantages and challenges'. *Journal of Cleaner Production*. 2016, vol. 137, pp. 1573–87.

[36] Roper D. A., Good B., McCauley R., *et al.* 'Additive manufacturing of graded dielectrics'. *Smart Materials and Structures*. 2014, vol. 23(4), pp. 1–9.

[37] Kennedy J. 'The influence of additive manufacturing processes on the performance of a periodic acoustic metamaterial'. *International Journal of Polymer Science*. 2019, pp. 1–11.

[38] Bours J., Adzima B., Gladwin S., Cabral J., and Mau S. 'Addressing hazardous implications of additive manufacturing: Complementing life cycle assessment with a framework for evaluating direct human health and environmental impacts'. *Journal of Industrial Ecology*. 2017, vol. 21(1), pp. 25–36.

[39] Hague I. C. and Dickens P. 'Implications on design of rapid manufacturing'. *Journal of Mechanical Engineering Science*. 2003, vol. 217(1), pp. 25–30.

[40] Fictiv. *2019 State of Hardware Report* [online]. 2019. Available: https://www.fictiv.com/ebooks/2019-state-of-hardware-report. [Accessed 9 February 2021].

[41] Newman S. T., Zhu Z. C., Dhokia V., and Shokrani A. 'Process planning for additive and subtractive manufacturing technologies'. *CIRP Annals*. 2015, vol. 64(1), pp. 467–70.

[42] Quinlan H. E., Hasan T., Jaddou J., and Hart A. J. 'Industrial and consumer uses of additive manufacturing: A discussion of capabilities, trajectories, and challenges'. *Journal of Industrial Ecology*. 2017, vol. 21(1), pp. 15–20.

[43] Deloite. *Challenges of additive manufacturing. Why companies don't use additive manufacturing in serial production* [online]. 2019. Available: https://www2.deloitte.com/content/dam/Deloitte/de/Documents/operations/Deloitte_Challenges_of_Additive_Manufacturing.pdf. [Accessed 9 February 2021].

[44] Monzon M. D., Ortega Z., Martinez A., and Ortega F. 'Standardisation in additive manufacturing: Activities carried out by international organisations and projects'. *The International Journal of Advanced Manufacturing Technology*. 2015, vol. 76, pp. 1111–21.

[45] Liu J., Gaynor A. T., Chen S., *et al.* 'Current and future trends in topology optimisation for additive manufacturing'. *Structural and Multidisciplinary Optimization*. 2018, vol. 57, pp. 2457–83.

[46] Guessasma S., Zhang W., Zhu J., Belhabib S., and Nouri H. 'Challenges of additive manufacturing technologies from an optimisation perspective'. *International Journal for Simulation and Multidisciplinary Design Optimization*. 2016, vol. 6(9), pp. 1–13.

[47] Zhou M. Y., Xi J. T., and Yan J. Q. 'Modelling and processing of functionally graded materials for rapid prototyping'. *Journal of Materials Processing Technology*. 2004, vol. 146(3), pp. 396–402.

[48] Jones R., Haufe P., Sells E., *et al.* 'RepRap the replicating rapid prototyper'. *Robotica*. 2011, vol. 29, pp. 177–91.

[49] Tuttle G. G. *3D Printing of metamaterials with conductive elements*. PhD thesis [online]. 2018. Available: https://island1.uncc.edu. [Accessed 9 February 2021].

[50] Ngo T. D., Kashani A., Imbalzano G., Nguyen K. T. Q., and Hui D. 'Additive manufacturing (3D printing): A review of materials, methods, applications and challenges'. *Composites Part B: Engineering*. 2018, vol. 143, pp. 172–96.

[51] Moreau C. *The state of 3D printing by Sculpteo*. 2020 Edition [online]. Available: https://www.sculpteo.com/es/ebooks/state-of-3d-printing-report-2020. [Accessed 9 February 2021].

[52] Berman B. '3-D printing: The new industrial revolution'. *Business Horizons*. 2012, vol. 55(2), pp. 155–62.

[53] MarketsAndMarkets [online]. Available: https://www.marketsandmarkets.com/Market-Reports/3d-printing-market-1276.html. [Accessed 9 February 2021].

[54] Bertoldi K., Vitelli V., Christensen J., and Van Hecke M. 'Flexible mechanical metamaterials'. *Nature Reviews Materials*. 2017, vol. 2, pp. 1–11.

[55] Wong Z. J., Wang Y., O'Brien K., *et al.* 'Optical and acoustic metamaterials: Superlens, negative refractive index and invisibility cloak'. *Journal of Optics*. 2017, vol. 19(8), pp. 1–13.

[56] Sklan S. R. and Li B. 'Thermal metamaterials: Functions and prospects'. *National Science Review*. 2018, vol. 5(2), pp. 138–41.

[57] Wohlers. *Additive manufacturing state of the industry reports* [online]. 2017. Available: https://wohlersassociates.com/state-of-the-industry-reports.html. [Accessed 9 February 2021].

[58] 3DPrintingMediaNetWork [online]. Available: https://www.3dprintingmedia.network/the-global-additive-manufacturing-market-2018-is-worth-9-3-billion. [Accessed 9 February 2021].

[59] Frost & Sullivan. *Global additive manufacturing market, forecast to 2025* [online]. Available: https://www.researchcosmos.com/reports/additive-manufacturing-market/1170635782. [Accessed 9 February 2021].

[60] Annunziata M. *Manufacturing-as-a-service platforms: The new efficiency revolution* [online]. 2019. Available: https://www.forbes.com/sites/marcoannunziata/2019/05/13/manufacturing-as-a-service-platforms-the-new-efficiency-revolution/#21ed512657fd. [Accessed 9 February 2021].

[61] Lu B., Li D., and Tian X. 'Development trends in additive manufacturing and 3D printing'. *Engineering*. 2015, vol. 1(1), pp. 85–9.

[62] US Department of Energy. *Additive manufacturing: Building the future*. Office of Technology Transitions [online]. 2019. Available: https://www.energy.gov/sites/default/files/2019/07/f64/2019-OTT-Additive-Manufacturing-Spotlight_0.pdf. [Accessed 9 February 2021].

[63] Tanaka T. and Ishikawa A. 'Towards three-dimensional optical metamaterials'. *Nano Convergence*. 2017, vol. 4(34), pp. 1–9.

[64] US Department of Energy. *Additive manufacturing: Pursuing the promise*. Energy Efficiency & Renewable Energy [online]. 2012. Available: https://www.energy.gov/eere/amo/downloads/additive-manufacturing-pursuing-promise. [Accessed 9 February 2021].

[65] Design World [online]. Available: https://www.designworldonline.com/engines-with-3d-printed-parts-lift-ge-partners-over-31-billion-in-paris-air-show-orders. [Accessed 9 February 2021].

[66] CSIRO [online]. Available: https://www.csiro.au/en/research/technology-space/energy/Photovoltaics/Printed-Solar-Cells. [Accessed 9 February 2021].

[67] Automobil Industrie [online]. Available: https://www.automobil-industrie.vogel.de/additive-drives-wir-denken-den-elektromotor-neu-gal-945211. [Accessed 9 February 2021].

[68] ETMM [online]. Available: https://www.etmm-online.com/3d-printed-engine-is-20-percent-lighter-a-998651. [Accessed 9 February 2021].

[69] Wrobel R. and Mecrow B. 'Comprehensive review of additive manufacturing in construction of electrical machines'. *IEEE Transactions on Energy Conversion*. 2020, vol. 35(2), pp. 1054–64.

[70] Wu F. and EL-Refaie A. M. 'Toward additively manufactured electrical machines'. *IEEE Transactions on Industry Applications*. 2020, vol. 56(2), pp. 1306–20.

[71] Wrobel R. and Mecrow B. 'Minimization of winding AC losses using inhomogeneous electrical conductivity enabled by additive manufacturing'. *2020 IEEE Energy Conversion Congress & Exposition, Conference*, Detroit, MI, USA, pp. 1–8, 2020.

[72] Lopez-Fernandez X. M., Ertan B., and Turowski J. *Transformers. Analysis, Design, and Measurement*. Boca Ratón, FL: CRC Press, Taylor and Francis Group; 2012.

[73] Magisettya R. P. and Cheekuramelli N. S. 'Additive manufacturing technology empowered complex electromechanical energy conversion devices and transformers'. *Applied Materials Today*. 2019, vol. 14, pp. 35–50.

[74] French D. M., O'Loughlin J. P., Hoff B. W., Maestas S., Lepell D., and Montoya T. 'Dielectric strength testing of 3D printed plastics and application to a high voltage transformer pulser'. *2015 IEEE Pulsed Power Conference (PPC)*, Austin, TX, USA, pp. 1–6, 2015.

[75] Javora R. and Sekula R. *Method of producing parts of a transformer*, EU Patent EP 3 671 777 A1 [online], 2020. Available: https://data.epo.org/publication-server/document?iDocId=6274875&iFormat=0. [Accessed 9 February 2021].

Conclusion

A retrospective glance at the body of the book is presented here with the aim to highlight the main results but also pinpoint possible future developments.

Chapter 1 presented an overview of materials and computers. Materials have a paramount importance in technology. This is true not only now but also has been since ancient developments, and throughout history of mankind, new materials and new substances have been capable of starting new technological advances. In parallel, but more recently, we have built new powerful computers which expanded, to an unpreceded extent, our ability to predict the behaviour of our devices and ultimately to design themselves. The merging of these two developments into computer-designed materials promises the emergence of innovative materials and devices which we can only vaguely imagine today what they may be able to do.

Design optimisation is a challenging field of research, but also a focused attempt to equip the designers and users of electromechanical devices with efficient, reliable and relevant tools. In Chapter 2, we reviewed the state of the art and identified the most promising new developments. Several modern aspects of the design process were emphasised, underlined by the concepts of hierarchical design, 'no free lunch theorem', kriging-assisted optimisation, sensitivity analysis, and robust design. Projections of possible future developments were also offered. There is a huge literature on the subject and a healthy pool of available algorithms, but the ultimate choice always needs to be carefully considered.

In Chapter 3, after a review of the main methods of optimal shape design, an algorithm of freeform optimisation of dielectric design was developed and used to identify the best geometry of a shielding ring electrode for a high-voltage circuit breaker. Subsequently, the issue of technology-related limits affecting the practical realizability of an optimal geometry was addressed; in this respect, additive manufacturing (AM) was presented as a technology virtually capable of creating surfaces of any shape. In fact, this unique feature has greatly enhanced the class of achievable prototypes allowing for the implementation of shapes and geometries identified through numerical methods of design optimisation. In particular, most popular techniques of AM, e.g. fused deposition modelling and selective laser sintering, have been presented in a comparative way. Eventually, advantages and drawbacks of the various techniques have been highlighted with reference to recent literature and against a real example, i.e. a prototype of electromagnetic lens based on the concept of dielectric metamaterials.

Novel electric motors find broad applications across multiple sectors of industry and Chapter 4 discussed recent innovations and future trends, including application of new materials to improve the efficiency and introduce new technologies. The demand for innovative electric motor topologies is driven particularly by e-mobility. AM with 3D-printing will enable designing, prototyping and manufacturing of electric motors to become easier, faster, reliable and economical.

Shape design is a multi-physics problem and many approaches are possible; two of them were implemented and illustrated by case studies, while three nature-inspired optimisation algorithms (genetic algorithm, particle swarm optimisation and cuckoo search algorithm) were discussed in some detail.

Another approach involved a numerical experiment using finite-element analysis elucidated by the cogging torque minimisation of a brushless DC motor. New frontiers for innovative electric motor topologies, new materials and novel manufacturing processes by exploiting 3D printing are awaiting future developments.

Chapter 5 presented the state of the art and perspectives for the development of ferro- and ferrimagnetic materials. The chapter also reviewed the physical properties of magnetic materials currently available on the market and new materials under development. Manufacturing methods were described, and examples of the application of magnetic materials in electromagnetic and electromechanical devices were given. Moreover, the frontiers and challenges for future applications were discussed. In the last part of the chapter, new manufacturing technologies of ferro and ferrimagnetic materials were presented.

Chapter 6 covers selected aspects of the modelling, design, optimisation and fabrication of metamaterials based on gradient dielectric structures and split-ring resonator arrays. In the first part, a synthesis method for gradient metamaterials was presented and its application to a flat focusing lens studied. The second part focused on meta-surfaces comprising conductive resonant elements. It discussed full-wave and equivalent circuit modelling of resonator arrays and details of applications embracing a flat lens for magnetic resonance imaging and an artificial magnetic conductor. The last part of the chapter discussed the 2D/3D printing of effective dielectric media and proposed a method to improve its accuracy. The follow-up research will undoubtedly include tuneable meta-surfaces.

Finally, in Chapter 7 it was shown that both AM and metamaterials encompass a very broad range of research and applications. It is evident that AM will shift from being a disruptive technology, used only by innovators and for prototyping, to become a common method for core production. The aerospace and automotive industries are in the forefront of promising application fields for AM, while metamaterials are expected to have an impact on energy harvesting (both solar electric and solar thermal), high-power long-distance power transmissions and wireless charging. Moreover, AM could potentially be employed in construction of all integral components of an electrical machine assembly saving time and transportation costs. Finally, it is predicted that the total 3D printing market will continue to approximately double in size for the period 2020–2023, while the global metamaterials market is expected to grow by 36% in the period 2017–2025.

The main message and overall conclusion of this book is that – in the age of technology directly affecting our lives and generally the society – setting up an automated design centre in a manufacturing environment is no more an extravagant and ambitious goal; it is in fact a sound investment that academic and industrial institutions should be encouraged to pursue.

Eventually the book illustrates several examples of automated optimal design exploiting new materials, which aims at making the design of electromagnetic devices easier and, hopefully, more enjoyable.

In memoriam

Prof. PhD, DSc, Eng. Barbara Ślusarek

Professor Barbara Ślusarek, one of the authors of this book, was born in Wrocław, Poland, in 1946. She passed away on 19 February 2021 in Warsaw.

She graduated from Warsaw University of Technology in 1971 with MSc Eng in Mechanics. In 1997, she was awarded a PhD in Electrical Engineering by the same university, having submitted her thesis entitled ‘Bonded magnets for small electrical machines’. In 2002, she received a degree of Doctor of Science in Electrical Engineering from Wrocław University of Science and Technology, following a publication of her monograph ‘Nd–Fe–B-bonded magnets’. In 2014, she received the Title of Professor, conferred by the President of Poland.

Upon completing her studies, she worked in the Institute of Precision Mechanics and, since 1975 until her last days, in the Łukasiewicz Research Network of the Tele and Radio Research Institute in Warsaw, where she headed the Magnetic Materials Department. She was particularly interested in permanent magnets and soft magnetic materials made of powder and their application in electric motors and other electromagnetic devices.

She was the Principal Investigator of 11 research grants from the Polish National Centre for Research and Development and Polish National Science Centre. She also participated in research grants in the fifth, sixth and seventh European Union Framework Programmes.

In her long and fruitful academic career, Barbara published about 250 research articles in periodicals and conference proceedings, many of them in prestigious journals. She wrote two monographs and was an author or co-author of about 20 patents.

She received many awards for her inventions, amongst others in Brussels Innova fair and Concours Lepine in France fair.

She supervised two PhD students and was a reviewer of many PhD and habilitation theses.

She was a member of Polish Society of Theoretical and Applied Electrical Engineering, and also APMI International – Powder Metallurgy Society.

Until her last days, while struggling with a disease, our friend Barbara was devoted to the work on Chapter 5 entitled 'Frontiers and challenges of new ferromagnetic materials' for this book.

Index

www.ingramcontent.com/pod-product-compliance
Lightning Source LLC
LaVergne TN
LVHW011847200926
848484LV00002B/410
* 9 7 8 1 8 3 9 5 3 3 5 1 8 *